An Introduction to the Meteorology and Climate of the Tropics

An Introduction to the Meteorology and Climate of the Tropics

J F P Galvin

WILEY Blackwell

Registered Office
John Wiley & Sons, Ltd, The Atrium, Southern Gate, Chichester, West Sussex, PO19 8SQ, UK

Editorial Offices
9600 Garsington Road, Oxford, OX4 2DQ, UK
The Atrium, Southern Gate, Chichester, West Sussex, PO19 8SQ, UK
111 River Street, Hoboken, NJ 07030-5774, USA

For details of our global editorial offices, for customer services and for information about how to apply for permission to reuse the copyright material in this book please see our website at www.wiley.com/wiley-blackwell.

Library of Congress Cataloging-in-Publication Data

Galvin, Jim (J. F. P.)
   An introduction to the meteorology and climate of the tropics / Jim Galvin.
      pages   cm
   Includes bibliographical references and index.
   ISBN 978-1-119-08622-2 (cloth)
1. Tropical meteorology.   2. Tropics–Climate.   I. Title.
   QC993.5.G25 2016
   551.50913–dc23

                         2015020802

A catalogue record for this book is available from the British Library.

Cover image: Pacaya-Samiria Reserve, Reserva Nacional Pacaya-Samiria, Maranon River, Rio Maranon, Loreto Region, Maynas Province, Amazon Basin, Peru
(The cover image is from Getty Images, number: 505330107)

Set in 10.5/12.5pt Times Ten by SPi Global, Pondicherry, India

1   2016

Dedication

To Grace, Hannah and Cecilia

Contents

About the Author

J F P Galvin has spent his career with the Met Office, most of it working as an observer or forecaster of the weather in the UK and Mediterranean. He has had a long-held fascination with tropical weather and its different character.

He gained a Master's Degree in Applied Meteorology at the University of Reading in 2000, and much of the inspiration for this book has its origins in a module on tropical meteorology, studied as part of the course.

This book is updated, revised and extended from a series published in the RMetS journal *Weather*, of which he is now editor. The original papers were written in part as a guide for forecasters in the World-Area Forecast Centre at the Met Office, where he is a specialist meteorologist.

Between 2000 and 2006, he served as Honorary Photographic Editor on the Editorial Board of *Weather* and is a prolific writer on meteorology. Recently he has assisted in the production of a forecasters' guide for West Africa (Parker & Diop-Kane, 2015).

He is married with two girls, and greatly enjoys photography, travel and reading novels.

Preface

In this book I describe various aspects of tropical weather and climate as it is understood today. Each chapter includes a description of the effects of tropical weather and there are several case studies included to show the effects of tropical weather.

Written as an introduction (based on a series published in the RMetS journal *Weather*) with an inter-disciplinary text, the emphasis is on observational science. A number of very good texts are available for those who wish to go into greater depth, including the mathematics of the tropical atmosphere; in particular, I recommend the books *Climate Dynamics of the Tropics* by Hastenrath (1991) and *The Climate and Weather of the Tropics* by Riehl (1979).

The text is aimed at readers with an interest in – and at least some knowledge of – meteorology or physical geography. It is intended as a background for students beginning their examination of tropical weather and climate. Readers may not be familiar with some of the diagrams discussed in the text and a guide to these appears as Appendix 4.

I have approached the subject as a weather forecaster, as much as possible, with the needs of fellow forecasters in mind. Computers are increasingly accurate in their analyses and prediction of weather patterns, in particular in the first few days (Met Office 2013a) and carry out mathematical calculations to formulate forecasts to a high precision. In general, the forecaster – or even the researcher – needs to have little more than a basic grounding in the mathematics peculiar to the weather of the tropics. In general, this is covered in foundation courses for atmospheric physics (e.g. McIlveen 1992) and so will not be revisited here.

It is far more valuable to the meteorologist to know about general patterns of weather and climate, as well as analysis within the tropics – most of which are very different from those of the high latitudes. With this knowledge it is possible to add detail to the predictions from (global) models with a coarse grid length, thus adding value to the low-resolution numerical forecast product. It is also possible to assess the likelihood of some longer-range numerical forecasts. Nonetheless, there are many aspects of tropical weather that can be explained in the same way in the higher latitudes: the variation of temperature with height, the effects of continentality (or the converse, the oceans) and the effects of the radiation balance, all of which are described by universal physical equations. The atmosphere is, after all, a single entity with essentially the same constituents. For this reason, where appropriate investigations have been carried out, I have included references to the meteorology of the higher latitudes.

Acknowledgements

Special thanks go to Anglosphere Editing Limited, whose dedicated copy-editing work greatly improved all aspects of this book.

David Membery, Richard Young, Nigel Bolton, Chris Jones, Mariane Diop-Kane, Rowl Twidale, David Hopkins and Joe Tennant provided helpful assistance.

Thanks to Nige Emery, who prepared Figure 4.4, and Clive Jones of the Met Office, who supplied Figure 5.1.

Climate data for many world cities are available at http://www.bbc.co.uk/weather/world/city_guides/index.shtml, as well as in Pearce and Smith (1984).

The descriptions of soil types are as given in the *Encyclopædia Britannica* and *The Concise Oxford Dictionary of Earth Sciences*, from Encarta and through the US Department of Agriculture.

With thanks to Julian Heming, Penny Tranter, Richard Young, Ken Horn, Andy Dexter, Ralph Cooper, Glenn Greed, Nigel Bolton, Tony Gillard, Mel Collier, Dick Francis, Nick Grahame, Chris Tyson, Dai Naylor, Steve Manktelow, Phil McGarry, Ian Deavin, Rick Robins, Fiona Smith, Clive Wilson, Ian Black, Neil Higginson, Paul Arbuckle, Jay Merrell (all Met Office), Chris Sear (DEFRA), Danuta Martyn, Doug Parker (University of Leeds), Andreas Fink (Institut für Geophysik und Meteorologie, University of Cologne) and Sue Brown (formerly RMetS) for their useful information and correspondence.

Wikipedia was a useful source, having made available many references from the World Wide Web, in many cases very soon after significant weather events.

Some elements of the text drew on figures in the *Britannica Atlas* (1989 edition) and *Philip's Universal Atlas of the World* (2005 edition).

David Membery supplied information for Chapter 7 and for various other parts of this book; Chris Jones supplied information for the preparation of Chapter 6.

Images were kindly supplied by National Aeronautics and Space Administration (Figs 2.1, 2.2, 2.3, 2.4, 13.5); National Oceanic and Atmospheric Administration/Earth System Research Laboratories (Figs 3.1, 3.2), RMetS (Figs 3.5, A3.1), Renault F1 team 2006 (Fig. 4.3), © Crown copyright (Met Office) (Figs 4.5, 5.1, 7.4, 9.8, 10.1, 10.4b, 10.5, 10.6, 11.5, 11.5a, 11.6, 12.6, 12.7, 15.2, A5.1, A5.2, A5.4, A5.6, A7.1, A7.4, A7.7), © EUMETSAT (Figs 4.6, 7.1, 7.7, 15.2, 15.3, A7.6), NERC Satellite Receiving Station, Dundee University (Figs 5.2, 5.4, 8.4, 8.5, 8.7, 10.2, 10.3, A7.6), © Richard Young (Fig. 6.3), © Barbara Pettigrew (Fig. 6.6), © Nigel Bolton (Fig. 6.11), National Aeronautics and Space Administration/Goddard Space Flight Center (Figs 6.12, 10.4, 13.4, 14.2, 14.1, 14.3), © Grant McDowell/Naturepl.com (Fig. 7.2), © Sue Wilson (Fig. 7.3), European Centre for Medium-Range Weather Forecasts (Fig. 8.2), National Aeronautics and Space Administration Rapid Response Team

(Figs 8.6, 8.13b), © University of Wyoming, Department of Atmospheric Science (Figs 8.8, 11.5b, A5.3, A5.5, A7.2, A7.5, A7.8), © Bureau of Meteorology (Figs 8.13a, 8.14), © Russell White (Fig. 8.15), World Meteorological Organization (Figs 9.1, A1.1), © University College London, courtesy Mark Saunders and Adam Lea (Fig. 9.4), US Navy Historical Center (Fig. 9.7), National Oceanic and Atmospheric Administration/National Climate Data Center (Figs 12.1, 12.2), American Meteorological Society (Fig. 12.4), Courtesy Freie Univerität Berlin (Fig. 12.5), National Oceanic and Atmospheric Administration (Figs 13.5, 15.4), J Wiley & Sons (Fig. 14.4), Servicio Meteorológico Nacional, Mexico (Fig. 15.1), Instituto Nacional de Tecnica Agropecuria (Figs 15.5, 15.6), World Health Organization (Fig. 16.1), RMetS (Fig. A4.1), © Grace Galvin (Fig. A3.4), © Faye Davies (Fig. A3.14), courtesy B J Burton, accessed at http://www.woksat.info/wwp.html (Fig. A5.7). The author produced all other images.

Julian Mayes, formerly Editor of *Weather*, accepted the series on which this book is based and has encouraged me throughout.

1

Setting the Scene

1.1 Introduction

Much of what we know about the weather has been focused on mid-latitude weather systems, first because most early researchers came from rapidly developing western Europe and eastern North America, and second because of the risks and consequences of weather systems prevalent in these zones. However, although there are simple non-scientific descriptions of weather events from the tropics and tropical climates going back hundreds or thousands of years, it is only since the late 1960s that much scientific research has been carried out within the tropical zone, although observational networks were established in many populous areas as early as the 19th century. What we know of the weather (and, to some extent, the climate) of the tropics remains limited and has typically focused on severe weather events, such as tropical revolving storms (Emanuel 2005) or data from a limited range of observing stations. This is partly because upper-air networks, in particular, are thinly spread with weather balloons launched rarely more than once per day. Of necessity, desert regions have few observing stations. Weather observations are often the first 'casualties' of economic hardship (see Appendix 1).

However, many factors of the day-to-day weather are important in the tropics, not least

for aviation and public safety. It is also important for holiday-makers to know what to expect. In recent years, tropical resorts have become readily available and popular for their warmth and sunshine, but travellers often know comparatively little about seasonal weather variation or the effects of exposure to sunshine in the tropics.

For instance, the primary purposes of forecasting for aircraft operations in the tropics are safety and the maximization of efficiency for the benefit of passengers and aircraft operators. The most accurate and appropriate forecasts will achieve this goal using a mixture of numerical weather prediction products, observed data and good forecasting knowledge. It is the effects of the weather, in other words its outcomes, which must be considered.

In order to maintain observational networks and co-ordinate the exchange of atmospheric and hydrological data and numerical forecasts, the World Meteorological Organization (WMO) – an agency of the United Nations (UN) – has established various programmes, not least the World Weather Watch (WMO 2013). This agency has more members worldwide than any other UN agency, emphasizing the importance of meteorology and hydrology. The research carried out as part of the World Climate Research Programme since the early 1970s is very important in allowing us to

An Introduction to the Meteorology and Climate of the Tropics, First Edition. J F P Galvin.
© 2016 John Wiley & Sons, Ltd. Published 2016 by John Wiley & Sons, Ltd.

understand many of the processes and associated weather of the tropical zone (Gates & Newson 2006). Knowledge continues to grow through more recent research programmes, such as Tropical Ocean Global Atmosphere (TOGA), which investigates the important links between the tropical ocean and the global atmosphere (Fleming 1986). It is clear that the tropics have an important effect on weather systems throughout the globe, providing much of their energy.

1.2 What do we mean by 'the tropics'?

In order to examine tropical weather and climate we need to define what we mean by 'the tropics'. Although most of us have some concept involving warmth, the definition is not straightforward and it is possible even for meteorologists and climatologists to have different views of what constitutes the tropical zone.

The most commonly used definition of the tropics is the zone within which the sun is directly overhead at some time during the year: the zone between the Tropics of Cancer and Capricorn (23.45°N and 23.45°S, respectively). However, a weather-related definition, rather than the elevation of the sun at midday, is probably more useful to the weather forecaster.

1.2.1 Climatological methods

The range of temperature is often large over land in the tropics. This is because the high sun brings high daytime temperatures, but the loss of temperature by long-wave radiation overnight is also large, the rate of loss of radiation increasing exponentially with absolute temperature (Stefan's law). Because condensation and a high water-vapour content reduce the loss of temperature, its range is also governed by humidity, so the mean daily temperature range is greatest across the deserts (15°C) and lower in the humid zone, in particular close to the oceans. However, even in humid coastal

areas and on mountains the range is rarely less than 5°C. Temperature varies little over the ocean surface, where most solar radiation is absorbed rather than re-radiated, although daily variation is larger in many (but not all) parts of the tropics than it is at higher latitudes.

We could define our 'tropical' zone as, say, the region within which the mean temperature is above some nominal value, say, 20°C, throughout the year. However, this has the disadvantage of excluding even relatively modest high ground and, for part of the year, continental areas that are relatively close to the equator. Whilst this failing could be corrected by adjusting the nominal value to a specific level,[1] this is not the most satisfactory method as it can include high mountain areas well north and south of zones normally considered as tropical.

Figure 1.1 demonstrates the difficulties of using mean temperature as the main method of definition of climatic zones. It is readily apparent that some of what most meteorologists or climatologists would call the tropics – albeit along the poleward extremes – is classified by Köppen as 'temperate' (zone C), largely due to altitude or the presence of cool air over the coasts of cool oceans. His scheme defines the tropics in two zones: A, wet climates and B, dry climates. However, this results in a narrowing of the tropical zone to within 25° of the equator in places and an expansion as far poleward as 50° (in the dry zone) elsewhere. One result of this classification scheme is to place northern parts of the monsoon zones in the temperate zone and the dry (often cold) deserts of central Asia in the tropics.

A geographical method could be to use the limits to growth of a particular characteristic crop: bananas or fruiting pine trees might be considered ideal crops in this respect. This approach is useful in many ways, especially if adjustment is made for areas that are too dry to grow the crop, such as deserts. However, some parts of these dry lands are susceptible to frost (which would kill sensitive plants, such as

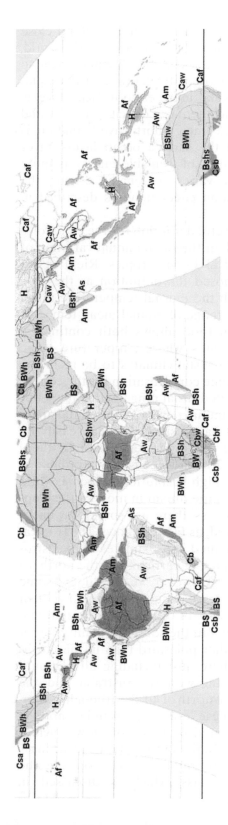

Figure 1.1   Climatic zones of the tropics: Af/Am, humid tropical; Aw, savanna; BS/BSh/BShw, tropical steppe; BWh, tropical and sub-tropical desert; Cf, warm temperate with no dry season; Cs, warm temperate with dry summer; Cw, warm temperate with dry winter; Cb, temperate climate with cool summer months; H, highlands; stippled, modification due to altitude (using the system devised by Wladimir Köppen (McKnight & Hess 2000)). The effect of high ground has a profound influence on climate in the tropics, particularly above about 2 km. The black lines indicate latitudes 30°N and 30°S.

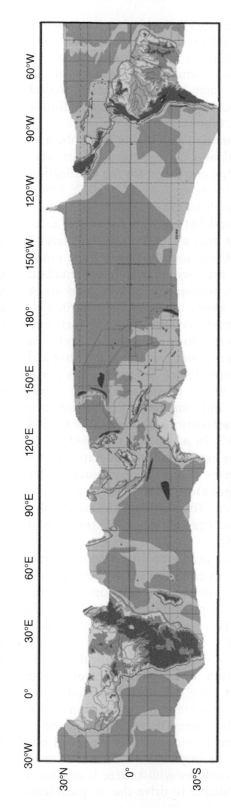

Figure 1.2   Area fitting Riehl's scheme for the definition of the tropics. From all available data, areas where the diurnal temperature range exceeds the annual range of mean temperature are included in the tropical zone. Some notable peculiarities can be seen: much of Arabia, India and South-East Asia, locally as far south as 10°N, is excluded, but some cool-water coasts of the Americas and Africa are included. Much of the mountainous tropics would also be likely to be excluded, as high ground has a low diurnal temperature range (a few degrees), exceeded by the seasonal mean-temperature change. (This is further complicated by the difference in the climatology of valleys, which have a high diurnal range, and exposed mountaintops, where the diurnal range is low.)

banana bushes) and the cold of high mountains would also exclude their growth (rather than latitude constraints). However, both crops are grown in many sunny frost-free parts of the extra tropics, such as south-western coastal Cyprus, so perhaps an agricultural indicator has too many limitations?

A simple method is to divide the globe into fixed tropical and extra-tropical zones. This method is often employed for the verification of numerical forecasts (WMO 1982; Fuller 2004). One such form divides the globe into two equal-sized halves with somewhat arbitrary dividing lines at 30°N and 30°S (marked on Fig. 1.1). Conveniently, this latitude range includes the area within which near-surface winds are predominantly easterly and all of the climatic zones that can be regarded as tropical: humid, savanna, semi-desert and tropical desert. However, part of some of these climate zones, notably the tropical desert and semi-desert, frequently lies north or south of 30°, in the margin of the extra-tropical zone and some zones that do not have tropical characteristics extend equatorward of 30°N and 30°S. In addition, as we will see in Chapter 2, the area between these latitudes on average receives a surplus of incoming solar radiation (insolation) over that lost by long-wave radiation. This surplus is carried poleward into a more readily defined extra-tropical zone (Fig. 2.5).

The zones of predominantly westerly winds make incursions equatorward of these lines of latitude, particularly in winter. In order to keep within a zone of predominantly easterly winds at most levels, a narrower north–south zone must be used.

We might also define the tropical zone as the area within which there is more solar energy (short-wave radiation) received at the surface than that emitted from it (long-wave radiation). The energy balance is discussed in Chapter 2 and is an important aspect of our definition of the tropical zone. However, even this has its problems: whilst there is an excess of energy available to drive the 'tropical heat engine' (Pidwirny 2013; Lindsey 2009) close to the equator all the way around the world, areas of relatively cold ocean extend near to the equator (within 10° latitude of the equator), in particular in the eastern Pacific Ocean. In these areas there is a deficit of energy and all of the incoming short-wave radiation is absorbed in the surface ocean layers. Over land, the excess of available energy is more evident, but varies seasonally, driving the major monsoon circulations, as described in Chapter 8.

A more useful definition for the climatologist is based on the small annual variation of climate typical in the tropics. Riehl (1979, Ch. 2) proposed the definition as the area within which the diurnal temperature range exceeds the range of annual mean temperature. This method allows both continental (inland) climates, where temperature range is high, and coastal climates to be included, as there is some compensation for the combination of high diurnal range with a tendency for a larger annual range of continental areas. It has great value, since data can easily be sorted using this definition, although, theoretically at least, data needs to be available at a relatively high resolution, which is little more than a pipe dream in most of the tropics. Nevertheless, there are some drawbacks to this method, as Fig. 1.2 shows. The tropical area thus defined approximates to the astronomical tropics, in particular in the northern hemisphere. However, it includes some extra-tropical areas, such as a small part of the cool coast of California around San Francisco, and in the southern hemisphere it extends further poleward than in the northern hemisphere, just including all of South Africa, coastal western Australia and the Chilean coast north of 40°S. Although accommodating these areas, it excludes notable areas dominated by monsoon flows, including much of India and the south of China – areas that are rarely, if ever, outside the tropical air mass, which is discussed in section 1.2.2.

1.2.2 The tropical air mass

Day to day, the weather forecaster usually needs something more closely related to the day-by-day weather, without reference to seasonal variations.

Using the current state of the atmosphere in depth, treating tropical air as a single air mass, it is possible to define the tropical zone on a daily basis. The characteristics of the tropical troposphere are discussed in the following sections. The temperature difference between the tropics and middle latitudes causes a jet stream to develop at the poleward limit of the tropics. This sub-tropical jet stream (STJ) is a generally broad belt of winds, often extending across a latitude range of 10° or more. It has a core close to 30°N and 30°S and has little high-amplitude wave development along it (see Fig. 4.1). The area between these jet streams has a tropospheric depth characteristic of the tropical zone. This depth allows us to define the periphery of the tropics, even when the STJ weakens or is absent, as occurs in summer in both hemispheres and is usually the case in the northern hemisphere summer. Furthermore, the equatorward edge of the STJs coincides (albeit weakly) with the transition from westerly lower tropospheric winds to poleward and easterly trade winds nearer the equator.[2] The subtropical jet stream is described in more detail in Chapter 4.

Although use of the STJ as the northern and southern limits of the tropics means that the tropical zone extends north of 40°N over Asia during the northern summer, it is appropriate, since the air to the south of it retains tropical characteristics. Thus, to provide consistency, in this book some areas poleward of the mean latitude of the STJ will be discussed, since these areas are often within the meteorological tropics. For instance, it allows inclusion of areas frequently affected by troughs in the STJ and upper troposphere (Chapters 4, 10 and 11). The incursion of extra-tropical air has important effects on the weather of the tropics, in particular in the monsoon zones, as well as within upper-tropospheric troughs, notably in winter.

1.3 The geography of the tropics

To understand the weather of the tropics we need first to understand the geography and its controls on the movement of weather systems.

Two main factors affect the development and motion of weather in the tropics: the distribution of land and sea and the effect of mountain chains on this motion (Chapter 11). Between 30°N and 30°S the proportion of land to sea is about 1:4. There is much less land in the tropics than at higher latitudes. Not only does this generate the great monsoon flows of the tropical zone (Chapter 8), but it also provides the moisture that drives the tropical heat engine. With much more land there would be significant differences in the tropical weather systems and the tropical winds. The diurnal and seasonal heating and cooling of land are much greater than that of the sea, generating weather systems on a variety of scales. In the tropics there is more land north of the equator than in the southern hemisphere; this brings more rain to the north of the equator than to the south.

The tropical zone also has a disproportionately large amount of high ground, including parts of the Himalayan and the South-East Asian mountains, much of the Andes and all of the Central American sierras. These mountains help to guide and block the motion of weather systems. However, they also generate their own weather.

These effects will be described in detail in succeeding chapters.

1.4 The tropical troposphere

The atmosphere is composed of a number of layers and almost all weather occurs in the lowest of these: the troposphere. In general, temperatures fall with altitude within this layer, although locally there are layers of warmer air overlying cooler, forming shallow layers where temperature rises (inversions). Both the overall

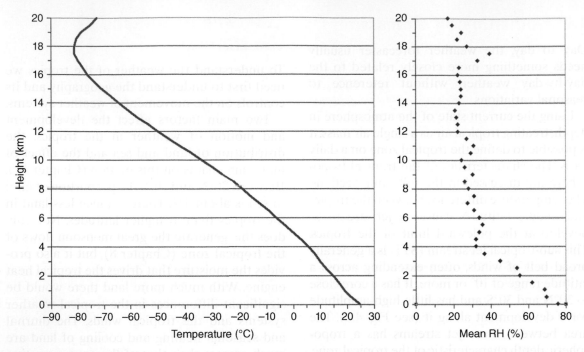

Figure 1.3 The mean change of temperature and humidity with height in the tropics from a selection of radiosonde ascents made across the range of tropical weather regimes. RH, relative humidity. In practice there is considerable variability in these figures: the profiles are much drier over the deserts and during the winter monsoon, where the upper troposphere tends to be somewhat cooler than in the humid zone. However, the height of the tropopause – the level at which temperature stops falling quickly, marking the transition into the stratosphere – is remarkably constant (unlike in the higher latitudes), near 17 km.

fall in temperature and the presence of inversions are necessary components of the formation of cloud and rain, particularly in the tropics. The tropopause is at the top of the troposphere and above it is a near-isothermal layer, the stratosphere. Convective cloud locally reaches into the lowest part of this layer, but with temperatures no longer falling quickly with height, the cloud can no longer rise, as its density becomes equal to (or below) that of the surrounding environment. The cloud tops readily evaporate because of the very low humidity of much of the stratosphere. The temperature and humidity profiles of the tropical troposphere and lowest part of the stratosphere are shown diagrammatically in Fig. 1.3.

Within a single air mass, temperatures (and humidity) vary most in the layer just above the surface: the boundary layer. The balance of incoming and outgoing radiation changes the temperature of the surface and convection carries these temperature changes into the lower layers of the troposphere. By day the development of instability carries warmth to a much greater depth by convection than overturning can when the air is cool and stable by night. The greater the instability, the more heat can be carried to depth away from the surface. However, the entrainment of heat (and moisture) becomes comparatively small above about 2000 m and the boundary layer is generally considered to exist below that level, even by day.[3] By night the boundary layer may be little more than a few tens or hundreds of metres deep.

On a broader scale, radiation is the main source of energy exchange in the atmosphere. Although most outgoing long-wave radiation is emitted from the earth's surface in response to heating by the sun, the atmosphere also radiates long-wave radiation to space. Most atmospheric radiation is emitted by the troposphere, proportional to its mean temperature. As there must be a balance of outgoing and incoming radiation, where the lower troposphere is warm, the upper troposphere and lower stratosphere are cold (and vice versa). To accommodate this, the tropopause must be high where the air is warm and lower where it is cool. Thus, the height of the tropopause may be used as a marker of a particular air mass (although cooler air may make incursions at low levels, so that a particular air mass may not extend all the way to the surface from the tropopause).

The height of the tropopause varies little in the tropics, but changes as the mean temperature of the troposphere changes and although there are significant differences in weather within the tropical zone, there is only a single tropical air mass. Nevertheless, differences in mean tropospheric temperature, humidity and dynamics cause some variation within the air mass, so that the highest tropopause is generally found close to the equator. The troposphere rarely extends above about 18 km and is most often close to 17 km altitude. Within the tropical air mass, height gradients are usually small, but increase somewhat near the sub-tropical jet streams. Indeed, the association of the STJ with a tropospheric depth[4] of about 15 km provides a definition of tropical air.

Despite its relatively uniform depth for much of the year, some variation occurs with the changing of the seasons and these variations are notably marked in the northern hemisphere summer. Between May and September, the tropopause is higher over northern Africa, and southern and eastern Asia than it is close to the equator. The intense warming of these land masses causes the troposphere to expand and the tropopause occasionally reaches a height of 18 km or more over northern India, Nepal and Tibet. Smaller expansions occur over Australia, southern Africa and South America during the southern summer. The expansion is a key element in the development of the summer monsoon circulations described in Chapter 8.

The density of tropical air is low because it is warm, so the heights of pressure levels are high in the deep tropical troposphere and the thickness between fixed (standard) levels is high. Globally, this is usually measured in terms of the height difference between a pressure of 1000 hPa and 500 hPa. This height difference generally has a minimum around 580 decametres[5] (dam) along the northern and southern boundaries of the tropical zone and may reach 590 dam or more.

The height of the freezing level can also be used as a marker for tropical air. Although night-time frosts sometimes occur at lower levels on high ground within tropical air, the free-air freezing level is never below 4000 m in the tropical air mass and is usually found between 4500 and 5500 m.

The great disadvantage of defining the tropics by means of an air mass is that its area changes day by day, so that, locally, we must discuss extra-tropical air that affects areas in winter to include areas under the influence of tropical air in summer. However, this is an inclusive method, including all areas, such as the monsoon zones, which remain within the tropical air mass in winter despite having cool (or even cold) air near the surface. There is comparatively little expansion and contraction of the tropical air mass (between the sub-tropical jet streams) from winter to summer. The waves in the STJ are comparatively modest and move slowly, upper troughs of sub-tropical air tending to fill as they move.

1.5 Climate and population in the tropics

On average, population density is low in the tropics, typically around 10 km^{-2}. However, in India, Bangladesh, southern China and parts of South-East Asia the population density is much higher, reaching 100 km^{-2} or more.

Away from rivers, the hot-desert environment is not able to support large populations and tropical rainforest is a dark foreboding environment with trees that are difficult to clear unless heavy machinery is available. Thus populations are found mainly along coasts or rivers, where transport has been available for centuries or millennia. It is near the mouths of the larger rivers that the largest cities are found.

The savannas are moderately populated, the land easy to clear for agriculture and road building. However, in some areas the population density is close to the ability of the land to support it. In Africa and parts of South America this presents a problem, since populations have grown rapidly during the late 20th century and the rainfall is not sufficiently reliable in these areas for there to be confidence that populations can survive without a major risk of drought and famine.

The degree of urbanization, by contrast, varies considerably from continent to continent. In South America it is generally above 50%, the proportion in Venezuela above 90% and in Australasia 80%. However, in much of Asia and Africa it is below 45%; Ethiopia, Uganda, Malawi and Nepal have fewer than 15% of their populations living in towns. This stresses the relative importance of agriculture in these countries, even those that are rapidly industrializing, such as India and China. However, there are significant variations from country to country, as well as within countries.

1.6 Question

List the main factors determining climate, considering why the tropics differ from the higher latitudes.

Notes

1 In meteorology we can use the standardized physical properties of air to describe air masses, which may be compared or grouped. In this case

the observed (or interpolated) mean temperature could be assumed to be its equivalent at, say, the 1000 hPa level (i.e. very near the surface). The change in altitude would compress and warm the air at a standard rate of 0.0098 km^{-1}, so that air originally at 3000 m with a pressure of 695 hPa and a temperature of 0°C in the Andes of Brazil would be warmed to 28.5°C as the pressure rose to 1000 hPa. Similarly, air at 25°C with a pressure of 1025 hPa in Death Valley (85 m below sea level) in the western desert of the USA would cool to 22.5°C as its pressure reduced to 1000 hPa.

2 A careful examination reveals that the core of the STJ usually lies above the margin of the cooler air of the sub-tropics, which, as is the case along the frontal zones of the high latitudes, undercuts the warmer lower-density air. However, the 'true' tropical air is found all the way to the surface close to the equatorward margin of the jet-stream belt (which is usually several hundred kilometres wide).

3 There are some problems with this somewhat arbitrary approach. Areas of high ground, such as inter-montane plateaux, will develop their own boundary layers. If the altitude of the plateau is near or above 2000 m, the top of the boundary layer will often be much higher than the broadly accepted level given here. Although mountains also affect the temperature of the air in contact with them, the effects are generally rather small, with temperature changes affected more by advection than the exchange of radiation.

4 In winter, as cool sub-tropical (sometimes polar) air undercuts the tropical air mass, a tropopause may be found below the level of the STJ and, indeed, a second jet-stream core (a cut-off from the polar-front jet stream) may be found below this secondary tropopause.

5 Atmospheric height and thickness are, in fact, measured in geopotential metres, which allows for the difference in potential energy in air at different distances from the centre of the earth, geopotential being a measure of the work required to lift a fixed mass of gas to a new altitude. Geopotential height (Z) uses gravity to calculate altitude, so that where acceleration due to gravity g = 9.79 ms^{-2} (the value near sea level at the equator), true height h = gZ/0.979 m.

2

The Energy Balance and the Dynamics of Weather in the Tropics

2.1   The tropical 'heat engine'

Radiation emitted by the sun supports all life on earth and drives all the world's weather systems (Lindsey 2009; Pidwirny 2013). Averaged through a whole year, the tropics receives far more of this radiation than the higher latitudes, and the heat absorbed at lower latitudes (especially over land) is the main source of the energy that generates global weather. At the top of the atmosphere, a surface perpendicular to the sun receives an average of 1360 Wm^{-2}, but this varies a little through the year due to the elliptical orbit of the earth (Kiehl & Trenberth 1997; Strangeways 2011). However, not all of this energy is available to heat the earth's surface (Fig. 2.1). Variations in the elevation of the sun, cloud cover and suspended particles in the atmosphere reduce the daily average at the earth's surface to 342 W m^{-2} (5395 MJ m^{-2} yr^{-1}), although in the tropics the average is around 700 W m^{-2} (11 GJ m^{-2} yr^{-1}).

Across the globe, the mean temperature is determined by several factors: the area-averaged amount of solar energy arriving at the top of the atmosphere (Fig. 2.2), absorption by the atmosphere, reflection and re-emission. Averaged through the year, most energy is received in the tropics, where the midday sun is directly overhead on one day of the year (two days at the equator). However, as the sun's apparent position migrates north and south through the year, the angle of the sun at midday is as great at 47°N on 21 June as it is at the equator. In response, the zone of tropical weather moves north and south through the year (although the migration follows, rather than accompanies, the sun's apparent motion). Through the year, however, the top of the tropical atmosphere receives much more radiation than the high latitudes.

Through the year, there are large differences in the amount of radiation received at the top of the atmosphere in the high latitudes, but the difference is modest near the equator (Fig. 2.2). This is not only because of the sun's elevation, but also because the angle of the earth's axis to the sun lengthens the days in summer in the high latitudes and shortens them in winter.

2.2   Absorption, reflection and apparent solar elevation

The most important effect on incoming solar radiation is reflection. Cloud reflects up to about 210 W m^{-2}. Land surfaces with little or no

(a)

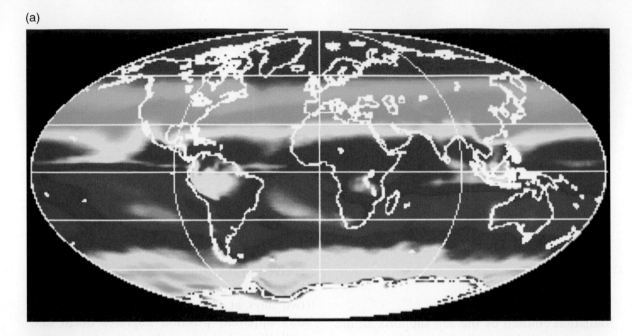

(b)

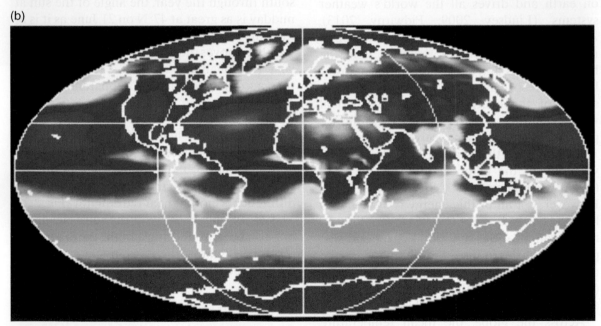

Figure 2.1   Average insolation at the earth's surface, 60°N–60°S: (a) January 1984–1991, (b) July 1983–1990. High values occur in the sub-tropical oceans of the summer hemisphere where the air is dry and clear of dust or cloud. Colour range: blue, ~50 Wm⁻²; cyan, ~100 Wm⁻²; green, ~150 Wm⁻²; yellow, ~200 Wm⁻²; red, ~250 Wm⁻²; magenta, ~300 Wm⁻²; white, ~350 Wm⁻². Tropical mean, ~ 280 Wm⁻². © NASA Surface Radiation Budget Project.

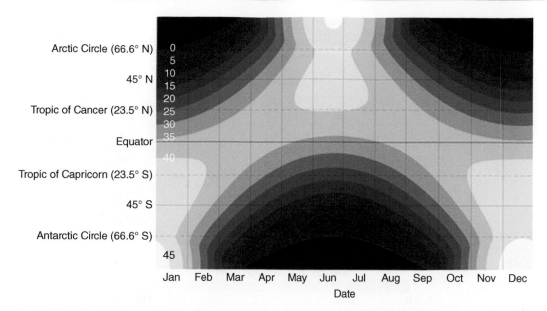

Figure 2.2 The annual variation of solar energy received each day at the top of the atmosphere across the globe (MJ m⁻²). Note that a little more radiation is received in tropical latitiudes north of the equator than to its south, although in most respects the areas encompassed by each contour in each hemisphere are very similar. Between the Tropics of Cancer and Capricorn, the total is about 32 MJ m⁻² ± 9 MJ m⁻², whereas at 45°N and 45°S the total is about 26 MJ m⁻² ± 16 MJ m⁻² (Kiehl & Trenberth 1997). © NASA/Robert Simmon. See also Fig. 3.5.

vegetation or snow cover also have a high reflectance (albedo). Values reach about 180 W m⁻² in the world's deserts and snow-covered mountain chains, as shown in Fig. 2.3. The average reflectance by clouds and the surface in the tropics is about 100 Wm⁻².

Even though the sun emits radiation with a short wavelength and the atmosphere is relatively transparent at these high frequencies, the atmosphere absorbs a surprising amount of energy, in particular in the tropics. In this zone, amounts of water vapour, ozone and dust – the main absorbers of insolation – are present in relatively high proportions. However, the depth of atmosphere through which the sun's rays must travel to reach the earth's surface is the main overall determinant of absorption. Although the high solar angle near the equator throughout the year means that the path length is small between about 30°N and 30°S, the tropical troposphere is much deeper than that of the high latitudes. (The effect of the deep atmosphere and the high concentrations

of absorbers near these latitude lines is evident in the deficit of insolation at the Tropic of Capricorn compared with that at 60°N between early May and early August in Fig. 2.2). Absorption averages about 80 W m⁻² in the tropics.

The apparent migration of the sun north and south is the final determinant of radiation received at the surface. This shows strong seasonality, with differences growing rapidly away from the equator.

2.3 Emission from the surface

In response to insolation, the surface emits radiation, but most re-emission is in the infrared (long-wave) spectrum. The emission is proportional to temperature (Stefan's law), so is greatest in the tropical zone and is broadly in balance with the amount of radiation received

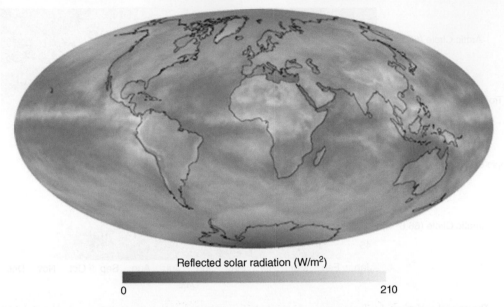

Reflected solar radiation (W/m²)

0   210

Figure 2.3   Reflected solar radiation, September 2008, observed by CERES. Near the equator clouds reflect a large proportion of sunlight, as do the tropical deserts and snow-covered mountains. The reflectivity of snow is very evident from the high peaks of the Himlaya and Andes mountains. Although the oceans typically have a low reflectivity at low solar elevations, the sea surface may reflect a high proportion of insolation. © NASA/Robert Simmon.

at the surface. However, the process is complex: cloud is a good reflector of long-wave radiation and greenhouse gases – particularly water vapour – absorb strongly in the infrared, so that little radiation is lost to space in cloudy areas and variations in the gaseous components of the atmosphere affect the radiation balance (Strangeways 2011). The net emission from the surface in January and July is illustrated by Fig. 2.4.

Comparison of Figs 2.3 and 2.4 reveals a surfeit of radiation in many parts of the tropics, whereas there is a deficit in the high latitudes (Fig. 2.5). This surplus drives a 'heat engine' that moves heat from the tropics to the higher latitudes. The additional heat drives convection and evaporation, generating global-scale air motion – the Hadley cells. These in turn drive warm surface water from low latitudes onto the western margins of the high-latitude continents, as well as the eastern margins of tropical continents, and pull cold water along the western coasts of tropical

lands and eastern coasts at high latitudes (Chapter 5).

The combination of latitude and continentality (distance from the sea) with sea temperatures and cloud cover determines climate (Chapter 3).

2.4 The radiation balance and the tropical zone

As we saw in section 2.2, there is a surplus of energy from the sun within the tropics and a deficit at high latitudes. The latitude average at which the annual radiation surplus becomes a deficit (~40°) is another way to define the tropical weather zone (the one that covers the greatest range of latitude, so that the weather of the extremities of this zone is dominated by extra-tropical air masses throughout the winter. Locally, some northern or southern extremities of this zone are not within the tropical air mass at any time of the year (see Chapter 4.)

(a)

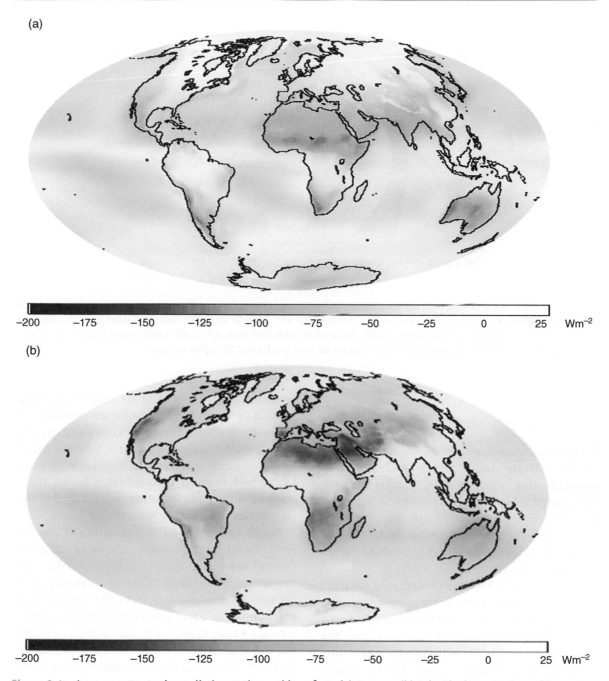

(b)

Figure 2.4   Average net outgoing radiation at the earth's surface: (a) January, (b) July. The largest values of long-wave radiation loss occur where surface temperatures are highest, such as from the tropical deserts. Cold surfaces emit small amounts of long-wave radiation and net emission is low in cloudy areas. Courtesy of NASA Surface Radiation Budget Project.

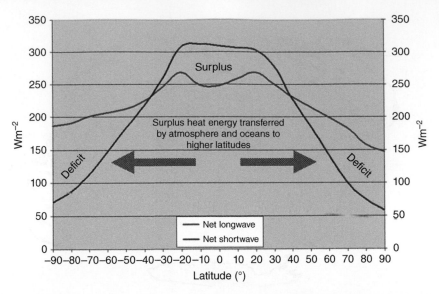

Figure 2.5   Annual zonal means of net short-wave and net long-wave radiation. The lines representing incoming and outgoing radiation do not have the same values. From 0° to 40° latitude incoming solar radiation exceeds outgoing terrestrial radiation and a surplus of energy exists. The reverse holds true from 40° to 90° latitude and these regions have a deficit of energy. As a result, energy must be transferred from the tropics to higher latitudes.

2.5   The dynamics of weather systems in the tropics

In the extra-tropics, weather systems are driven by the strong temperature and humidity gradients that lead to density differences – baroclinicity. However, the tropical air mass may be regarded as barotropic (i.e. horizontally homogeneous), so these large horizontal density differences are unable to drive large-scale weather systems.

Instead, winds and their associated weather systems are driven by vertical motions. The most important of these vertical motions is as a result of both heating by the sun and the convergence of air from the north and south. Along this line of convergence, where there is sufficient moisture, the converging air masses can ascend through much of the troposphere as a line of large convective cells close to the equator. This upward motion is most vigorous over the western Pacific Ocean and the mid-Atlantic regions. Although this is the most significant and vigorous vertical motion, other factors contribute to the ascent of air, most notably the heating of the air by solar radiation (insolation), as discussed above. This causes the air to expand so that the troposphere reaches a height maximum where there is a combination of both insolation and upward convection. Clearly, there must be some compensating motion so that the air can continue to rise from the surface.

North and south of the equator, within the zone of warm high-pressure cells, there is gradual descent as a result of the development of a 'convective ridge' and insolation near to the equator. As this air moves away from the equator, the rotation of the earth causes it to turn eastward, forming a zone of anti-cyclonic rotation throughout the troposphere, within which a balance of flow causes gradual descent.[1] Although gradual, these zones cover many thousands of square kilometres, providing the air that moves towards the equator[2] that will, once again, ascend as it converges near the equator.

The winds of the tropical zone resulting from these vertical motions and the weather systems that form where air moves towards the equator are discussed in more detail in Chapters 3, 4 and 5.

The reliance on moist convection to drive tropical weather systems is the main complexity. The individual components of moist convection are towers rarely more than a few tens of kilometres across, so they are hard to predict (Chapter 11) and their effects vary widely over small areas. Even when they form larger-scale systems (squall lines or mesoscale convective complexes), as is often the case where subtle variations in air motions and air characteristics exist, the life of these systems is short, especially compared with most mid-latitude baroclinic systems, making prediction beyond a very short (nowcasting) time scale a challenge to the weather forecaster and the atmospheric numerical modeller. These systems are described in Chapter 10.

2.6   Questions

1. What effects might there be on the radiation balance if the axis of the globe was not at an angle?
2. Why does the amount of radiation received and emitted differ in the northern and southern hemispheres?

Notes

1   This descent is enhanced by the gradual cooling of the air by radiation as it moves towards the poles.
2   Some of the air that descends through these anti-cyclones moves poleward, helping to provide some of the energy that forms the mid-latitude weather systems in zones where the mean sea-level pressure is low.

3
Winds, Temperature and Weather in the Tropical Zone

3.1 Winds

Within the tropics winds are often relatively light, in particular at upper levels. Over the Atlantic and much of the Pacific these are westerlies throughout the year. Over the western Pacific, Indian Ocean and Africa, high in the troposphere, there are easterlies close to the equator (Fig. 3.1). At low levels, trade-wind flows predominate (Fig. 3.2), originating in the sub-tropical high-pressure systems (the areas of the 'doldrums') centred near 30°N and 25°S. North-easterlies in the northern hemisphere converge near the equator with south-easterlies in the southern hemisphere, providing the additional forcing necessary for deep convection within the tropics and forming a belt of convective cloud: the inter-tropical convergence zone (ITCZ). However, significant reversals in the low-level wind flow occur over and around the tropical continents during the change from winter to summer. The effect of mountains and strong convective motion over the tropical land masses is significant. The trade winds are usually comparatively light over land covered in trees or sheltered by mountain ranges, particularly in India and Africa. However, variations in these winds produce important climatic variations. The Amazon Basin allows the trade winds ready access, in particular the north-east trades, which may cross the equator. The shelter provided by the mountains of Venezuela and the borders of Guyana, Suriname and French Guiana with Brazil brings a comparatively dry climate to much of the north of the Amazon Basin, locally extending to within about 2° of the equator. Much of this area is savanna and there is semi-desert across the border of Venezuela and Brazil.

Summer monsoon circulations have westerlies at low levels,[1] but middle- and upper-tropospheric easterlies strengthen as the troposphere warms and deepens to form an upper high over the continents, on the equatorward rim of which there is a steep temperature and height gradient. The greatest strengthening occurs as the equatorial high migrates away from the equator and deepens. Close to the equator these winds may reach jet-stream strength in a shallow layer near 15 km. This jet stream is found only about 1500 m below the tropopause. During the northern hemisphere summer these winds stretch all the way from South-East Asia to Africa's Gold Coast, although the main activity and highest speeds are generally across the southern tip of India, where speeds occasionally reach 60 m s^{-1} and there is a local minimum over eastern Africa (Atkinson 1971). A similar, but weaker, jet stream forms over New Guinea and Indonesia in response to the warming of Australia during the southern summer.

An Introduction to the Meteorology and Climate of the Tropics, First Edition. J F P Galvin.
© 2016 John Wiley & Sons, Ltd. Published 2016 by John Wiley & Sons, Ltd.

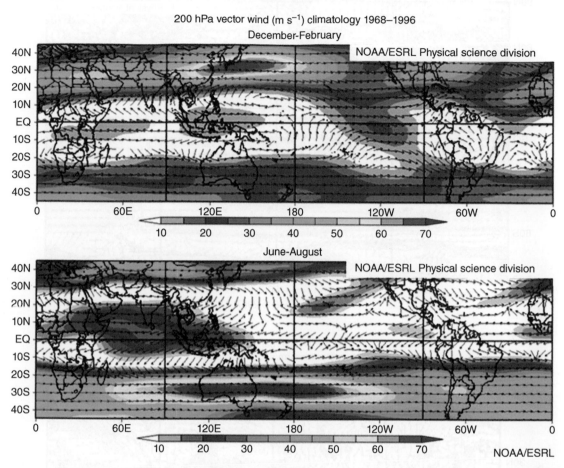

Figure 3.1 The mean wind flow in the tropical upper troposphere, near 12 km altitude. Easterly winds are normal near the equator, whilst the flow is variable, often very light 15–20° from the equator, marking a transition to the westerlies further north and south, which reach a maximum in the STJ, near 30°N and 25°S. There is considerable seasonal variability in the strength of the easterlies close to the equator in association with the monsoon circulations of southern Asia, West Africa and Australia. Courtesy of NOAA ESRL.

The easterly winds at high levels diverge north and south away from the equator.[2] North and south of 15° latitude they become westerlies (Fig. 3.1). The upper westerlies may reach jet-stream strength to form the STJ along the poleward edge of the tropical air mass, close to 30°N and 30°S, as described in section 1.2.2. The STJ is present throughout the year in the southern hemisphere, but has large speed changes between winter and summer in the northern hemisphere. In the northern winter its speed may reach 110 m s⁻¹ or more over east Asia and the western Pacific. However, its speed is rarely more than 50 m s⁻¹ in high summer with a mean closer to 25 m s⁻¹.

At medium levels near 4 km altitude, winds of moderate strength often form wave trains, known as easterly waves, which form in response to temperature differences and convective development. These have a strong association with severe weather, notably the summer mesoscale weather systems of the monsoon, including the squall lines of Africa and parts of the Pacific Ocean (Persson 2000; Leroux 2001).

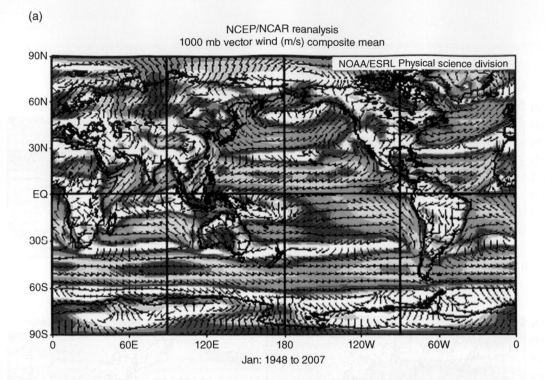

(a)

NCEP/NCAR reanalysis
1000 mb vector wind (m/s) composite mean

Jan: 1948 to 2007

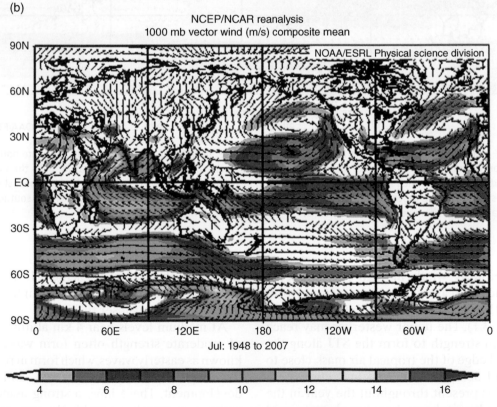

(b)

NCEP/NCAR reanalysis
1000 mb vector wind (m/s) composite mean

Jul: 1948 to 2007

Figure 3.2   The mean wind flow near the surface: (a) January 1948 to 2007 and (b) July 1948 to 2007. Easterly winds are normal near the equator, whilst, as at high levels, the flow is variable, often very light 15–20° from the equator, marking a transition to the westerlies further north and south in the middle latitudes. These mid-latitude winds show considerable variation between winter and summer, as do the flows associated with the monsoons (Chapter 8). Courtesy of NOAA ESRL.

Over the Atlantic, easterly waves are intimately associated with the development of tropical revolving storms (McIlveen 1992).

The low value of the Coriolis force close to the equator presents a difficulty in the assessment of wind speed and direction (Godbole & Shukla 1981), so streamline analysis, rather than conventional pressure analysis, is generally used in the tropics, as described in Box 3.1.

Box 3.1 The effects of latitude on geostrophic balance

Over much of the globe, wind speed and direction are the result of a balance of forces, the most significant of which are the pressure-gradient force and the Coriolis force, which are in equilibrium where the flow is straight and parallel to the equator (Holton 1979). However, this relationship breaks down close to the equator, where the Coriolis force becomes very small since the air is subject to little rotation around the earth's axis.

Nevertheless, winds remain in reasonable balance until they reach latitudes of about 6° and even then momentum usually carries the wind in the direction it was moving when in near-geostrophic balance. Bigger problems for the analyst include the distance between isobars, which become very large in the tropics, such that barometric errors (and those of correction to sea level) make standard analysis of pressure gradient, based on mean sea-level pressure (or geopotential height), almost impossible. As a result, most analysis in the tropics uses streamlines.

If we define the tropics as the area between the sub-tropical discontinuities, the tropical atmosphere is barotropic, so horizontal temperature differences are small. Furthermore, the apparent progress of the sun from east to west across the sky causes a pressure difference as the air is heated and cooled by incoming and outgoing radiation. The wavelength of this disturbance is 12 hours and its amplitude is approximately 2 hPa at the surface (Table 3.1); the amplitude is larger than in other latitude zones or changes due to the development of all but the most vigorous tropical weather systems.

Over many inland areas the use of mean sea-level pressure is also inappropriate

Table 3.1 Typical pressure changes (in hPa) due to the heating and cooling of the atmosphere in the tropics

Correction to the observed pressure necessary to allow for diurnal variation		Average component of the 3-hourly barometric change due to the diurnal variation	
	0°–10°N or S		0°–10°N or S
Local time	10°–20°N or S	Local time	10°–20°N or S
0000	−0.6 −0.5	0000–0300	+1.1 +1.0
0100	−0.1 −0.1	0100–0400	+0.9 +0.9
0200	+0.3 +0.3	0200–0500	+0.3 +0.3
0300	+0.7 +0.7	0300–0600	−0.5 −0.5
0400	+0.8 +0.8	0400–0700	−1.2 −1.1
0500	+0.6 +0.6	0500–0800	−1.5 −1.4
0600	+0.2 +0.2	0600–0900	−1.5 −1.3
0700	−0.4 −0.3	0700–1000	−1.0 −0.9
0800	−0.9 −0.8	0800–1100	−0.2 −0.4
0900	−1.3 −1.1	0900–1200	+0.7 +0.6
1000	−1.4 −1.2	1000–1300	+1.5 +1.3
1100	−1.1 −1.0	1100–1400	+1.8 +1.7
1200	−0.6 −0.5	1200–1500	+1.9 +1.6
1300	+0.1 +0.1	1300–1600	+1.4 +1.2
1400	+0.7 +0.7	1400–1700	+0.7 +0.5
1500	+1.3 +1.1	1500–1800	−0.3 −0.2
1600	+1.5 +1.3	1600–1900	−1.0 −1.0
1700	+1.4 +1.2	1700–2000	−1.5 −1.4
1800	+1.0 +0.9	1800–2100	−1.6 −1.5
1900	+0.5 +0.3	1900–2200	−1.4 −1.1
2000	−0.1 −0.2	2000–2300	−0.8 −0.6
2100	−0.6 −0.6	2100–2400	0.0 +0.1
2200	−0.9 −0.8	2200–0100	+0.8 +0.7
2300	−0.9 −0.8	2300–0200	+1.1 +1.0

since the correction of observed pressure to sea level causes errors. Thus, where geostrophic balance needs to be measured, the height of the 925 or 850 hPa surface should be used and even then, this should be on a broad scale.

Vertical motion occurs on a range of scales throughout the tropics. Much of the rainfall of the tropics is from the rapid convective ascent of moist air, but convection and rainfall are suppressed where there is large-scale descent. Whilst the effects of broad-scale descent are most evident in the desert and semi-desert lands north and south of the humid tropical zone, some areas near the equator also experience descending air for at least part of the year. This vertical motion was first described by Sir Gilbert Walker (McIlveen 1992) and his name was given to the system that is usually associated with ascending air over the western Pacific and the tropical Atlantic with corresponding descent over the western Indian Ocean and the eastern Pacific. Variations in the Walker circulation are closely linked to El Niño and the Southern Oscillation, discussed in Chapters 4 and 12.

Combined with variations in orography over land, and differences between areas of land and sea, the Walker circulations help to form distinct areas of weather associated with variations in surface wind regimes. This is particularly notable in East Africa, where surface winds are associated with a descending limb of the Walker circulation. Easterly winds at the surface are blocked in their westward progress by the high ground around the East African Rift Valley, which reaches 2–3 km, with some peaks reaching 4.5 km (Fig. 1.3b indicates the significance of these altitudes). It is thus separate from the wind system to the west of these highlands and from the monsoon system of western Africa (section 8.8). The line dividing the eastern and western air masses at the surface is called the Congo air boundary.

Named winds of the tropics are given in Appendix 8.

3.2   Temperature

Day-to-day, temperature varies comparatively little in much of the tropical zone.[3] Over the oceans, the temperature of the sea surface largely determines temperature (although it is rarely identical). Over land, where temperature rises and falls significantly through the day, there is more variation in maximum and minimum temperature with season than between one day and the next. As the zone has a single air mass (except where troughs extend equatorward in winter), there is very little advection and maxima are determined largely by the mean temperature of the lower troposphere (Callen & Prescott 1982). Minima vary somewhat more, especially in coastal areas or close to high ground. As the condensation of moisture slows the fall of temperature, moist air cools much less than dry air, so minimum temperature varies according to whether the wind is onshore or offshore. In windy zones, mixing and entrainment minimize the rise and fall of temperature, although there is some variation and as temperatures fall a little so wind speed falls, allowing the temperature to fall a little more, although the diurnal variation in windy weather is rarely more than about 5°C.

Where cloud cover changes significantly with season, the temperature record reflects this. For instance, in the areas dominated by monsoons, maxima are lower and minima higher in the cloudy weather of summer than they are in the periods before and after the wet season. However, this is mainly a matter of accurate prediction of the arrival and clearance of the moist air, maxima and minima varying little day by day in the moist air mass or in the dry air.

Over mountains, the mean temperature shows rather more variability than at lower levels, especially in areas dominated by anti-cyclones. Variations of the height of the subsidence inversion of areas dominated by anti-cyclones can advect air of very different temperature onto the upper slopes, daily maxima and minima possibly changing as much as 10°C from one day to the next.

The greatest differences in temperature are with altitude. Mean temperature falls rapidly with height, the rate greatest by day close to the land surface, affected by both convection and wind speed. Strong winds carry heat away from the surface, tending to produce a lapse of temperature with height of about 10°C km^{-1}. However, above the near-surface layer, the lapse rate is lower; the effects of the surface becoming relatively small above about 1000–2000 m. By night, as the layer in contact with the ground cools, wind speeds decrease, convective entrainment decreases and the lapse rate with altitude falls (especially in the lowest layer of atmosphere, where it may be reversed). However, this tends to reduce the diurnal range of temperature on high ground, except where there is shelter, such as in a mountain valley. These sheltered locations may see the largest temperature ranges on earth, particularly if the air is dry and cooling is favoured by the development of drainage winds (Oke 1987).

3.3 The weather patterns and climates of the tropics

The tropical region experiences only gradual changes in weather patterns and variations are generally small, even between seasons. The main changes are between dry and wet seasons, marked by (i) the northward and southward movement of the ITCZ in the central (equatorial) portion and (ii) the winter incursion of cooler air at altitude near the poleward extremes. Even with this movement, there is almost no seasonal weather fluctuation within about 5–10° latitude of the mean position of the ITCZ. However, since most of the world's hot deserts have their equatorial flank within 20° of the equator, marked by transition to savanna vegetation (e.g. the Sahel of West Africa), seasonal variations can be large close to these latitudes.

Despite the relatively constant (or seasonal) weather types of the tropical zone, variability is an important factor across all of its climates. Annual and seasonal rainfall varies

both spatially and temporally, as discussed in Chapters 7, 8, 10 and 12. These variations are often related to local temperature changes.

As can be seen in Fig. 1.1, the predominant climates of the tropics are dry ones: the hot deserts, semi-desert scrub and expansive savannas. These lie towards the periphery of the tropical zone in regions where anti-cyclonic subsidence predominates and rainfall is either seasonal or ephemeral. The hot deserts are noted for a high diurnal range of temperature. A maximum of more than 40°C in places by mid afternoon may fall as much as 30°C by morning. This range is solely due to the effects of a dry atmosphere with little or no cloud and such a low vapour pressure that diurnal temperatures can vary greatly.

Over continental areas the periphery of the humid zone is dominated by monsoon wind regimes. These bring wet humid weather in summer and predominantly dry weather in winter. Characteristically, there is a seasonal reversal of wind at low levels. In the northern summer, south-easterly winds cross the equator and recurve to become south-westerlies (north-easterlies become north-westerlies in the southern hemisphere), bringing moist oceanic air across the equator into areas that are under the influence of dry continental easterly winds during the winter (Verbickas 1998; Fedorov 2002). West Africa, southern Asia and northern Australia all experience these monsoon reversals. Although still seasonal, the situation is more complex over the Amazon basin and Caribbean, where moist westerlies cannot become established due to the Andes–Sierra Madre mountain barriers. Thus the motion of convection is dependent on more complex changes in the atmospheric circulation and the influence of the Caribbean Sea, which warms and cools more than the neighbouring Atlantic Ocean.

Over the warmest ocean areas the tropics are characterized by the development of tropical revolving storms associated with strong lower-tropospheric winds and (perhaps more importantly) heavy rainfall.

Although a relatively narrow zone[4] (no more than about 1500 km wide, except in South-East Asia), it is the humid equatorial zone that many associate with the tropics. In this zone rainfall can be relied on year round, as the ITCZ is never far away. Maxima are generally limited to around 35°C[5] over land by the high water-vapour content of the air and minima are similarly restricted by cloudiness or the overnight formation of dew, mist or fog. Over the sea, temperature changes little day by day. Thus, throughout the tropics the diurnal range of temperature is relatively small. Nevertheless, this climate is uncomfortable for most humans, who find it difficult to lose excess body heat in these conditions. Indeed, this energy-sapping weather is used as a test by élite British army and Royal Marines regiments, which carry out part of their training in Brunei.

Where seasonality is the main effect on rainfall, savannas predominate. In these areas of extensive grassland, annual evaporation exceeds precipitation and trees grow only in stunted groves. These areas are home to relatively large populations in some parts of the world and agriculture is critically dependent on the summer rains (both locally and to re-charge river flows), so any reduction or failure of seasonal rainfall often causes notable famines, especially in recent years. Some areas that have some seasonality, but a predominantly maritime climate, such as the northern Caribbean, have an intermediate climate with extensive forest, as well as grassland. Elsewhere, steppe surrounds the arid deserts. Here few plants can grow, but there is sufficient rainfall or run-off to support agriculture and moderate-sized populations. Included in this climatic zone are the highlands of much of Arabia.

However, within each climatic zone there are important variations due to orography, latitude and longitude. Some of the world's hot deserts receive most of their (meagre) rainfall in summer, others in winter – here the variation is mainly by latitude and altitude, with equatorward regions having a summer rainfall peak. Examination of the mean annual rainfall in the tropics reveals that there are significant differences on a broad scale within climatic zones, as shown in Fig. 3.3. In general, the desert areas generally see less than 200 mm yr⁻¹,

mm	
Over 2000	
1000–2000	
500–1000	
250–500	
Under 250	

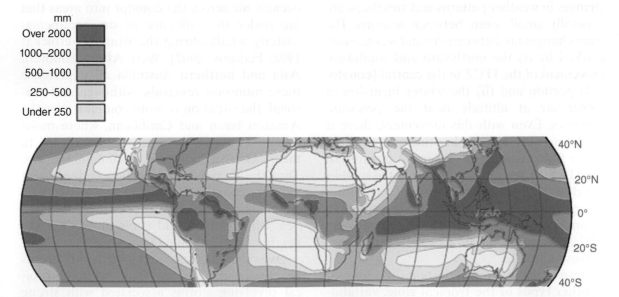

Figure 3.3 Mean annual rainfall (mm) in the tropics. Local orographic effects are not evident at this scale, although a rainfall maximum in excess of 2000 mm year⁻¹ is shown over the southern Himalayas.

although Australia's dry interior is defined by annual rainfall less than 600 mm. The monsoon zones see between about 1000 and 2500 mm yr^{-1}, occurring in summer. The semi-deserts and oceanic areas under the influence of the sub-tropical anti-cyclones have a total rainfall between 200 and about 1000 mm yr^{-1}. Areas under the influence of upper-tropospheric troughs (see Chapters 4 and 10) see about 1000 mm yr^{-1} and the ITCZ experiences totals above 1000 mm yr^{-1}, with some areas seeing more than 3000 mm yr^{-1}. Orographic effects add further local detail.

High ground has two main effects: it lowers the mean temperature (although nights may be less cool if the mountains are modest and the observation is not in a valley) and increases the rainfall, increasing the likelihood of precipitation by 'forced' convection or convergence. For instance, copious rainfall is generated by the Ethiopian Highlands and India's Western Ghats. These highlands have a relatively equable climate within the almost universally hot tropical zone and their climate is markedly different from that of their surroundings.

Along the western margins of Africa and the Americas there are cool currents as a result of the upwelling of cool deep water under the influence of the trade winds. The effect is most marked along the south-west coast of South America, as well as the north-western and south-western coasts of Africa. This brings a surprisingly cool and usually dry climate north to only a few degrees south of the equator, although this climatic zone, comparatively

equable for the population, is narrow, extending only a few tens of kilometres inland. Populations are comparatively high in these zones, largely due to the excellent fishing available as a result of the upwelling of plankton-rich deep waters.

Heating and cooling have a significant effect on air pressure throughout the tropics, the pressure falling in response to diurnal heating and rising in response to nocturnal cooling. As pressure rarely changes due to synoptic-scale weather systems in the tropics, the diurnal changes are significant, as shown in Table 3.1. Changes that differ from these tend to indicate the motion of weather systems, so that large falls of pressure overnight, or even small rises (~+0.2 hPa) in the middle of the afternoon, for instance, are likely to indicate weather deterioration.

3.4 Clouds and fog in the tropics

The difference between the high latitudes and the tropics is related to the great depth of the tropical troposphere. Convective clouds are predominant in the tropics, whilst the amorphous altostratus and nimbostratus genera, formed by gradual mass ascent, are rare. However, the increase in the depth of the troposphere means that the range of height at which each species is seen is greater than at higher latitudes. The WMO recommends the range of cloud height given in Table 3.2 (WMO 1956). Appendix 3 describes the nine cloud genera and their composition, as seen in the tropics.

Table 3.2 The variation of range of cloud-base heights (in m) in tropical, middle latitude and polar air masses

Étage	Polar regions	Middle latitudes	Tropical regions
High (Ci, Cs, Cc)	3000–8000	5000–13000	6000–18000
Middle (Ac, As, Ns*)	2000–4000	2000–7000	2000–8000
Low (St, Cb, Cu, Sc)	0–2000	0–2000	0–2000

* Nimbostratus frequently has a base that extends down into the low-cloud étage and a top that is frequently well into the high-cloud étage.

All stratocumulus clouds[6] in the tropics are at a temperature above 0°C and it is altocumulus and altostratus (as well as occasional nimbostratus) that occupy the level of transition to supercooled water. The tops of the deepest medium-level clouds may contain a significant amount of ice as the temperature of the cloud falls below about −20°C. This temperature occurs at an altitude between 8 km and, locally, 9 km. Clouds formed predominantly of ice are not found below 8 km in the tropical air mass.

Cumulonimbus clouds always have tops extending into the middle levels and frequently into the high-cloud étage. This is the usual state in the tropics, where cumulonimbus clouds rarely have tops below 12 km. In hot-desert regions, where there is sufficient moisture, cumulus or, if there is sufficient moisture at high levels, cumulonimbus clouds may form around the time of maximum temperature.

Precipitation from cumulonimbus clouds is usually heavy. These clouds, resulting from air with a high humidity mixing ratio ($r \geq 20$ g kg^{-1} in the ITCZ), can produce large amounts of precipitable water and are the source of thunderstorms in the ITCZ. Thunderstorms are frequently seen in association with the deep convection of the tropics. However, they are more common near the poleward edge of the ITCZ, where instability and convective available potential energy (CAPE) values (measures of the ability of buoyant pockets of air to rise) are larger than nearer to the equator, especially over the continents (see Fig. 3.4). Despite the presence of cumulonimbus, which often produces hail, much of the tropics is simply too warm for all but very large hail to reach the ground. Nevertheless, it occurs in places where wind shear (the change of wind velocity with height) and vorticity (a measure of the rate of rotation of the air, proportional to its rate of ascent) are conducive. Waterspouts and tornadoes may also form, but only where there is sufficient (positive) vorticity.

From towering cumulus clouds, precipitation is generally slight to moderate. Nevertheless, this rainfall (or, over high mountains, snowfall) locally forms the majority of annual precipitation.

Stratus and fog are characteristic of parts of the tropics. Stratus with hill fog is frequently seen in the early morning over tropical woodland, in particular the tropical rainforest, following overnight cooling. Here, transpiration from the trees in a very moist environment assists its formation.

Fog and stratus are also features of cooler parts of the tropical oceans and are frequently seen along the west-facing coasts of northwestern Africa, southern Africa and South America. In these areas much of the meagre precipitation available for plants to grow is deposited from wet fogs.

Early-morning fog is relatively common on the coasts of the Arabian Gulf, where sea breezes frequently bring moist air inland during the afternoon, humidifying the air and raising its fog point. Extreme summer cases (observed in Doha and Bahrain), when sea-surface temperatures in the Gulf may reach 35°C, may bring fog with visibility less than 100 m around dawn with air temperatures and dew points above 30°C. These fogs feel very moist, almost suffocating, making it difficult for humans and animals to lose excess heat. As such, they can be dangerous to life in a very different way from fogs in the middle latitudes.

Layer clouds in the tropics are usually the product of convection and are often relatively thick as a result. Most of these clouds of convective origin are of genera altocumulus or stratocumulus. However, there are some exceptions and altostratus or nimbostratus sometimes form, usually in association with cyclonic disturbances or where air masses converge over high mountain ranges. These will be discussed later in the series. Hours of steady precipitation may fall from these deep layer clouds.

Altocumulus castellanus (or floccus) clouds are also characteristic of tropical air. Principally, they form in response to high-level cooling. This occurs in two ways. First, there is the dynamical cooling associated with advection, often ahead of upper troughs. This is usually

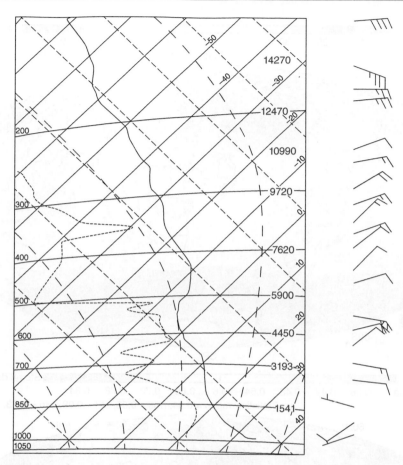

Figure 3.4 Radiosonde profile for Niamey, Niger, in Africa's Sahel at 1200 UTC on 30 July 2006.[7] This ascent is absolutely unstable and has a convective cloud base near 850 hPa, in contrast to ascents typical of the ITCZ. It shows a typically high level of CAPE (1273 J kg^{-1}) and θ_w falling with height. (Appendix 4 includes a brief introduction to temperature–entropy diagrams.)

evident above the 700 hPa level (near 3 km). Second, there is the long-wave cooling of the atmosphere during the evening, causing a decrease in stability at the cloud tops. Castellanus development may prolong the life of cumulonimbus through the night. In some cases the destabilization is associated with instability at lower levels, which feeds into the unstable medium-level clouds.

Any unstable cloud with a top that becomes glaciated is reclassified as cumulonimbus. Thus the base of these cumulonimbus clouds, originally altocumulus castellanus, may be at 4000–5000 m or more. Most thunderstorms that develop over the hot deserts of the tropics have

such high bases. Precipitation from these high-based cumulonimbus clouds usually evaporates before reaching the ground and strong down-draughts can be the result. Figure 3.5(a) shows the mean cloud cover of the tropical zone and Figs 3.5(b)–(d) its effects on the radiation budget, assessed as part of the International Satellite Cloud Climatology Project (ISCCP). The lack of cloud over the tropical deserts (<50%) is particularly notable, as is the persistence of cloud along the ITCZ (>60%, even over land, despite its migration in the monsoon zones and its variability through the day). Many areas of high ground are also cloudy, even those of the dry tropics, although this is not apparent

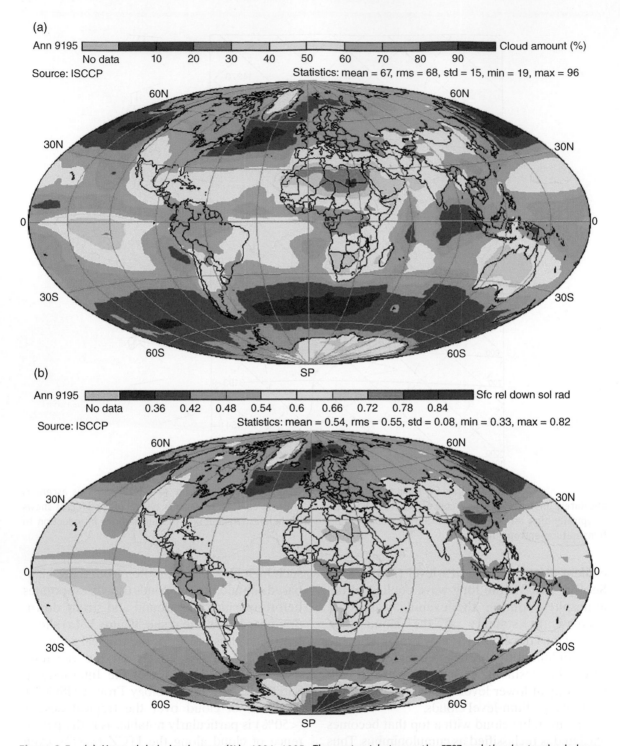

(a)

(b)

Figure 3.5 (a) Mean global cloud cover (%), 1991–1995. The contrast between the ITCZ and the dry tropics is large. The mean average cloud cover of the monsoon zones is also notable. (b) The ratio of solar radiation received at the surface to that at the top of the atmosphere. These values of the effective transmittance of the atmosphere provide an impression of the effects of cloud. A notable effect is that of high ground, where a much greater proportion of radiation reaches the surface.

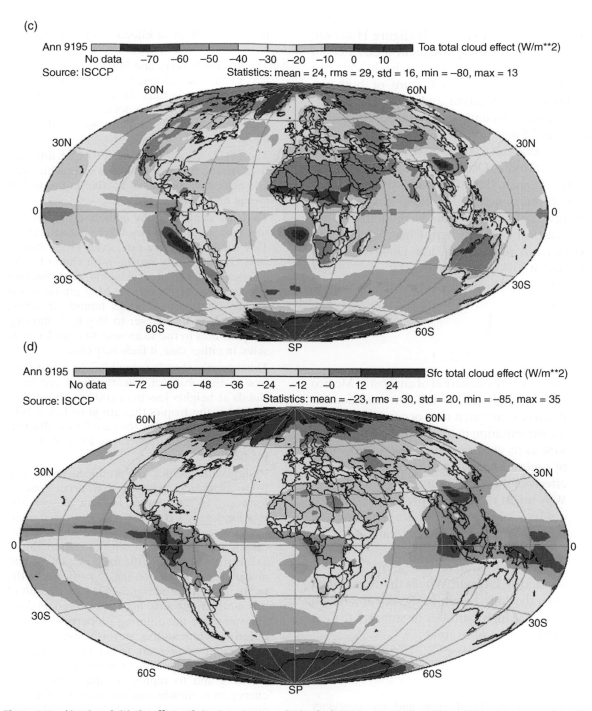

(c)

Ann 9195 | Toa total cloud effect (W/m**2)
No data −70 −60 −50 −40 −30 −20 −10   0   10
Source: ISCCP   Statistics: mean = 24, rms = 29, std = 16, min = −80, max = 13

(d)

Ann 9195 | Sfc total cloud effect (W/m**2)
No data −72 −60 −48 −36 −24 −12 −0   12   24
Source: ISCCP   Statistics: mean = −23, rms = 30, std = 20, min = −85, max = 35

Figure 3.5 (*Continued*) (c) The effects of cloud on the net radiation budget (cloudy–clear) at the top of the atmosphere (W m^{-2}). Negative values predominate, showing that in general clouds reduce the net radiation and so cool the planet. Only over the tropical deserts is the effect near zero and there is a large overall cooling effect in the cloudy zones of around 40 W m^{-2}. Note the strong effect of cloud over southern China, which reduces incoming radiation by more than 60 W m^{-2}. This is discussed in section 8.5. However, in some parts of the tropics the effect is positive. (d) The effects of cloud on the net radiation budget (cloudy–clear) at the surface (W m^{-2}). The effects are similar to those at the top of the atmosphere. These data are from satellite observations collected as part of the ISCCP (from Rashke et al. 2005). rms, root mean square; std, standard deviation.

in most areas at the scale of this figure. However, there is a large semi-permanent area of cloud over much of China and this is clearly seen in Fig. 3.5(a). This is a feature of both the Asian winter monsoon and a tropical upper-tropospheric trough, as discussed further in sections 8.5 and 11.4.

The cloudiest tropical zone stretches across the central Indian Ocean, Indonesia and Malaysia to New Guinea. Here the cloud is a result of convection and much of the cloud cover is high. However, gloomy conditions predominate off the coasts of Peru, northern Chile, Namibia and Angola, where extensive low cloud covers an average of 70% or more of the sky.

Cloud types and species are given in Appendix 3.

3.5 Questions

1. Examine the likelihood of air frost in Mexico City (altitude ~2240 m) and consider the times of year that it may occur. (Hint: examine the variation in altitude across the city as well as its elevation, the general orography of central Mexico and the possible climatic effects of these variations.)
2. Why might radiosonde profiles or synoptic observations for a single site sometimes be unrepresentative of the general state of the troposphere?
3. Revisit the section on the variation of insolation with latitude. Why is there less effect from the sun's height in the sky and length of day at high latitudes in summer than at low latitudes in winter?

Notes

1 This is a simplified view and the necessary complexities of the summer monsoon flows here are discussed further in Chapters 5 and 8.
2 Wind speeds in this book are quoted in metres per second. However, wind speeds are most often observed or forecast in knots (sometimes in miles per hour or kilometres per hour): 1 m s^{-1} = 2 kn.
3 The months January and July are those of mean maximum and minimum temperature across most of the world since the rise and fall of temperature follows the sun north and south of the equator through the seasons with a delay of about 3 weeks.
4 The ITCZ is sometimes seen to split into two branches, each of which may be several hundred kilometres across. This is notable in the Indian Ocean and the western Pacific.
5 The 'rule' explaining the limitation of temperature assumes a boundary layer near saturation. This is comparatively rare, even in the humid zone. The humidity mixing ratio (hmr) locally reaches 20 g kg^{-1} in the humid zone, limiting temperatures. However, the May pre-monsoon season, for instance, brings humid air across India with an hmr closer to 16 g kg^{-1}, allowing temperatures to rise to around 40°C under clear skies. In either case, it feels very close!
6 This assumes that the cloud height is not relative to the observer. Clearly there are layer clouds at heights less than 2000 m above high ground in the tropics that are at or below 0°C. Typically, this would be the case for an observation made at an altitude above 2500–4000 m in the tropics.
7 Radiosonde profiles are produced using data from ascending balloons that relate energy (entropy) to temperature. As can be seen, the relationship is a complex one. There is a number of ways to portray the relationships; in this case a tephigram is shown. The temperature axis is rotated 45° and the fixed rate of dry adiabatic warming or cooling (Γ) intersects the temperature lines at right angles. As a result of this method, pressure lines are curved, making allowance for the differing properties of air at differing temperatures; nevertheless the basic relationship of pressure to height is preserved. In this way, it is possible to consider equal areas of the graph to represent equal amounts of energy, thus allowing the effect of the displacement or warming of air to be considered easily. Using the fourth axis, hmr, it is possible to consider the effects of displacement of parcels of air and the probability of condensation due to cooling or evaporation due to warming. This is discussed further in Appendix 4.

4
The Subtropical Jet Streams

4.1 The formation of jet streams at the margins of the tropics

One of the most notable aspects of the tropics is the strength of its temperature contrasted with the relative cool of the extra-tropics. The Hadley cells redistribute heat from near the equator to higher latitudes. There is, however, a marked discontinuity in the mean temperature of the troposphere close to 30°N and 25°S. The contrast generates a thermal wind,[1] the strength of which is proportional to the gradient of temperature difference between the tropical and extra-tropical air. For much of the year, this difference is sufficient to generate winds of jet-stream strength (>40 m s^{-1}).

Although it is formed by temperature contrasts and thus has some similarities to the polar-front jet stream (PFJ), the STJ also has many differences from its high-latitude cousin. First, waves in the flow usually have a small amplitude, so that in most areas the core of the jet stream remains within about 10° north and south of its mean latitude (Fig. 4.1). Second, the jet stream is usually associated with subsidence and the subtropical high-pressure belt, lying slightly to poleward of the line of highest pressures. Because of the subsidence, weather systems do not generally accompany the STJ and it is only where there is an incursion of cool air from higher latitudes and associated cyclonic curvature in the flow that weather systems can develop (Meteorological Office, 1994). These weather systems are discussed in section 4.2.

Like the PFJ, the STJ inhabits air on the warm side of the temperature discontinuity (i.e. the tropical air), but lies above cooler air near the surface, since there is a slope in the discontinuity between the contrasting tropical and sub-tropical air masses (Fig. 4.2) such that tropical air at the surface usually lies 600–1000 km equatorward of the jet-stream core. In both hemispheres, the core of the STJ is found close to the 200 hPa pressure level (near 12.5 km), with a height range between about 11 and 14 km (see also Fig. 3.1). The tropopause is at a height of 15–16 km at the poleward edge of the tropics, so the jet core lies between 1 and 4 km below the tropical tropopause. Wind speeds of jet-stream strength do not extend more than about 1000 m into the stratosphere, but may extend well below the core. At times, this may be 7 km or more, particularly in the southern hemisphere, where the PFJ and STJ can be found very close to one another during the winter months, increasing winds in the middle troposphere. In the northern hemisphere, the STJ reaches maximum speeds over east Asia during the winter. Here, speeds of 100 m s^{-1} are relatively common and maxima of 125 m s^{-1}

An Introduction to the Meteorology and Climate of the Tropics, First Edition. J F P Galvin.
© 2016 John Wiley & Sons, Ltd. Published 2016 by John Wiley & Sons, Ltd.

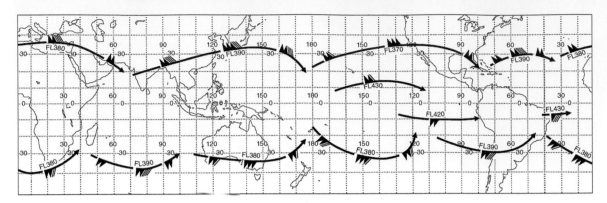

Figure 4.1   Typical wintertime flow of the STJ (both hemispheres). The core of the flow is shown with altitudes in hundreds of feet (International Standard Atmosphere) given opposite the fleches. Fleches show 50 kn (25 m s⁻¹) as pennants and 10 kn (5 m s⁻¹) as bars, with the head and tail of the flow at 80 kn (40 m s⁻¹). The position of upper-tropospheric troughs can be seen clearly.

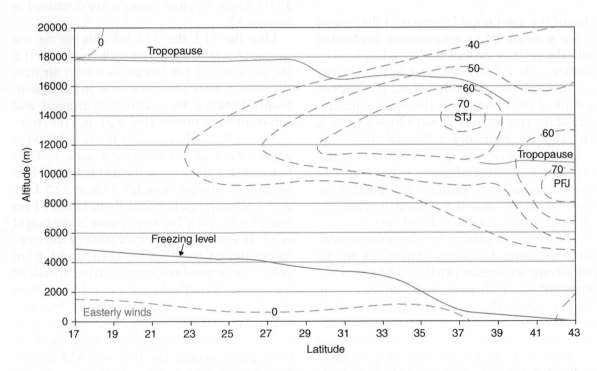

Figure 4.2   Cross-section of the northern hemisphere STJ at 1200 UTC on 15 February 2006 over east Asia, based on radiosonde observations over east China, Korea and Japan. Westerly wind-speed contours (— — — —) are shown in m s⁻¹. The core speed is close to 75 m s⁻¹, associated with the axis of the subtropical high-pressure belt at 37.5°N. To the south of the core the highest speeds of the jet stream are at a lower level than the core. In this case tropical air at low levels lies a long way south of the jet-stream core, south of about 26°N, and this is probably the reason for the lower tropopause to the north of 28°N. The shallow trade winds are also evident below about 1500 m. At 42.5°N the PFJ can be seen with a core speed of 75 m s⁻¹, lying below and to the north of the STJ. The polar stratospheric jet stream can be seen above the PFJ, extending above 20 km.

have been observed at times close to Japan in response to the cooling of the Asian landmass.

In the southern hemisphere, where the STJ is more or less continuous around the globe throughout the year, speeds are generally more modest, but can reach 100 m s^{-1} or more when there is a large temperature gradient between the tropics and the southern polar region.

The STJ is often a broad flow and the contour of wind speeds greater than 40 m s^{-1} (the definition of jet-stream speed used for high-level aviation) often spans about 10° (1100 km) equatorward and 5° (550 km) poleward of the core (Fig. 4.2). This is much broader than the flow of the PFJ and means that a very large volume of the upper atmosphere is affected by these strong winds.

In both hemispheres, the STJ weakens during the summer as the temperature contrast between the tropics and high latitudes decreases. The northern hemisphere STJ becomes discontinuous with maximum speeds rarely more than 60 m s^{-1}.

4.2 Weather associated with the subtropical jet stream

Although the waves in the flow of the STJ have relatively small amplitude, there are semi-permanent troughs in the flow (Krishnamurti 1961; Leroux 2001). These are located close to 70°E, 170°E, 90°W and 20°W in the northern hemisphere, and 40°E, 120°E, 170°W, 100°W and 30°W in the southern hemisphere, with a variation of about ±20°. The orientation of the troughs is skewed and their longitude decreases towards the equator. The eastward motion of these troughs is slow, and at times they may be stationary. However, small changes of position may be linked to small-scale features within the STJ, as well as wind-speed changes, and these may be sufficient to develop areas of severe weather.

Another trough is frequently seen in the northern summer flow near 120°E, while the flow of the STJ is north of the Himalayas. This brings long periods of cloudy skies and local heavy rain from deep convection along the eastern coast of China, as well as Korea and Japan. The system is most active in spring and early summer (Suzuki & Hoskins 2005), when these rains are known as the cherry-blossom 'baiu' in Japan and the plum rains in China (Ding & Hu 1988), since they begin when fruit trees are in flower.[2] The baiu brings a welcome relief from the mainly dry weather of the winter (Tomita et al. 2003). In autumn, as the trough develops, cloudy skies bring some relief from the heat of summer. In both cases it is humid. A notable rainfall event associated with this trough made for exciting driving at the FIA Formula 1 Grand Prix in Shanghai on 1 October 2006 (Fig. 4.3).

Within these upper-tropospheric troughs (discussed further in section 11.4), the flow is ageostrophic (i.e. there is a component of flow out of balance with the pressure gradient force) and the cross-contour (confluent) flow ahead of the trough provides dynamical uplift (Meteorological Office 1997, Ch. 8). Combined with a convergent flow at low levels, this can provide the additional forcing required for the development of deep cumulonimbus clouds in the area ahead of the troughs (Fig. 4.4). In these zones there is a weakening of the subtropical high and there may be cyclonic curvature (occasionally, a low-pressure system) in a westerly flow in the lower troposphere. Thus frontal zones may develop and although their character is more convective than most mid-latitude fronts, their layer clouds mainly the product of the spreading out of convective clouds. These cloud systems are near-stationary, although cloud cells usually move gradually in the direction of high-level winds.

In the southern hemisphere land areas typically affected by these troughs and frontal disturbances include the southern coast of South Africa, the Solomon Islands, southern Australia, New Zealand, islands of the south-east Pacific and the east coast of Brazil. Japan, island nations in the north-west Pacific, the Canary Islands, Morocco

Figure 4.3 The wet start to the Chinese Grand Prix in Shanghai on 1 October 2006. © Renault F1 team 2006.

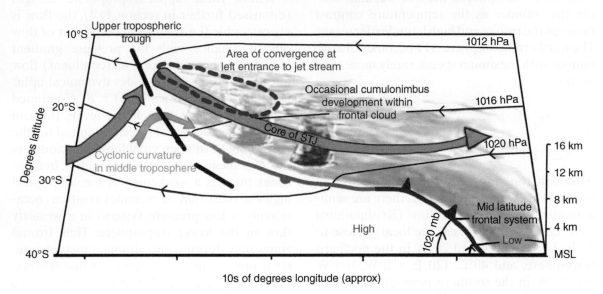

Figure 4.4 In the southern hemisphere convergence at the left entrance to the STJ, with mid-and upper-tropospheric cyclonic vorticity, associated with a tropical upper-tropospheric trough, provides the forcing to release convective potential energy. Low-level (easterly) winds (←) are warm and moist. Note: Only the core of the STJ is indicated; westerly winds are present above about 4 km.

and, occasionally, the Sahara are affected in the northern hemisphere (Fig. 4.1). It is notable that the trough over south-west Asia, which is present only during the winter, produces little severe weather, since anti-cyclonic subsidence is prevalent in the lower troposphere. It is mobile mid-tropospheric troughs that are the main cause of severe weather outbreaks here. These are described in Chapter 7.

The severe weather described above occurs mainly during the winter months, in particular early in the season, when cool air aloft moves equatorward over warm oceans.

As the northern summer begins, the STJ makes its greatest change of track over Asia. First the semi-permanent trough over southwest Asia moves east and fills, then it moves to the north of the Himalayan mountain barrier and, for much of the summer, until the autumn seasonal reversal, it is found near 40°N across Asia, thus encompassing all of the Tibetan plateau.

This has an interesting consequence for Himalayan mountaineers. Much of the mountain chain is under the influence of the STJ in winter, although the core of the jet stream lies to the south of the highest peaks. Over the main Himalayan ranges and the Hindu Kush, climbing is frequently perilous and may be impossible. Indeed, wind speeds may reach 50 m s^{-1} or more at the level of Mount Everest. However, during the seasonal reversals wind speeds fall and the best times for climbing are in May and October. In May, temperatures are rising and the STJ moves away north; in October some summer warmth remains before the high-level westerlies return.

It might be thought that there would be suitable weather on the Himalayas throughout the summer season. However, once the STJ moves well to the north, the warmth of the Tibetan plateau forms an upper-tropospheric high-pressure area; on the southern (Himalayan) flank of this there are strong easterlies for much of the time. The south-west monsoon also brings poor weather when the foothills of the highest mountains in the world are under the influence of relatively warm tropical air, so despite slightly higher temperatures, monsoon-related convection also brings unsuitable weather! The warmth of the tropical flow brings another risk: that of avalanche. Warming, in the order of 10°C, occurs as tropical air spreads onto the southern flanks of the mountains, temperatures rising over a matter of weeks to melt lying snow (which may lie at altitudes below 3000 m at the end of the winter). Once the snow and main avalanche risk have gone, there is a lower risk to mountaineers.

The development of troughs in the STJ is also linked to the development of low-pressure systems close to the equator. Pressure tends to fall on the eastern equatorward flank of a jet stream accelerating away from the equator, causing surface convergence and ascent. Where sea-surface temperatures are sufficiently high and other factors conducive, some of these depressions may form into tropical revolving storms. These are discussed in Chapter 9.

4.3 Folds and bifurcations in the flow

Although most of the waving flow of the STJ has only small amplitude, the situation over the Pacific Ocean, where the warmest air is usually confined to a narrow band near the equator, frequently causes the STJ to split into two widely separated streams. This development is most marked in the winter.

When this bifurcation occurs, one stream is found close to 15°N or 15°S, its core near 13 km, and the other (faster) stream near 40°N or 35°S, its core altitude around 11 km. Where the flow splits, a Z-shaped fold forms in the northern hemisphere flow and an S-shaped fold forms in the southern hemisphere flow. These folds are found at 10–30° latitude, close to the 180th meridian. Although this is the most common area for folds to form in the STJ, they sometimes occur elsewhere. Over land in particular, cooler air aloft can enhance deep convection as the poleward side of the jet stream is brought unusually close to the equator, over warm moist air in the boundary layer.

The equatorward branch of the flow becomes part of the Pacific cell of the zonal (Walker) circulation described by Newell (1979). This flow extends downwards into the middle troposphere, although it rarely reaches the surface, where the predominant flow remains from the east. The entrainment of moisture from the ITCZ allows cloud to develop, particularly in the northern hemisphere, where the bifurcated flow moves poleward. The development of this cloud brings significant rainfall to the Pacific coasts of northern Mexico and California in a system known as the 'pineapple express'.

Except where monsoon flows develop, upper tropospheric winds are westerly above the equator and form a series of cells over South America, the Indian Ocean and the central Pacific. Each is spurred by temperature differences in the tropical oceans.

Although the flow is relatively modest in the contorted flow close to the equator (usually no more than 50 m s^{-1}), there is often considerable deceleration of the flow into the fold, which is associated with curvature and divergence, as well as convergence into the equatorward branch (Bysouth 2000). (This is particularly true in the northern hemisphere, where the strongest flow of the STJ is usually found close to Japan.) The result can be moderate (occasionally severe) turbulence (Fig. 4.5), which almost always occurs in clear air. The development of one of these troughs was a particularly notable cause in an investigation by Turner and Bysouth (1999). Close to the equator, the STJ is also only in weak geostrophic balance, so small-scale confluent and diffluent flow may be expected, resulting in overturning and local pockets of light to moderate clear-air turbulence (CAT).

4.4   Clear-air turbulence

Clear-air turbulence is caused by changes of wind speed in the horizontal and vertical. The factors controlling turbulence are complex and include speed, momentum, shear, deformation (related to anti-cyclonic curvature), convergence and temperature (density) changes (Ellrod & Knapp 1992). CAT is usually observed in association with high wind speeds and a high rate of shear.[3] It may affect flights at high levels and is a major cause of discomfort and disruption to passengers in flight. It is rarely significant in the tropics, but does occur at times in association with the STJ (Roach & Bysouth 2002). Table 4.1 shows the percentage

probabilities of a flight encountering turbulence for every 100 km of flight within the tropics. It is usually accepted that moderate or, occasionally, severe turbulence should be forecast where the indicated probability of CAT is greater than about 6×10^{-7} m^{-1}.

Although the likelihood of CAT is greatest where momentum is greatest – close to the jet stream core – the greatest wind shear usually occurs near the top of the jet stream, around wind speeds of 40 m s^{-1}. Thus we can see that shear alone is insufficient in calculations of the likelihood of turbulence.

Given the high altitude of the STJ, air density is relatively low, so the effects of acceleration and deceleration are relatively modest (Meteorological Office 1994). CAT is most likely close to the core of the jet stream and below it, on the cold side. However, the effects of turbulence may also be seen on the warm side of the STJ, in which case the turbulent motion is most likely above the level of the jet-stream core and may extend hundreds of kilometres into the tropical air (Galvin et al. 2011).

As the ageostrophic component of the flow in the STJ is usually small and accelerations are gradual, its flow is not generally turbulent. However, the horizontal wind shear on the poleward edge of the flow is often large and, especially where there is anti-cyclonic curvature, moderate turbulence may occur (Asnani 1993). This is almost always in clear air, since the STJ is only associated with weather systems in limited areas. Tests have revealed that the turbulence indicator devised by Brown (1973) is a good predictor of turbulence along the STJ. This index does not include a calculation of small-scale vertical motion, but this is a benefit in the tropical area, which is characterized by convective ascent, even though some element of turbulent motion may be omitted.

The depth of the flow of the STJ tends to bring its lower part into contact with several mountain ranges, notably the Andes in the southern hemisphere, and the Rockies and the Himalayas in the northern hemisphere. This interaction generates CAT close to the level of

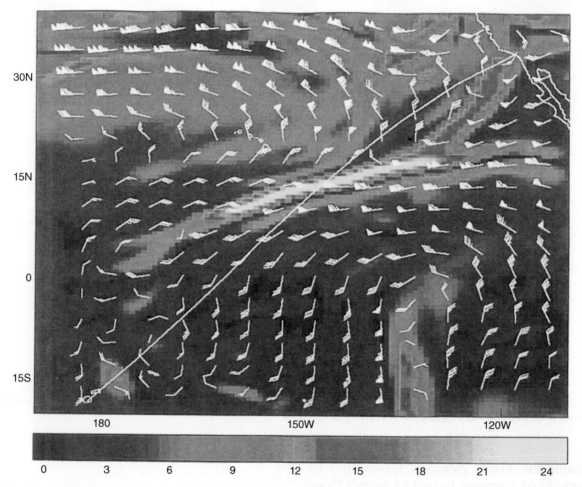

Figure 4.5   Wind speed and direction at 200 hPa and values of Brown's indicator of clear-air turbulence (10^{-1} s^{-1}) over the eastern equatorial Pacific at 1800 UTC on 24 February 1999 (Bysouth 2000) (3×10^{-1} s$^{-1} \approx 1 \times 10^{-1}$ m^{-1}). The white line indicates the path of a flight from Fiji to Los Angeles that encountered turbulence for much of the journey on that day. © Crown copyright (Met Office).

Table 4.1   Percentage frequency of CAT within the tropics (Atkinson 1971)

Area	Turbulence category (%)			
	None	Light	Moderate	Severe
0–30°N	91.3	6.0	2.7	0.1
Indian Ocean (0–15°S)	96.0	2.7	1.3	0.0
Africa (south of 15°N)	94.8	3.7	1.4	0.1
All zones	91.9	5.5	2.5	0.1

the peaks, in particular those aligned across the flow. The highest mountain ranges that lie across the flow of the STJ in winter are the Hengduan Shan of southern China, which reach 4000 m, and the Andes of South America, which reach around 5500 m. Turbulence is very likely to occur to the east of these ranges, especially as there is a generally easterly flow below the level of the mountaintops, against that of the STJ, as described in more detail in Box 8.1. This turbulence, which may be within cloud

Figure 4.6   Satellite image of ribbons of medium- and high-level cloud associated with wind shear over north-west Africa at 0315 UTC on 12 February 2013. Courtesy of EUMETSAT.

during the winter monsoon season, may be considered a serious hazard to aircraft flying at medium levels. Where large-scale gravity waves are generated by the stable flow over these mountains, moderate turbulence may occur throughout the troposphere and into the lower stratosphere.

Associated with the STJ, but separate from it, further areas of tropical CAT are also observed at times. These may be identified by their high-cloud signature, as in many cases they occur where wind speeds are not above 40 m s⁻¹. Ribbons of cirrus may be seen and may persist for days within areas where there is a large horizontal wind shear (Fig. 4.6). This often occurs within upper troughs previously associated with a cold front, the deep cloud of which has dissolved. These have been noted in particular over the Arabian Sea, close to the wintertime trough in the STJ.

4.5   Questions

1. Bearing in mind that the pilots of long-distance aircraft file a flight plan in which they state the air lane that they wish to use and the altitude at which they plan to fly, consider whether the figures in Table 4.1 might be an underestimate.
2. Explore journal and internet sources of severe weather that may be linked to the STJ.

Notes

1   A thermal wind is generated by a difference in temperature, its strength proportional to the gradient and magnitude of the difference. Because of the rotation of the earth, the thermal wind (as any other wind not at the equator) does not flow direct from hot to cold air, but is

deflected, so that in the northern (southern) hemisphere the cold air is on the left-hand (right-hand) side. Its flow is in balance with the thermal gradient and changes in the thermal gradient, so that it blows approximately parallel to lines of equal temperature. In the case of the STJ there is usually some ageostrophic flow, although it can be regarded, in general, as a thermal wind in balance between the heat of the tropics and the relative cool of the extra-tropics.

2 Keen botanists will notice that the difference in name between the spring rains in China and Japan is subtle. Cherry and plum trees both belong to the genus *prunus* and have many similarities.

3 Shear has the units s^{-1} and turbulence may be observed in either these terms or as a probability over a distance of flight (m^{-1}). The former is more useful as a method of forecasting turbulence, using numerical weather prediction data, although empirical CAT indicators are usually given in probabilistic terms.

5
Synoptic-scale Weather Systems

5.1 Introduction

Most of the cloud in the tropics has its origin in convection, forming as warm moist air ascends through comparatively cool air. The tropical atmosphere is almost homogeneous and so has few significant gradients of density or pressure, it is largely barotropic[1] (effectively a single air mass). Where gradients of wet-bulb potential temperature[2] (θ_w) occur they are usually small and are more important in the vertical than the horizontal. In the tropics, if the surface is warm, convection can be expected and, with sufficient moisture, where large-scale descent is absent, deep convection may occur.[3]

Nevertheless, because vertical gradients of density are small and much of the tropical air has, at some time or another, been involved in large-scale subsidence, layer clouds also form and are a common component of tropical systems, spreading out from the cumuliform towers of convective clouds. However, stratocumulus and altocumulus cloud layers generally occur as relatively thin layers, spreading out from cumulus or cumulonimbus clouds. Rarely is any of these cloud layers more than about 1500 m deep.

5.2 Convection in the tropics

5.2.1 Deep convection

Deep convection can be revealed by instability indices applied to observed or forecast atmospheric profiles (George 1960; Jefferson 1963; Bradbury 1977) and shallow convection is readily evident in low-level temperature lapse rates. These reveal that most of the tropical zone is essentially convective, although variations occur on diurnal, latitudinal and seasonal scales, as well as with altitude. In some areas, convection is relatively shallow, confined to the lower layers of the troposphere (section 5.2.2), and it is significant differences in the temperature and, to some extent, the humidity of the upper troposphere that determine the depth of convection to higher levels. Deep convective clouds have cold tops and produce precipitation by the Wegener–Bergeron–Findeisen (ice) process, described in section A3.1.1. Heavy rain, sometimes accompanied by hail and thunder if the cloud deepens to become cumulonimbus, is associated with these deep clouds.

Because of the comparatively limited instability of tropical air, there is a fine balance between

An Introduction to the Meteorology and Climate of the Tropics, First Edition. J F P Galvin.
© 2016 John Wiley & Sons, Ltd. Published 2016 by John Wiley & Sons, Ltd.

shallow (locally dry) convection in the near-surface layer and deep convection. As presented in Chapter 3, there are factors limiting deep convection, as well as those promoting it. In areas where large-scale factors favour deep convection, local factors favouring development include high humidity (high θ_w), particularly in the boundary layer, high surface temperature, a high level of CAPE and a modest change of wind with height. Convection is inhibited by layers of stable or dry air, although the former can assist in the build-up of energy that may be released as the temperature rises to its maximum (Ludlam 1980; Scorer & Verkaik 1989; Bader et al. 1995; Grant 1995; Young 1995). The process is particularly important over land and a technique to use these factors has been developed recently to forecast pre-monsoon thunderstorms in Calcutta, often the most damaging in north-east India (Chaudhuri & Middey 2009).

Although the accurate forecasting of deep convection is important for agriculture, the general public and tourism, the development of deep convective cloud and the layer cloud that forms from it are possibly of greatest significance to aviation. This is due mainly to the potential for icing, turbulence, large hail or lightning from deep convective clouds. Whilst local forecasters are concerned with tropospheric development and resultant weather within a limited area, long-distance aviation forecasts are concerned only with developments at higher levels, above 3 km. Only large-scale weather systems are picked out for their significance to passengers and pilots. Areas of convective or layer cloud must be at least 150 km across to feature in these continental-scale forecasts. Convection is of greatest significance where deep convection is either embedded within layer clouds or closely spaced so it would be difficult to avoid (Meteorological Office 1994).

5.2.2 Shallow convection

Although deep convection is predominant in the humid tropics, some areas have cloud and rainfall brought predominantly by relatively shallow convection. Cumulus cloud-tops that reach an altitude between about 4 and 7 km in the tropical air mass have 'warm' tops with temperatures in the range +3 to −10°C and cannot produce precipitation by the Wegener–Bergeron–Findeisen process. However, cloud liquid–water content is often high enough to allow rain droplets to form by agglomeration (section A3.1.1). The random coalescence of cloud drops may be sufficient to allow small drops to fall and collect more drops, eventually large enough to fall as rain. These raindrops are often relatively small and precipitation from these clouds is rarely more than slight and always of rain. Some climate zones in the tropics have most of their (meagre) rainfall from these clouds, notably the equatorial Atlantic, characterized by the climate record from Ascension Island.

5.2.3 Cloud development over the oceans

Relative to the atmosphere, the oceans are cool in much of the tropics (Fig. 5.1) and only in a belt near the equator, in particular on the western sides of the Atlantic and Pacific Oceans, does it become relatively warm. In the trade-wind zone (Figs 1.3 & 3.2), the modest temperatures of the surface are carried into the lowest part of the troposphere by mixing to form a well-developed but cool boundary layer. Above this well-mixed layer there is a temperature inversion, where the temperature rises by several degrees in only a few hundred metres ascent. Above this inversion, the air is dry and subsided, associated with the near-equatorial ridge system and upper-tropospheric convergence, which causes the air to descend. The inversion is at a height of about 1500 m (but with large variability) near 30°N and 30°S, but overturning in the brisk trade winds allows cumulus and stratocumulus to form over the oceans (Box 5.1). In some places there may be large amounts of cloud, although for the most part the stratocumulus cloud layer is well broken, with shallow cumulus the

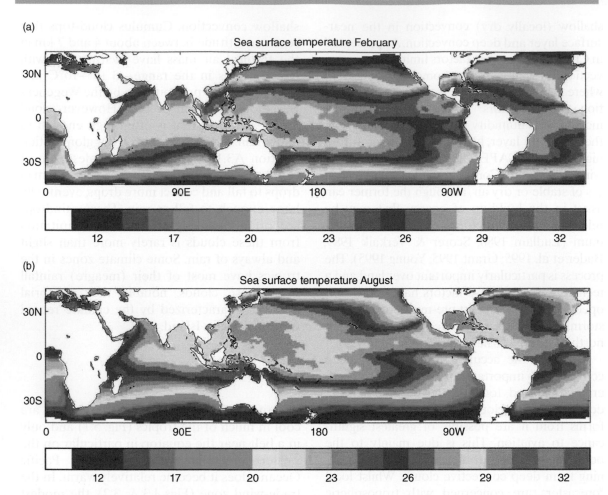

Figure 5.1   Tropical sea-surface temperature (°C) in (a) February and (b) August from the Met Office HadISST dataset. © Crown Copyright (Met Office).

Box 5.1   *Overturning in the atmosphere*

In the atmosphere, the meaning of 'overturning' is large-scale turbulent mixing. This most often occurs when there are strong winds in the boundary layer.

Surface friction causes the wind speed near the ground (up to about 900 m above the surface) to be slower than and backed in relation to the gradient wind speed. Turbulence is the result, and the ascent and descent of the air forms a well-mixed temperature profile, provided the surface is not warm compared with the air in contact with it. The well-mixed air has neutral stability.

The overturning also carries moisture from the surface to the top of the boundary layer, where saturation may occur, particularly over the ocean. If there is a stable layer at the top of the boundary layer (a frequent occurrence in much of the tropics), areas of stratocumulus cloud may result (perhaps with poorly formed ragged cumulus clouds below them).

characteristic cloud type. However, the equatorward motion of the trade winds progressively (but slowly) warms the boundary layer and convection over warming seas deepens the moist boundary layer, which reaches a depth of around 3 km at the edges of the humid equatorial zone. Within this humid zone cloud development is of a generally greater extent, reaching the middle and upper troposphere.

When an air mass passes over a large sea or ocean, even where sea temperatures are rising along the track of the air mass, instability (in particular the need for a superadiabatic[4] lapse rate above the surface) is very limited. In part this is due to entrainment in the mixed flow and sensible heating of the cool boundary layer by convection. The mechanism that allows large-scale deep convection is the convergence of airstreams in the boundary layer, combined with divergence in the upper troposphere above the meteorological equator (ME), the line along which the trade winds from north and south meet (Fig. 3.2).[5] Even along the ME the air temperature at the surface must be sufficient to generate convection. In general, the sea-surface temperature

necessary for deep convection is 25–26°C (Emanuel 1988). The resulting areas of deep convection are frequently several hundred kilometres wide and areas may be thousands of kilometres long, as shown in Fig. 5.2. The formation of the most vigorous convection occurs where the air mass source is unstable to greater depth or there is additional convergence in the form of so-called 'westerly wind bursts'. Surface winds with a westerly component have crossed the equator and their effects are revealed by their frequent association with the development of tropical depressions and cyclones (Verbickas 1998), which are discussed in Chapter 9.

As a result of the limited stability and pre-existing (though shallow and modest) near-isothermal layers or temperature inversions in the middle troposphere, layer cloud often forms, spreading out from the convective towers where the environment is sufficiently moist.

As the ocean surface warms during the day, convection develops to become most vigorous during the afternoon and evening. Over the warmed surface there is a time lag in the development of deep convection, bringing the most

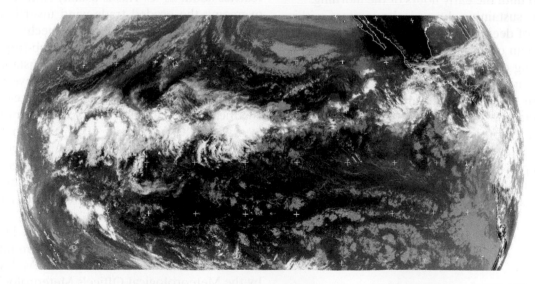

Figure 5.2 Infra-red image from GOES-W in geostationary orbit over the eastern Pacific Ocean at 2100 UTC on 14 November 2006. The inter-tropical convergence zone is evident between the equator and about 10°N as large-scale cold-topped cloud masses hundreds of kilometres north–south and thousands of kilometres west–east. Courtesy of University of Dundee Satellite Receiving Station.

copious rainfall overnight (Leroux 2001, Ch. 12). Both the heating of the ocean surface and the resulting atmospheric circulation drive the diurnal modulation over the oceans.

5.2.4 Forecasting cloud development over land and sea

The diurnal rhythm is much more marked over land.[6] Very few deep convective clouds can be seen (except in coastal areas) before midday, but because the warming of the land surface is greater than that of the sea, within an air mass unstable cloud develops to greater depth over land by late afternoon. Thus cloud tops generally reach a higher level over the continents during the evening than is the case over the sea. The development of deep cloud late in the day emphasizes the relatively modest instability present in the humid-tropical zone. As temperatures rise to around their maximum, the lower troposphere is seeded as cumulus clouds develop. Cooling of the troposphere during the afternoon and evening allows convection to deepen and cloud tops may rise to 14 km or more by evening. Deep convection is often sustained until the early hours of the morning.

The sustaining dynamical processes within areas of deep convection is such that development can occur in a more-or-less steady state, allowing cloud areas to persist for some time, either in one area or as a steadily moving system. However, all areas of cloud vary through the diurnal cycle, at least to some extent, spurred by surface-temperature changes, as well as changes in the atmosphere in depth. However, large-scale changes in dynamics occur only slowly in the tropics, as discussed in Chapters 4, 10 and 11.

5.3 The inter-tropical convergence zone

In the humid tropical zone near the equator, the convergence of the trade winds usually forms a discontinuous band of deep convection – the ITCZ – associated with the ME (Fig. 5.2). However, in some areas the trade winds converge without generating significant cloud. This is because in the ITCZ deep convection can only occur where warm air can ascend through a weakened sub-tropical inversion, usually in the presence of divergence in the upper-tropospheric easterlies, which causes dynamical ascent, increasing the depth to which convective clouds may rise. The variations in development of cloud along the ITCZ can be seen in Fig. 5.2.

The position of the ITCZ moves north and south with the seasons in response to the movement of the sun and the resultant waning and waxing of the sub-tropical high-pressure belt. However, its mean position is about 5° north of the geographical equator over the oceans (see Figs 5.1 & 3.2). Over the continents, the line of convergence is distorted (Fig. 5.3) as the land surface warms in summer and the ITCZ becomes poorly defined where a series of monsoon circulations forms. These will be discussed in Chapter 8.

Close to the ITCZ the depth of the convective boundary layer is about 3000 m and θ_w reaches about 21°C. This is usually sufficient to overcome the eroded trade-wind inversion in the presence of dynamical forcing mechanisms.

The feed of cooler air from the sub-tropical high-pressure belt ensures sufficient instability to maintain the ITCZ. As a result, convection from the surface to form cumulonimbus clouds is critically dependent on surface temperature, even within the meandering ITCZ. Nevertheless, in the presence of upper-level divergence, cumulonimbus convection is vigorous once the critical temperature is reached. Thus, over tropical seas almost all deep convection occurs along the ITCZ.

Although convective in origin, the ITCZ is not a continuous 'wall' of cumulonimbus clouds. Indeed, a brief survey over Malaysia by the Meteorological Office's Meteorological Research Flight suggests that cumulonimbus towers are rarely less than 20 km apart (Zobel & Cornford 1966). However, some of the most

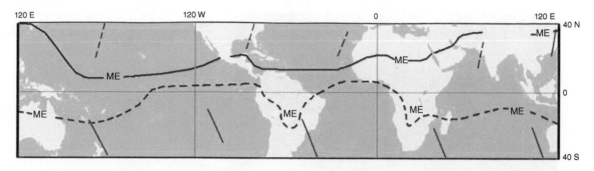

Figure 5.3 Typical seasonal variation of tropical upper tropospheric troughs (blue) and the ME (red): - - - - January; ——— July. The ME can be seen to respond to the movement of the sun and is particularly distorted over the world's major landmasses in summer. The buckling of the ME over South America during the southern hemisphere summer reflects the fact that westerly winds at low levels cannot cross the Andes mountain barrier. ME, meteorological equator.

vigorous systems, such as tropical depressions or cyclones and mesoscale convective systems, are likely to have clusters (or lines) of cumulonimbus somewhat closer together.[7]

The layer cloud that characterizes much of the ITCZ is a result of convection, so disperses over a matter of hours in areas where convection is absent. However, it is much longer lasting than the convection that forms it.

5.3.1 The inter-tropical convergence zone over the oceans

Over the Indian Ocean the ITCZ often splits into two parallel bands of cloud to the north and south of the equator. Between them at these times, subsidence and slightly lower sea-surface temperatures prevent deep convection. The northern branch is most active in the south-west monsoon season, when it becomes broad and may be difficult to identify as a line of convergence. The southern branch is most important during the north-east monsoon. Either of these branches may spawn tropical depressions or cyclones as 'easterly waves' in the middle troposphere, with a period of about 5 days move along them. This small-amplitude north–south motion in easterly winds is discussed in sections 3.1, 9.2, 9.4, 9.5, 9.8, 10.2 and 10.4. One such cyclone developed over the Bay of Bengal on 20 June 2007 and is pictured over southern India in Fig. 5.4.

Over the warmest oceans – the western Pacific in particular, where sea-surface temperatures are above 28°C throughout the year (Fig. 5.1) – are the areas where most deep convection occurs. In general, the ITCZ is observed to be broadest over these zones (Fig. 5.2). However, even here the balance of vigorous deep convection and atmospheric stability, which is reinforced by descent between convective towers, remains delicate. The main consequence is the development and relative persistence of extensive layers of altocumulus, which spread out from cumulus and cumulonimbus clouds beneath modest temperature inversions.

As Fig. 5.1 shows, where the oceans are cooler due to upwelling along western coasts, convergence in the oceanic gyres and the equatorward motion of ocean currents, convection is limited, often absent, so that cloud along the ITCZ is at best discontinuous in these areas.

In the northern winter, east–west variations across the Pacific basin are amplified. In general, the eastern Pacific becomes cooler as the California Current strengthens. More energy is lost from the reduced winter insolation and values of diabatic heating become strongly negative. This means that much energy is lost by radiation, the sun providing insufficient energy to restore the loss. The result is a reduction, or absence, of convection and the weather is drier in the period December–April. A similar

Figure 5.4   Infra-red satellite image from IODC (Meteosat-7) showing isolated convection typical of the northern summer monsoon over South-West Asia at 1800 UTC on 22 June 2007. Cumulonimbus cells can be seen over the Ethiopian Highlands, the 'Asīr Tihāmah and Hadhramaut of Yemen, the Jabal Akhdar, the Zagros Mountains, the ranges of the border between Afghanistan and Pakistan, and the foothills of the Himalayas. Coastal convergence has also formed large cells near Abu Dhabi and over Gujarat and Rajasthan, India. The large cirriform shield with embedded convection over southern India is the remains of a tropical revolving storm that had formed over the southern Bay of Bengal two days earlier and which later caused serious floods in coastal Pakistan. The grey altocumulus cloud layer over coastal Somalia has a temperature around 0°C. This is where the boundary layer has been cooled over upwelling water offshore. Convection is inhibited and cumulus clouds have spread out to form an extensive, but broken, layer. Courtesy of University of Dundee Satellite Receiving Station.

situation is observed over the South Pacific gyre for much of the year.

In the western Pacific, as in the Indian Ocean, the ITCZ splits to form two branches, divided by the South Equatorial Current, which is relatively cool (Fig. 5.1), so that convection occurs to the north and south of it. The northern branch of the ITCZ, at about 5–10°N, is discontinuous and convection is generally weak during the northern winter, whilst that in the south, at about 10–15°S (but closing with the northern branch near New Guinea), strengthens in the southern summer. When the south-west Pacific is at its warmest, the ITCZ may be seen as a broad north-west/south-east

zone from north of the Solomon Islands to the south-west of Fiji, where it often becomes incorporated into cloud associated with a weak semi-permanent upper trough. At these times the northern branch of the ITCZ may weaken and break up across the central Pacific, despite sea temperatures above 27°C for much of the year (Fig. 5.1), as the convergence of trade winds moves into the southern hemisphere. The southern discontinuous branch rarely extends east of about 120°W.

In the southern summer, the ITCZ moves south and generally lies between 10°S and 15°S over the Indian and south-west Pacific Oceans by February, although its movement south is much less over the Atlantic and eastern Pacific, where it typically remains in the northern hemisphere. This is due mainly to the relatively cool currents of the south-east Pacific and South Atlantic. This modest southward migration of the ITCZ in these areas demonstrates the comparative warmth of the North Equatorial Current, the effect of the relatively cool Equatorial Counter Current and the modest instability of the tropical air mass. However, the position of the ITCZ over the North Equatorial Current (between 5°N and 10°N) also aids convective development. Where stronger winds carry more momentum in the south-east trade winds across the equator, the effect is most significant. The change of direction of the Coriolis force causes these winds to become south-westerly, forming 'westerly wind busts' that directly oppose the north-east trades. There is more discussion of the development of westerlies near to the equator, associated with the Madden-Julian Oscillation, in section 12.3.

5.3.2 Seasonal changes over the north of Africa and south-west Asia

The effects of sea temperature on convection vary markedly with season along the coasts of Africa and south-west Asia.

The north-western and south-western coasts of Africa are areas where upwelling brings cold water to the surface (Fig. 5.1) due to the offshore motion of the sea surface in response to the trade winds. The water moves equatorward and is gradually warmed, but at certain times of the year close to the equator the ocean surface is sufficiently warm to generate deep cumulonimbus clouds. However, dry weather predominates in many coastal areas of Africa.

In West Africa, the distortion of the ITCZ during the summer monsoon causes it to become the so-called inter-tropical front (ITF, described in Box 5.2). This is a diffuse zone, not co-incident with the convergence of low-level winds. Deep convection lies to the south of the ITF, where the 850 hPa θ_w is between 22°C and 25°C.

Over the Gulf of Guinea, strong south-westerly winds overturn the ocean surface in summer, cooling it. The cooling is greatest along the Gold Coast, east of Cape Three Points, Ghana where there is upwelling due to offshore motion of the Guinea Current (Holton 1979, Ch. VI). The cooling of the waters of the Gulf of Guinea is reinforced by upwelling along the west coasts of Namibia, Angola, Congo and Gabon, south of the equator, under the influence of the south-east trade winds. By late summer, the temperature of the Gulf of Guinea cools to around 24°C. As the ocean cools, deep convection becomes progressively confined to the land (Leroux 2001, Ch. 2). Along the Gold Coast there is a rainfall minimum in August and September, and fogs are common, directly related to the cooling of the ocean. Indeed, an annual average of only 715 mm of rain falls in Accra due to the weakened monsoon convection over this cool-water coast. This compares with an annual average of 1339 mm in Abidjan, Ivory Coast. The local climate differences in the West African monsoon are discussed further in section 8.6.

The atmospheric circulation with lighter winds in winter causes much less overturning than in summer and so seas are warmer, the sea surface typically reaching 28°C or more, some 4–5°C above its minimum (Fig. 5.1). The warming of the Gulf of Guinea reinforces the southward motion of the ITF, a form of large-scale land breeze developing the north-east trade wind. The ITF becomes aligned with the coast soon after the apparent position of the sun moves into the southern hemisphere and now lies approximately below the (weakened) area of divergence in the upper atmosphere. Thus, convection is more likely to occur over the Gulf of Guinea and through equatorial Africa from late October to April than from April to early October.

In the east of the continent, the processes that form deep convection are even more complex. Forced uplift over high ground

Box 5.2   The inter-tropical front in monsoonal West Africa

The definition of the inter-tropical front (ITF) is problematic, in that there has been considerable disagreement about its definition as well as its name, the latter being largely dependent on historical usage within colonial meteorological services. None of the names given is truly appropriate during the summer monsoon season, since although it is a convergence zone between dry and moist air masses, its line at the surface has no weather directly associated with it.

The ITF has a structure and resultant weather that differs from that of the ITCZ. The distortion of the ME is so great that the line of convergence at low levels it is not coincident with convergence of the mid-tropospheric easterlies or the divergent easterly equatorial jet stream. In terms of moisture, the vertical structure has some resemblance to a front with deep convection present only on its equatorward side (discussed in section 8.6). The structure of the ITF also has some similarities to that of a sea-breeze front, although on a scale an order of magnitude larger. In high summer, the position of the ITF at the surface may be more than 5° displaced from the higher-level ME, with a slanting moist zone descending to the surface from around 3000 m (Leroux 2001, Ch. 5).

Layer clouds do not mark this zone of changing moisture content and its slope is reversed from that normally expected in a mid-latitude frontal zone. This is because the temperature difference between the moist and dry air masses is large and air-density differences are typically around 3%. Unlike a mid-latitude front, therefore, the baroclinic zone based on θ_w is reversed. The warm air lies on the low-θ_w side and the front slopes upwards to the south from the northern edge of the gradient of boundary-layer θ_w. Furthermore, the normal 'rules' of advection (winds veering and increasing with height indicating the progress of warm air) apply. Winds back with height, whether or not the ITF is advancing or in retreat. In terms of temperature, this northward movement is cold advection, although there is moisture advection, as normally associated with a warm front.

The easterlies to the north (above south-westerlies at low levels) are so hot and dry that the airstream is effectively stratified. It is only when the moist air is sufficiently deep and air at higher levels is comparatively moist that convection can occur.

The monsoon in China tends to develop a similar form to that in West Africa as it moves north, late in the summer.

releases conditional instability, which reaches a maximum during the northern summer. The south-easterly monsoon winds that cross Ethiopia lose much of their moisture on the country's highlands, which lie between about 5°N and 15°N. Thus, the ME lies to the north, across northern parts of Sudan and Chad at the surface, but with air to the south much less moist than westward from Niger (Fig. 5.4). Eritrea, despite its high ground, is in the rain shadow of the Ethiopian Highlands and is one of the driest countries on earth.

Although the monsoon trough crosses southern parts of south-west Asia in summer

(Fig. 5.3), large-scale convection is generally limited along it here. This is partly because mid-level convergence remains as much as 1000 km to the south of the ME, but also because of the relative dryness of the atmosphere in depth. As a result, significant convection is usually restricted to the mountains of Yemen, Oman and south-western Saudi Arabia (Al-Maskari & Gadian 2005), although isolated thunderstorms do occur in other parts of the peninsula (Ghulam & Dorling 2005), but rain from these rarely reaches the ground. Convection is also seen over the mountains of Iran, Afghanistan and Pakistan, locally bringing rain or snow, the latter to be expected above about 4700 m in summer. Nevertheless, many parts of south-west Asia may remain dry, as shown in Fig. 5.4. In winter, southern parts of south-west Asia are usually dry under the influence of the north-east trade winds.

5.3.3 Convection over the southern hemisphere continents

Over southern Africa in summer there is similar distortion of the ITCZ to that in the northern hemisphere. Here, most of the rainfall during the southern summer falls on the Kenyan, Tanzanian, Zambian and Zimbabwean highlands, leaving some central-southern African areas, as well as the coastal plain, comparatively dry.

At the same time, over Australia periodic monsoon rains move inland from the north coast, following the ME south (Fig. 5.3).

In South America, a form of monsoon development occurs as the sun moves south, bringing moist air at low levels well to the south of the convergent flow in the middle troposphere. This will be discussed further in section 8.9. Deep convection also occurs preferentially on ground with an elevation above about 1000 m, in particular on the slopes of the Andes of Ecuador and Peru (see Fig. 5.3). This is due to the enhanced warming of higher ground by day, itself due to a decrease in absolute humidity with altitude. (Humidity is an important factor the sub-tropical high pressure and dry, limiting the rise and fall of temperature in the equatorial zone, as will be discussed in Chapter 11).

5.4 The depth of convective clouds

The depth of convection is much greater in the tropics than it is for most of the year at higher latitudes. In part, this is due to the greater depth of the troposphere, but it is also because the atmosphere is barotropic and, in the ITCZ,

relatively moist. Although in higher latitudes it is relatively common to observe cumulonimbus clouds that reach, or overshoot, the tropopause, they rarely reach the tropopause in the tropics, even though deep convective clouds frequently reach 13 km or more where the cloud-top temperature is around –60°C. This is because subsidence at the top of the troposphere may warm and dry 1000–3000 m of air below the tropopause. Table 5.1 shows the mean altitude of convective inhibition levels (CILs; where most convective uplift ceases) assessed from observed radiosonde profiles on 31 October 2005, along with the corresponding tropopause heights over each continent. Deep convection rarely reaches more than a few hundred metres above the CIL, except in the most vigorous convective areas, since the amount of CAPE (the difference in energy between an ascending parcel of moist air and the environment) is relatively small.

Strong convection through the day may cause the inhibition level to rise, so that ultimately a new tropopause forms at an altitude lower than before since the temperature at the cloud top is lower than that of the pre-existing environment above the CIL.

Despite the inhibition of convection, some updraughts are sufficiently strong to reach the stratosphere and a significant mass of air is carried above the tropopause in the equatorial zone. This is important worldwide, since tropical

Table 5.1 Mean observed tropopause heights and convective inhibition levels in the tropics, 0000 UTC on 31 October 2005

Region	Tropopause height (m)	Convective inhibition level (m)
North Africa	16000	13600
South-West Asia	16600	13600
India	15300	13900
South-East Asia	16600	13900
Australia	16900	13600
Pacific Islands	16300	13300
Central and South America	15600	14300

air contains a high proportion of ozone, which is formed from oxygen in strong sunlight and is carried away from the tropics to form part of the layer that protects life from harmful short-wave radiation.

5.4.1 Forecasting the depth of cloud using satellite imagery

Forecasters can use pre-existing cloud-top temperatures as a way of estimating the likely tops of deep convective cloud. This is particularly useful in forecasts for up to about 24 hours ahead. Since there is only a small amount of CAPE available for ascent – in particular over the oceans and in the lower troposphere of the equatorial zone, where the humidity is high – 'ready reckoners', based on the tropical air mass (as described in Chapter 1), give a reasonably good guide to the height of cloud tops, using their cloud-top temperatures, recorded by satellite (Table 5.2). The height of the freezing level varies between about 4500 and 5000 m in the tropical zone (but 5500 m or more near the centre of a tropical cyclone and locally over the continents during the summer), causing variations in cloud-top height.

5.5 Layer clouds and shallow convection

Although oceanic and bordering coastal areas in the tropics (away from the cool-water coasts of the continental western margins) are characterized by shallow cumulus development, the high absolute humidity of the tropical atmosphere confers high liquid-water content to convective clouds. By the process of agglomeration, slight showers may fall, bringing much-needed rainfall from 'warm' clouds with cloud-top temperatures close to freezing.

Even in the near-equatorial areas of deep convection by no means all of the rainfall is from cumulonimbus clouds. Rain from layer clouds is most often seen near the poleward margins of the ITCZ, as well as locally over high ground. Cumulus with associated altocumulus and altostratus bring significant rainfall, which may be prolonged. In some areas this slight precipitation is more important than the heavy downpours from deep cumulonimbus clouds.

In winter, the onset of the north-east monsoon over southern China brings convective rain embedded in extensive layer clouds. This is discussed in section 8.5.2 and Box 8.1.

Over land, layer cloud develops following the initiation of large-cumulus convection, usually from late morning. Thus, although there may be broken remnants of layer cloud formed from cumulus and cumulonimbus clouds the previous day, a more continuous deeper layer forms, spreading out from convective towers, to reach a peak in the early evening. This cloud is maintained by the gradual loss of long-wave radiation from the atmosphere, which causes atmospheric cooling overnight. As the rate of loss is proportional to the fourth power of the temperature (Stefan's law), the cooling is greatest during the afternoon and evening (although the loss of sensible heat is limited while convection remains vigorous), so that thick layer clouds remain well into the night.

5.6 The effects of heavy rainfall in the tropics

Most parts of the tropics are subject to periodic heavy rainfall. In the humid tropics (near the ITCZ and in summer in the monsoon zones), where rain is frequent, heavy downpours falling

Table 5.2 The relationship of cloud-top temperature and freezing level height to cloud-top level (dam) in the tropics

0°C level (m)	−80°C	−75°C	−70°C	−63°C	−55°C
4500	1490	1430	1370	1310	1250
5100	1590	1520	1460	1400	1330
5700	1690	1620	1560	1490	1420

on saturated soils may be sufficient to have devastating effects. Even the large rivers of the tropics can soon fill, causing flooding in settlements bordering them. In many cases it is the poorer people who live closest to the rivers and their poorly constructed buildings are those most often flooded. Flash floods are also a feature of drier tropical climates, heavy rain soon overcoming valleys not sculpted over long periods to accept large quantities of rainfall in short periods.

In recent times of increasing population and forest clearance, the risks of these rainfall events have become very serious, especially where mud may be carried from hillsides many kilometres downstream. Thus, increasing effort is being applied to the control of these floods (Mathur & WMO Secretariat 2006). However, the deserts are also subject to flooding – indeed, this is one characteristic of the tropical-desert environment. Although many years may pass between events, when it occurs torrential thundery rain can sweep all before it, overtopping normally dry wadis. This leads to strange statistics. Six years' rain may fall in just a few hours, but it will be dry for years following the event! Chapter 7 is devoted to dry tropical environments.

5.7 Atmospheric teleconnections

Convection in the tropics is a key driver of world-wide weather. The Hadley and Walker cells redistribute heat and moisture, the former not only within the tropical air mass, but also to north and south of it. This effect is particularly marked when there is enhanced convection due to the Madden-Julian Oscillation (over the short term and in a spatially restricted sense) or El Niño (over periods of a year or more and usually world-wide). These phenomena are discussed in sections 12.2 and 12.3.

Indeed, the vertical and meridional motion of air from the tropics is connected, albeit weakly, to broad-scale weather throughout the extra-tropics The most notable teleconnection is that between the central Pacific and North America in the northern winter. An increase of 500-hPa height over the tropical Pacific usually causes an increase or decrease in 500-hPa height over north-western North America (James 1994). This direct effect has downstream effects elsewhere, in particular tending to reduce pressure over the North Atlantic and western Europe, this in turn altering the phase of the North Atlantic Oscillation (James 1994).

In addition, as we will see in Chapter 14, the Hadley cells redistribute the ozone formed in the lower stratosphere, reducing levels near the equator and replenishing the sub-polar stratosphere. As the broad-scale circulation in the troposphere changes, so does the redistribution of ozone and this may have health implications.

5.8 Questions

1. Why is convective inhibition important in determining the timing (and location) of precipitation in the tropics?
2. We saw in Fig. 1.3 that much of the tropical troposphere (above the near-surface layer) is dry. How does this dryness affect rainfall and sunshine amounts in the tropics?
3. Investigate the known effects of teleconnections on extra-tropical weather, considering in particular long-term rainfall variation and changes in the broad-scale circulation that can result.

Notes

1 Pressure depends only on density (and density on pressure) in a barotropic atmosphere, so that surfaces of constant pressure are also surfaces of constant density. Constant-pressure surfaces are also isothermal surfaces, so that the balanced geostrophic wind is independent of height because there is no horizontal thermal forcing. The motions of a rotating barotropic air mass

are thus strongly constrained. A barotropic flow is a generalization of the barotropic atmosphere: it is a flow in which the pressure is only a function of the density. A single air mass is broadly barotropic, although variations in the surface are always likely to have an effect, in both the horizontal and the vertical.

2 Wet-bulb potential temperature (θ_w), usually calculated at the 850 hPa level (close to 1500 m), is a good measure of air-mass characteristics, combining the effects of both temperature and moisture content in the near-surface layer. Forecasters usually assess θ_w from plotted radiosonde profiles, which are graphical plots, including lines of saturated adiabatic lapse rate, Γ_s. (Appendix 4 includes an introduction to temperature-entropy diagrams.) If we assume a mass of air is saturated and remains so as it is compressed to a pressure of 1000 hPa, its temperature will rise at a rate determined by its initial temperature and moisture content – Γ_s. In some countries and many areas of research other measures are used, such as the equivalent potential temperature (θ_e).

3 Large-scale descent strongly inhibits 'free' convection so that, even where there is an excess of energy available to a parcel of air rising from the surface, deep convection is rarely observed. The upper troposphere is usually dry and there may be a temperature inversion as a result of air-mass descent.

4 A superadiabatic lapse rate, where temperature falls with height at a greater rate than the dry adiabatic lapse rate (Γ, introduced in Appendix 4), is a requirement of free convection. Superadiabatic lapse rates are usually found only at the surface. However, in the presence of air of high absolute humidity, the superadiabatic layer need not be deep, so that surface temperatures need not be much more than 3°C greater than those at 300 m. By contrast, comparatively dry air needs a much greater fall in temperature in the near-surface layer for convection to be triggered.

5 Convergence is aided when the ITCZ lies north or south of the equator. Wind direction is partially determined by the rotation of the earth. The rotation decreases towards the equator, becoming 0 there, and as air crosses from one hemisphere to the other the sense of rotation of the earth reverses, causing the motion of the air to 're-curve', the westward component of motion becoming eastward. Thus convergence along the ITCZ (away from the equator) may be from almost diametrically opposing directions.

6 Although temperature usually rises by day, even under clear skies, this rise is largely due to a flux of heat from the soil until late morning. Insolation becomes the dominant factor in the rise of temperature around the middle of the day and it is this element of the input of energy that allows deep instability over the tropical continents. Before this time of day the input of energy is close to balance with outgoing longwave radiation (Lapworth 2009).

7 Mesoscale convective systems – clusters of cumulonimbus and associated layer clouds – also occur outside the tropics and are often seen in summer over Europe and North America.

6
Climate, Flora and Fauna

6.1 The relationship of climate to plants and animals

Following on from the introduction to tropical climates in Chapters 1 and 3, this chapter describes the major climates and aspects of their relationship to the plants and animals of the tropics, including humankind. A map of the climatic zones of the tropics appears as Fig. 1.1. The radiation balance was explored in Chapter 2 and Box 6.1 discusses the effect of the balance of incoming and outgoing radiation on the temperature at the earth's surface in the tropics. Convection and advection (the motion of air masses with a particular temperature profile) determine the temperature of the upper troposphere and the lowest part of the stratosphere.

In many places humans have altered the distribution of plant and animal species; in almost all areas they use plants and animals for food or labour. These changes in turn change the climate, as significant changes are made to the appearance of the environment. The zones discussed below are classified according to their natural vegetation, although settlement has often removed this.

The climates discussed include those that are only under the influence of the tropical air mass in summer, including those along its poleward limits: the Caribbean, the southern USA, northern Mexico, north Africa, south-west Asia, Tibet, central-southern China, the southern tip of Africa, northern Argentina, Uruguay, Bolivia, Paraguay, south-eastern Brazil, northern Chile and parts of southern Australia. In winter, these areas have weather dominated by mid-latitude weather systems or are dry.

The jungle teems with insects and these provide plentiful food for mammals and reptiles. Fruit-bearing trees also support some larger animals, including monkeys and apes. Amongst many insects, the mosquito is particularly well adapted to tropical rainforest. With it comes the malaria parasite, bringing one of many diseases that flourish in the warm, wet, equatorial environment.

6.2 Tropical rainforest

This is the archetypal environment of the humid tropics and often what is thought of as *the* tropical environment (Af in Fig. 1.1). However, there are many variations on a theme, based on seasonality of rainfall, distance from the meteorological equator and altitude.

In the non-seasonal tropics, areas of dense 'jungle' receive more than about 2000 mm of rainfall every year and day length varies little. Mean temperatures are between 22 and 28°C, with little variation throughout the year and a

An Introduction to the Meteorology and Climate of the Tropics, First Edition. J F P Galvin.
© 2016 John Wiley & Sons, Ltd. Published 2016 by John Wiley & Sons, Ltd.

Box 6.1　The difference in radiation balance in dry and moist atmospheres

As discussed in Chapter 2, the diurnal change in near-surface air temperature is determined mainly by the absorption of short-wave radiation at the surface (insolation) and the emission of long-wave (terrestrial) radiation by the surface. Over the course of a year, these are in balance, indeed there is very nearly a balance every day. However, various factors affect both insolation and the emission of long-wave radiation to space. Solar radiation may be reflected by clouds or dust, or it may be reflected by the surface (albedo) or absorbed by atmospheric gases.

Water vapour is the most important of the absorbing gases. Similarly, terrestrial radiation is reflected by clouds or suspended dust and is strongly absorbed by water vapour, carbon dioxide and other 'greenhouse' gases (Fig. 6.1).

Because warm objects radiate much more than cool ones, there is the potential for a great range of temperature in the warm environment of the tropics. However, this range is much more evident in the tropical deserts than in the humid tropics, as discussed below.

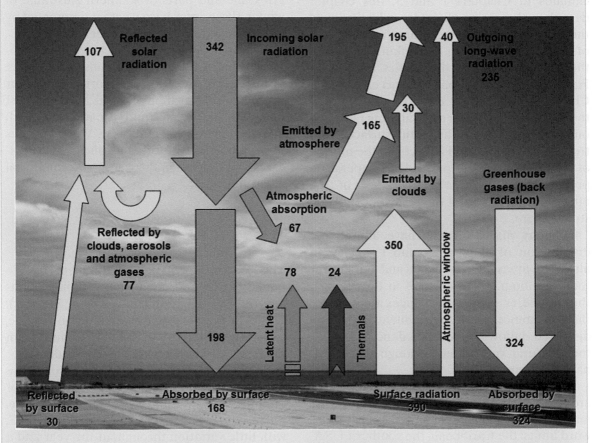

Figure 6.1　The global heat balance: mean values of energy exchange (Wm^{-2}). Note the important role of greenhouse gases (most importantly CO_2 and H_2O) in keeping temperatures in the habitable range for humans (global mean 15°C near sea level, but nearer 22°C in the tropics).

Albedo

The reflection of solar radiation by clouds and the surface is the most significant factor affecting the input and output of radiation in the atmosphere, so it affects surface temperature profoundly. Clouds reflect more than 25% of incident solar radiation and extensive low clouds may reflect more than 50%. Suspended dust also reflects a good deal of the solar radiation that falls on it (mainly because clay particles are flat and fly like wings). The surface also reflects sunlight and bare surfaces may be good reflectors. The total reflection of short-wave radiation to space is 33%, although the average is somewhat smaller in the tropics (due to the high elevation of the sun around midday). Variations are large between cloudy humid areas and dry (though sometimes dusty) regions. As much as 90% of solar radiation may be absorbed by the surface in dry areas, whilst in humid cloudy weather the amount may be as little as 25%, even in the tropics. Where most insolation is received, the temperature has the potential to rise most and where least is received, the temperature rises least (in the absence of warm advection, which is usually small in the tropics).

Upwelling long-wave radiation is reflected by cloud, greatly slowing the loss of radiation and consequent fall of air temperature.

Gaseous absorption

On average, the absorption of short-wave radiation by the atmosphere is very small and as the absorbed radiation is re-emitted as long-wave radiation, half of which is received at the surface, for most meteorological applications this absorption can be ignored. Nonetheless, where there is a high concentration of water vapour in the atmosphere, its effects may be noticeable.

Water vapour has a much more profound effect on the emission of long-wave radiation to space. On average, water vapour, carbon dioxide and cloud absorb about 95% of emitted terrestrial radiation, although re-emission accounts for as much as 59% of total global emissions to space. The absorption of terrestrial radiation slows the fall of surface air temperature from its mid-afternoon peak. Thus in the humid tropics (or near the ocean in dry areas), temperature falls slowly and the mean diurnal range is typically between 5 and 8°C. On any one day, the range rarely exceeds 10°C.* In the hot deserts, the mean diurnal range is between about 8 and 12°C, in summer sometimes reaching 20°C or more.

Convection

Convection has a very large effect on the rise and fall of surface temperature. Whilst convection is often thought of in terms of vertical motion of temperature in thermals, it is the wind that carries most heat to and from the surface, slowing the rise of temperature to mid afternoon and reducing the loss of heat from mid afternoon to dawn. Forced overturning of the air just above the surface in windy weather carries heat away from a warmed surface and towards a cooling one, all the time attempting to make the fall of temperature through the boundary layer equal to the dry adiabatic lapse rate (DALR; -9.8 K km^{-1}). However, as may be deduced, the transfer of heat by convection is not independent of the input and loss of radiation at the surface. The stronger the wind, the more efficient the transfer of heat, but there will still be a greater rise and fall of temperature in a windy cloudless dust-free environment than in a windy cloudy humid one.

The loss of heat is also greater in a humid environment where the atmosphere is unstable to great depth than in one where there is

*Near the minimum temperature, the fall in temperature may be further arrested by condensation of dew, fog or low cloud in a humid environment, which releases latent heat.

stability, even if the amount of water vapour in the atmosphere is (initially) the same. Deep convection carries heat (and moisture) to great altitude, in particular where convective cloud can rise to a great height. Thus the rise of temperature may be further slowed when deep convection commences.

All these factors need to be taken into consideration when forecasting the rise and fall of temperature. Calculations show that in a windless dust-free unvegetated stable environment with very low specific humidity, the temperature may rise and fall 20°C or more, occasionally bringing a maximum temperature of 50–55°C in summer in the tropical deserts. In contrast, with a vegetated windless environment near saturation, the absorption and insolation reach a balance at about 37°C, a theoretical maximum temperature for the humid tropics.

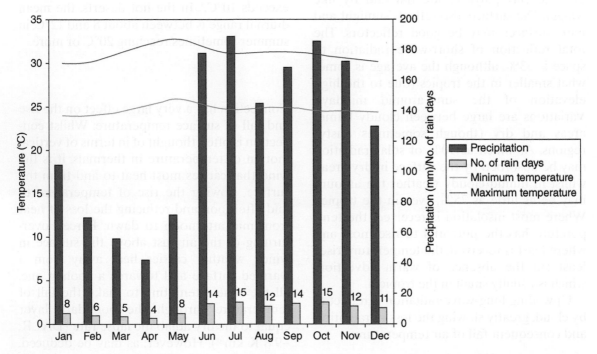

Figure 6.2 The climate of Mactan, Cebu, Philippines (10.3°N, 124.0°E, 24m) in the tropical rainforest zone. It is wet throughout the year, although monthly totals vary as the ITCZ moves north through the Philippines in June and July then returns south between September and November. Drier weather occurs as north-east trade winds make occasional incursions south between January and May. The number of rain days is high, in particular between June and December, when there is rain on more than half the days. However, the amount of rainfall is also related to factors described in Chapter 12. Note the small daily and annual temperature range, <10°C. Sourced at http://www.pagasa.dost.gov.ph/index.php/climate-nl.

daily range rarely more than 10°C. Climate data for a typical location, Mactan, Cebu, Philippines, are shown in Fig. 6.2.[1]

The main source of variability in the humid tropics is the alternation between El Niño and La Niña, linked to changes in the Walker circulation (Walker, 1923, 1924a,b; Reynolds, 2000). Whilst El Niño brings drier years, anomalously high rainfall in much of the Indo-Pacific region is linked to La Niña. Locally, La Niña may bring extreme rainfall, as was seen in January 2011 in Sri Lanka. More on the variability of climate in the tropics appears in Chapter 12.

The high humidity, near-constant high temperature and high degree of shelter usually provided by the forest possibly make this environment the most uncomfortable in the tropics. It is hard to lose heat by perspiration and work rapidly causes tiredness, despite the shade from trees. Nevertheless, the high productivity of the rainforest supports a large number of people. In many areas settlement is in the coastal zone or along rivers, although there is increasing settlement in formerly inaccessible forest. Two of the world's ten largest cities, Jakarta and São Paulo, lie within the tropical rainforest zone.

Trees are broad-leafed and evergreen; many have so-called 'buttressed' trunks (a result of the leached soils, see below). They often live hundreds of years, since there is no frost and year-round rainfall allows them to grow throughout the year. The leaf canopy is very dense in much of the zone, with lianas growing between and up tree trunks, so the jungle floor is relatively dark, even around midday. As a result, much of the forest floor has relatively thin vegetation, except near areas where more light can penetrate, such as along riversides, where trees have fallen or along coasts (Fig. 6.3).

Along the poleward edges of the forest zone, where rainfall is lower and the temperature range is greater, the forest thins and there is greater development in the under-storey.

The typical soil of the tropical rainforest is the latosol (FAO-UNESCO 1989). (A description of the soil types of the tropics is given in Box 6.2.) The soils are often well developed and up to 30 m deep, but are critically leached of fertilizing nitrate and phosphate. They are usually oxidized and have a shallow humus layer so that trees tend to spread their roots laterally rather than downwards to make use of humus nutrients. The leaching concentrates acid minerals, including iron and, particularly within in this zone of high rainfall, aluminium (Ellis & Mellor 1995). Interestingly, however, the forest canopy itself reduces the rainfall that reaches the ground, reducing leaching. Studies

Figure 6.3 Typical dense tropical rainforest at Latak waterfall, Lambir National Park, Sarawak. © Richard Young.

at the Centre for Ecology and Hydrology, Wallingford, suggest that the reduction is as much as 30% (Overton & Strangeways 2007).

Productivity is the highest of all the world's environments in these forests (Table 6.1).[2] The total productivity (plants and animals) reaches 4.5 kg Cm^{-2} yr^{-1}, and 50% of the world's species of plants and animals live in these forests. The low level of available mineral nutrients, however, has had an interesting effect: the diversification of plant and animal species, which must find ways to extract nutrient from the environment (Fothergill et al. 2006). This is mainly because growth occurs throughout the year.

On the leached soils, new growth is very dependent on the death of older trees, which can be decomposed in a matter of months or years in this high-temperature, moist environment. Thus beetles, as well as other insects,

Box 6.2   Soil types of the tropics

Alfisol Formed of clay with nutrient-enriched subsoil, typical of monsoonal and semi-arid regions. In monsoonal regions, however, it has a tendency to acidify when heavily cultivated, especially when nitrogenous fertilizers are used.

Andosol A highly porous, dark soil developed from volcanic rocks: ash, tuff or pumice. It typically occurs in highland areas.

Grumusol A brown, calcium-rich soil of dry environments. It forms over limestone or dolomite bedrock.

Latosol A soil rich in iron, alumina or silica formed in areas of tropical rainforest. The iron and aluminium enrichment is a result of high rainfall, which leaches the soil of soluble minerals, such as calcium and phosphorus.

Leptosol A very shallow soil (indicating little influence of soil-forming processes), often containing large amounts of gravel. It typically remains under natural vegetation, as cultivation makes it especially susceptible to erosion and desiccation.

Regosol A soil formed from unconsolidated silt or clay that may be of alluvial origin and that has a lack of a significant soil horizon because of a dry or cold climate.

Vertisol A soil containing immature (swelling) clays, commonly found in the wet or seasonally wet tropics where chemical erosion affects the bedrock, which is usually of volcanic origin. The soil is regularly overturned by wetting due to its high proportion of swelling clay.

Desert surfaces The surfaces of arid lands are composed of exposed bedrock outcrops and fluvial deposits, including alluvial fans, playas, desert lakes and oases. Bedrock outcrops commonly occur as small mountains surrounded by extensive plains. Where soils occur, they are poorly developed and shallow, although they may contain a high concentration of minerals since the low rainfall does not leach them. Where developed on limestone, they form *rendzinas*. However, they are also high in salts since any rainfall is evaporated rapidly. Deserts have highly specialized salt-tolerant natural vegetation.

Table 6.1   Typical productivity (measured as the fixing rate of carbon by plants/phytoplankton) in tropical environments (Fothergill et al. 2006; Lalli & Parsons 1993)

Climatic zone	Productivity (kg Cm^{-2} yr^{-1})
Tropical forest	1–3.5
Savannah	0.2–2
Arid zones	~0.3
Mountain	≤0.02
Ocean – barrier reefs	1.5–5
Ocean – coastal zones	~0.1
Ocean – continental western margins	~1
Ocean – subtropical gyres	<0.03

bacteria and fungi, which do not depend on sunlight but use undecomposed organic matter for food, are important inhabitants of this biome.

The archetypal tropical rainforest, however, is sensitive to temperature, humidity and rainfall changes. As a result, it soon gives way to other forms of tropical forest where temperatures are lower or where rainfall is seasonal. As a result, it is usually found within 200–500 m of sea level, where there is no long dry season. As a result, the main areas in which it is found are the islands and peninsulas of South-East Asia, the Congo and Amazon basins, and the Irrawaddy river system. Poleward of the main equatorial zone, upland areas of modest altitude may maintain dense evergreen forest because rain falls throughout the year.

At high altitudes trees are better adapted to changes in temperature. The fall in humidity with altitude discussed in sections 5.2.1 and 5.3.3 allows temperature to vary much more during the day, even though it remains above freezing to altitudes of 3000 m or more. On higher mountains, where occasional frosts risk damaging trees in leaf and which are usually above the humid tropical boundary layer, needle-leaf conifers predominate. At these altitudes there is often less rainfall available, placing greater stress on trees that are not adapted to steep slopes, thin soils and, at times, low temperatures.

6.3 Seasonal tropical forest

In the lower-lying tropics dominated by monsoons (Chapter 8), where there is sufficient precipitation to support extensive forest (an annual rainfall more than about 1000 mm; zone Am in Fig. 1.1), trees must withstand dry, periodically hot weather for a substantial part of the year. In the wet season, foliage must be able to withstand very wet humid weather.

In much of the monsoon zone there is great variability in forest cover, dependent not only on rainfall and the availability of soil water, but also soil-mineral availability, surface type and, to some extent, the presence of grazing animals.

Deciduous trees, such as teak, are well adapted to this regime, as they shed their broad leaves at the beginning of the dry season so are present in much of this zone, although evergreen species are also seen. The rate of carbon fixing is somewhat lower than in the predominantly wet rainforests, but the annual leaf fall allows the soils to be somewhat more fertile, with more nutrients returned to the soil, although the heavy rains of the monsoon wash them out for a proportion of the year. Seasonal forests are the climax vegetation of much of West Africa and eastern Brazil; parts of the Indian subcontinent, China, Central America and northern Australia.

Although these forests have an annual mean temperature range similar to that of the equatorial forests, the daily range is high during the dry season, reaching 12°C or more. At times inland the temperature may reach 40°C or more. Climate data for Chittagong, Bangladesh, reflecting this zone, are shown in Fig. 6.4.

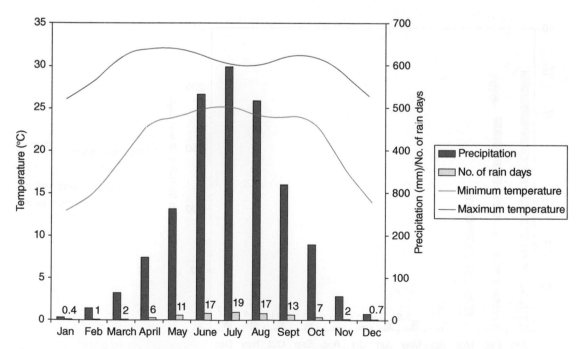

Figure 6.4   The climate of Chittagong, Bangladesh (22.3°N, 91.8°E, 6m) in the tropical deciduous forest zone. In this monsoon climate there is little rain between late October and early April. During this period potential evaporation exceeds precipitation, thus trees 'hibernate' through the winter and make use of the very high rainfall between June and September for rapid growth. Agriculture has replaced natural forest in much of this zone. The temperature range is high between November and March (>12°C), but comparatively low during the wet summer season.

Seasonal tropical forest supports a variety of plants as an under-storey; evergreen bamboo varieties are perhaps the most notable. The woodland is locally interspersed with grassland, in particular towards its poleward limit, and so also supports a wide variety of fauna, particularly larger animals of a great variety of species. The forest thins where the annual rainfall is less, reflecting the relatively large amounts of water required for trees to grow. The trees themselves, however, maintain a relatively moist environment, providing some protection from desiccation.

6.4 The savannas

Where rainfall is regular, but insufficient to support extensive forests – usually below an annual total of about 800 mm – grassland is the climax vegetation. The savannas (Aw and Caw in Fig. 1.1) thus mainly lie poleward of the monsoon forest zone; although in East Africa, altitude and shelter extend the zone across the equator. Although defined as grassland, trees are also a characteristic of much of this land, often occurring as small stands among large areas of grass. The zone usually has a pronounced dry 'winter' season and a wet 'summer' season characterized by large falls of rain in short periods.

The mean temperature range of these grasslands often exceeds 12°C, with the highest temperatures and largest temperature range occurring during the dry season, maxima reaching 35°C or more. Although frosts are rare, they sometimes occur in winter in upland valleys within this zone. Typical climate data for the savannas, recorded at Croydon, Queensland, Australia, are shown as Fig. 6.5.

The grassland characteristically supports many herds of large animals, but also a variety of smaller ones. Indeed, fauna abound: there is about 200 times the mass of animal life on the savanna compared with that in the tropical rainforest (Fothergill et al. 2006). The large

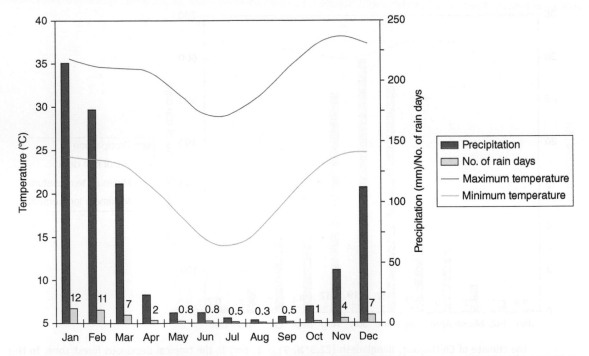

Figure 6.5   The climate of Croydon, Queensland, Australia (18.2°S, 142.2°E, 116m) in the savanna zone. The temperature range is high in winter (>12°C). Rainfall is plentiful in summer and has the potential to support highly productive agriculture, even though year-round potential evaporation is very high; winter is generally dry.

herds of animals, in their turn, support predator species in most tropical continents and have been a traditional source of food for humans. Reptiles often live well in this environment too.

Where there is plentiful water, there is a greater variety of species in the savannas and this is often from rivers that flow throughout the year in this climatic zone, even though rainfall is seasonal, with almost all rain falling in a wet summer period. Not only is this water important for vegetation, it also supports many animal species.

Winds can be strong over this zone, flat grasslands exerting little friction on the airstream. Dry trade winds predominate in winter, although the summer brings predominantly moist equatorial flow.

Although dry for a large part of the year with much less growth than more humid areas, in terms of carbon fixing rate the savannas deserve the title of the world's most efficient environment. Between 0.2 and 2 kg of carbon are fixed per square metre of land in each year (Table 6.1) from relatively scant resources by just a tenth of the biomass of the tropical rainforest (Fothergill et al. 2006), a rate up to about four times that of tropical rainforest.

The soil types of the savannas vary between vertisols, alfisols (Natural Resources Conservation Service 1999, 2006) and, in areas of limestone bedrock, grumusols. However, most have a form characteristic of the zone. Leaching reduces the organic and mineral content of the surface layer, but these soils are less leached than the latosols of the rainforest zone. They are formed both by leaching in the wet season and mineral differentiation during the dry season, when capillary action brings lighter dissolved minerals towards the surface. Much of the rainfall of the savannas occurs in short-period heavy downpours, increasing the potential for wash out, although the presence of grass cover helps to reduce the impact of the rain. Crops, as they grow, provide similar – or better – shelter, but when the ground is clear following the harvest, soils may be more open to leaching. However, the risks are small as the harvest usually occurs at the beginning of the dry season. Savanna soils can be hard to cultivate due to leaching and so favour pastoral farming, although arable farming is also very important in this zone, in particular where soils are rich in minerals and water is available for irrigation.

Fires are another characteristic of this zone, in particular towards the poleward edges. Following the dry season, vegetation may be desiccated and, particularly as more humid air begins to return, 'dry' thunderstorms occur, their lightning setting fire to trees (Fig. 6.6).

Figure 6.6   Blaze in stands of trees in the savanna of Australia's Northern Territory. The dry conditions make fire a near-constant hazard, Australian 'gum' trees easily burning. © Barbara Pettigrew.

However, the relatively large spacing of trees in many of the world's savannas (notably the gum-tree savanna of Australia) is adapted to the high risk of fire in this characteristically windy zone. Where trees are closely spaced, little vegetation grows on the forest floor and wind speed is reduced beneath the canopy top so that fire 'jumps' from tree top to tree top, burning the upper portion of the trees, whilst trunks and lower branches may be unaffected. These trees can recover from fires more readily than those that are more widely spaced (see also section 7.2.6).

Three of the world's ten largest cities, Mumbai, Delhi and Shanghai, have developed in these seasonally dry zones, reflecting the benefit to humans of seasonal climates with adequate rainfall during a relatively warm growing season.

6.5 Tropical deserts and scrublands

These environments (BSh and BWh in Fig. 1.1) cover more than one-third of the earth's surface and support a remarkably high population, despite their aridity. Throughout the year evaporation exceeds precipitation, which is generally below 300 mm yr^{-1} and more typically below 50 mm yr^{-1} in the desert plains. Rainfall is very variable, however: in some years there is relatively high rainfall locally, whilst in most years there is none. The usual lack of cloud cover and very low humidity (away from the cool-ocean coasts of Namibia, Angola, Chile and Peru, as well as, seasonally, those of Somalia, Yemen and Oman) bring a daily temperature range of 15°C or more. Sunshine totals may be more than 80% of the maximum possible. Many aspects of harsh dry environments are presented in Chapter 7.

Most scrublands along the equatorward fringes of the deserts see occasional monsoon rain. Many have their own peculiar environments and ecosystems. Ephemeral rivers, such

as the Okavango at the edge of the Kalahari, may not drain into the sea, but can support a variety of animals that must travel hundreds of kilometres to find water, following seasonal rainfall.

In many areas, especially over higher ground and in its rain shadow, the tropical forests and savannas give way to semi-desert scrublands composed mainly of stubby drought-tolerant trees and hardy grasses. This vegetation characterizes parts of southern and eastern Africa, India and the Australian outback.

Scrubland has mainly short-needle-leafed trees growing near ephemeral streams or in stands where sufficient water is available. Some hardy grasses are also seen, although there is a large amount of bare soil, especially at the edge of the deserts. In Africa and Australia, close to the furthest poleward burst of the summer monsoon, are scrublands (Fig. 6.7). In Africa, this land is known as the Sahel.

Rain, when it falls, is often intense and may be accompanied by hail. Indeed, it is estimated that much of the rainfall of the Sahel comes from mesoscale convective complexes, which are described in Chapter 8 (Gaye et al. 2005; Laing & Fritsch 1997).

Life must be very hardy to survive in the arid lands. Animals usually retain water or can make their own water metabolically. Nevertheless, all major types of animal are represented in desert or semi-desert environments and humans live in many tropical areas subject to drought. Perhaps surprisingly, two of the world's ten largest cities are in dry environments. The trading ports of Karachi and Lima were founded on rivers to supply plentiful fresh water from glaciers on neighbouring mountains to support populations of 10.8 and 8.9 million, respectively. As these glaciers have declined in recent years, however, there is a growing need to find alternative sources of drinking water (Aquino & Ford 2008; Asif 2008).

The climates of Lima, Peru and Aswān, Egypt are typical of the world's tropical deserts,

Figure 6.7 Chamber's Pillar, a sandstone outcrop with vertical sides in the western Simpson Desert, Northern Territory, Australia. This landform is typically seen in deserts and its slopes indicate a lack of regular rainfall.[4] However, meagre rainfall seeps gradually through the rock and supports the scrubby trees around its base.

as shown in Fig. 6.8. The scrubland (semi-desert) climate is illustrated by data from Timbuktu, Mali (Fig. 6.9).

Natural vegetation, where it can survive at all, is characterized by water-retaining plants. Other plants survive drought as dormant seeds that germinate rapidly following rain that may fall only one year in six. Characteristically, these are flowering plants and short-lived seas of colour are the beautiful result.

Cool-water desert coasts – in particular those of the Namib and Atacama – have an unusual vegetation, dependent on dew and fog in humid, cool air carried ashore.

Low rainfall and high evaporation make productivity low in the deserts (Table 6.1), although there is considerable variation, dependent on the variability of annual rainfall and potential soil fertility.

For much of the year these zones have relatively strong trade winds blowing across relatively flat land with little vegetation, although summer brings periods of moist monsoon flow poleward from their equatorward fringes (see Figs 8.4 & 8.5). The return of the trade winds may bring plagues of locusts across areas of seasonal growth, moistened in summer, whilst mosquitoes and the associated risk of malaria may occasionally be brought about by the moist equatorial flow of high summer (Chapter 16).

6.6 Mountain climates

Mountain climates are particularly important across the tropics; drainage of the rainfall they generate often helps to maintain populations on

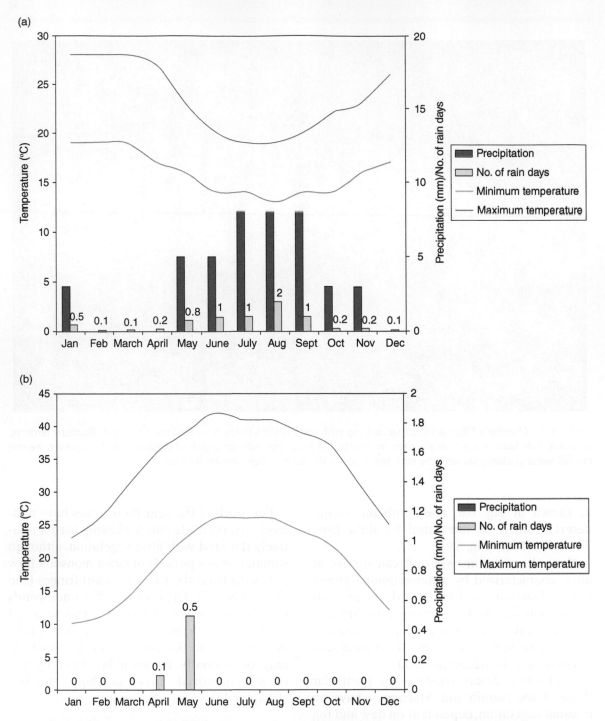

Figure 6.8   (a) The climate of Lima, Peru (12.0°S, 77.1°W, 13m) in the coastal Atacama Desert. Most rain falls in winter as upper troughs bring occasional mid-latitude disturbances from higher latitudes. Note the limit to maximum temperature through the summer, due to sea breezes from the neighbouring cool water of the Peru Current. (b) The inland desert environment at Aswân, Egypt (24.0°N, 32.8°E, 194m) in the eastern Sahara desert. No more than a trace of rain can be expected to fall here. Note that the rainfall scales in these diagrams are very much smaller than that in Figs 4.1, 4.5, 4.6 and 4.8. The difference in temperature range between the two desert environments is notable: large inland at Aswân (~15°C) and small at Lima (5–10°C), moisture from a cold ocean limiting both maxima and minima at the latter.

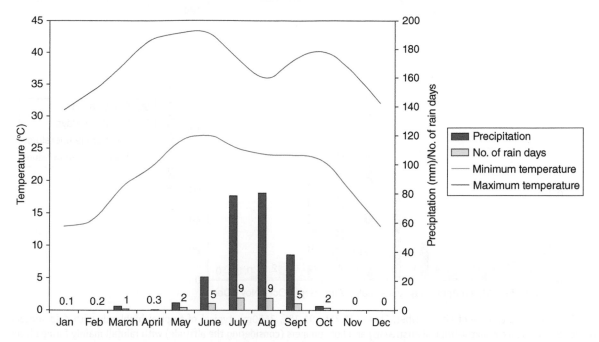

Figure 6.9 The climate of Timbuktu, Mali (16.7°N, 3.0°W, 264m) on the desert fringe of the Sahel. Almost all the rainfall comes from northward excursions of the summer monsoon from late June to early September. It occurs on only a few days during the summer, often in heavy bursts and about half comes from squall lines, associated with easterly waves (Laing & Fritsch 1997). The dramatic reduction in temperature range during the rainy season is notable.

dry, or seasonally dry, lands around them. These areas are stippled or designated H in Fig. 1.1.

One of the most extraordinary climates of the tropical zone, as it expands poleward in the northern summer, is that of the Tibetan plateau. Although much of the zone is a plain, its altitude brings it into the realm of mountains. Almost all the plateau lies above 4000 m and so even in high summer it remains cold, with surface temperatures barely above freezing at night and reaching more than 20°C by day. It is cold enough for the ranges of mountains (the Tangula Shan) that cross the plains to remain ice-covered throughout the year. Climate data for Lhasa, Tibet (at one of the lowest points on the plateau) are shown as Fig. 6.10.

At 4000 m above sea level the air is very thin, containing barely 60% of the oxygen available near sea level. Still more limiting for life is the lack of water. The plateau is sheltered on all sides: monsoon rains cannot cross the Himalayas and clouds from more northern latitudes deposit meagre rainfall mainly on the

Kunlun Shan, which forms a rim along its northern border. Apart from a few deep narrow passes, mountains reaching 7000 m or more surround the plateau. What little rain and snow that does fall mostly occurs in summer, mainly from isolated cumulonimbus clouds. With a base near 5000 m, convective clouds contain little precipitable water, even though they may be 11,000 m deep. The humidity mixing ratio for saturated air at 5000 m and 0°C is only 9 g kg^{-1}, the air density little more than half that at sea level. A lowland cumulonimbus cloud in the ITCZ, with a base at 750 m and 15,000 m deep, may contain five times as much water! Humans, plants and animals living in this environment must adapt to drought in the dry air and low oxygen levels, as well as to the cold.

Alpine grasses have adapted themselves to this environment, but there are very few trees (Fig. 6.11). The thin mountain soils, leptosols, andosols and regosols,[3] vary mainly in their mineral content and origin, but most remain

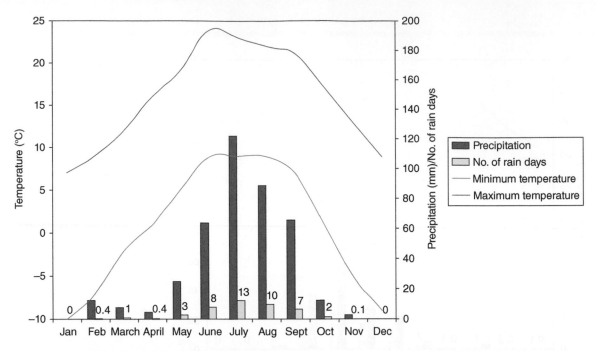

Figure 6.10   The climate of Lhasa, Tibet, China (27.7°N, 91.1°E, 3650m), typical of tropical highlands with cold winters (under the influence of extra-tropical airstreams) and hot summers (considering the altitude) with rainfall mainly formed from locally formed convection in summer. Many days are wet, but rain occurs over relatively short periods, mainly late in the day.

Figure 6.11   Vegetation of the trans-Himalayan mountain zone, essentially an upland cold desert (above an altitude of about 3500 m) with few trees and scant vegetation, Manang Himalaya, Nepal, October 1992. An effect of rural depopulation is visible as ruined buildings in the middle ground. © Nigel Bolton.

under their natural vegetation. In the tropics there are trees to near 4000 m (but this is highly variable) and thin grassland scrub at higher altitudes giving way to bare ice-shattered rock on the highest peaks. Where rainfall does not erode the soil, it may be rich in organic matter (Fothergill et al. 2006). However, herds of antelope and horses typify the natural fauna of the plateau. With such a short growing season and a desert environment, productivity is very low in Tibet and similar highlands.

Although the Tibetan plateau provides an interesting area to study, the tropics have many other mountain areas, such as the Ethiopian plateau. Relatively modest areas of high ground are likely to be more wooded than great plateau lands and benefit from a high rainfall, well above that of surrounding lowlands. The towering peaks of the great plateaux are so high that they are above most rain-bearing cloud. The greatest increase in rainfall is between about 1000 and 3000 m, the moisture content of the air reducing significantly above this level. High ground in the rain shadow of larger massifs remains almost as dry as its surroundings, but many dry or seasonal areas of the tropics between these heights have a significantly increased rainfall. Many of these areas of high ground are listed in section 11.3. Mexico City – one of the world's largest cities – at an altitude of about 2300 m, is founded on the relatively equable climate of the Mexican Highlands.

6.7 Tropical oceans and coasts

For the purposes of this book, the tropical oceans are found between the relatively steep temperature and salinity gradients that occur close to 30°S and 35°N of the equator (varying somewhat with season) and where the ocean surface has a temperature above about 22°C. However, the areas of upwelling cold water that lie along the western coasts of the continents, where temperatures are locally well below 22°C, are included (see Fig. 5.1).

Odd though it may sound, most tropical oceans are equivalent to the continental deserts. Their productivity is usually very low, since nutrients easily settle out. This is mainly due to the relatively light winds of much of the tropics, notwithstanding the moderate to fresh trade winds, which do ensure some essential minerals are retained in the surface layer.

The convergence and subsidence of surface water in the great ocean gyres, under the influence of the sub-tropical high-pressure belt (Lalli & Parsons 1993), along with the distribution of both rainfall and wind, ensure that most nutrients soon sink below the surface layer. The windier areas receive little rainfall and the wetter areas (the doldrums) have light winds. Under clear skies the intensity of radiation also reduces productivity near to the ocean surface (Table 6.1). This means that nitrogen, the main source of which is rainfall over the open oceans, is usually in short supply.

In the humid tropics, the drier season, dominated by trade winds, brings some increase in the maritime harvest, as does the weather following tropical storms (see Chapter 9). Large swell brings deep water to the surface and the tropical rainfall adds nutrients to the sea, some of it as run-off from the land.

Of course, there is a great variety across the tropical oceans. In coastal areas, there is a much greater supply of minerals, resulting from run-off in streams and rivers (Table 6.1). Most remarkable, however, are areas where there is little run-off, but a shallow continental shelf. Here, in the dry tropical ocean environment, coral reefs blossom. These are incredibly productive, despite the poor nutrient supply (Table 6.1). (The productivity of coral reefs is not reflected in chlorophyll concentrations (as shown in Fig. 6.12.) Indeed these animals, which grow together as a colony, depositing the coral mineral as they develop, are dependent for survival on warm, clear, still water with almost no suspended particulates. In the shelter of the corals live a variety of fish, shellfish and echinoderms (e.g. starfish, sea urchins).

An important exception to the usual paucity of fish in tropical oceans occurs in the areas of

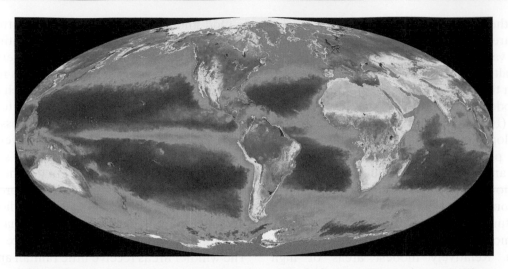

Figure 6.12 The global distribution of ocean chlorophyll concentration from satellite imagery, an indicator of productivity, September 1997–August 1998. Note the low productivity of oceanic gyres (blue shades <100 mg Cm^{-2} day^{-1}) compared with the productive areas of upwelling, in particular along the western coasts of Africa and South America, where values reach more than 400 mg Cm^{-2} day^{-1} (yellow-orange). The slight increase in productivity (~200 mg Cm^{-2} day^{-1}) near the equator in the Pacific and Atlantic Oceans (green shades) is not present in the Indian Ocean, where the ocean gyre is present across the equator and so convergent descending water is present across the equator. Courtesy of NASA-GSFC.

upwelling along the western coasts of the tropical continents. Trade winds blowing offshore cause cool water from the ocean deeps to rise to the surface, bringing with them a high nutrient supply, as indicated by Fig. 6.12. In this figure, the level of nutrients is proportional to the number of fish, since the nutrient feeds drifting plants (phytoplankton) and the phytoplankton form food for floating animals (zooplankton), which are available as food for fish. The west coast of South America, south of the equator, is particularly well-endowed. The waters rapidly deepen offshore, so that deep water can readily rise to the surface under the influence of the offshore trade-wind flow. The result is a cooling of about 6°C and fish stocks, including anchovies, pilchards and jack mackerel, are usually high. This coast, however, is periodically cursed. Close to the equator in northern Peru and Ecuador, the supply of upwelling water periodically decreases at a time when it would normally be expected to reach a peak in December and January. This is the El Niño phenomenon, which is discussed further in section 12.2. Without sufficient

upwelling, the nutrient supply is limited and fish stocks reduce – at times causing a failure of the ocean's harvest, typically around Christmastide. This gives the phenomenon its name – the Christ child.

Although the greatest stocks of fish are found along the South American coasts, those of the west coasts of Central America, southern Africa and North Africa are also important throughout the year (Table 6.1).

Periodically, there is upwelling along other tropical coasts – in particular those of the Gulf of Guinea, Somalia and Yemen, where the strong south-westerlies of the summer monsoon bring a good summer harvest from the Arabian Sea. Here, as in many parts of the world, fishing is a hazardous occupation, every bit as taxing as in the western European waters of winter with rough seas hindering the fishing effort.

A variation on tropical forest is found along its brackish coastal limit: the mangrove forest. Mangroves grow well semi-submerged in silt-laden waters and provide a defence against coastal inundation. The storm surges from tropical revolving storms can be much ameliorated by this dense, low-lying vegetation and its maintenance is regarded as a very important part of the protection of the prevalent coastal communities in areas affected by these storms. In addition, mangroves stabilize the foreshore and many can filter out poisonous chemicals from the water, helping to maintain its fertility.

The lack of rainfall in the oceanic deserts can make living conditions difficult on remote tropical islands, perhaps especially those that, because of their sunshine and dry weather, attract many tourists. Desalination remains a very expensive way to provide drinkable water, but can be the only way to provide sufficient for the population and local agriculture, even if tourists survive on imported bottled water. The climate of Praia, Cape Verde Islands (Fig. 6.13) illustrates the predominantly dry weather of the oceanic anti-cyclones, where relatively scant rain falls for only a few months of the year.

6.8 Climatic variability

In this chapter it has been necessary to divide the tropics according to 'typical' climate parameters, established by long-term means. However, there is, of course, great variety within these discrete zones and the margins of climate zones are, in truth, zones of transition. Climate change is also likely to impose changes to the limits of particular climatic regions, as discussed in section 12.6.

The climate of each country or province within the tropical zone (and at its perpiphery) is given in Appendix 6 and climatic data for a range of tropical locations can be found in *World Weather Guide* (Pearce & Smith 1984).

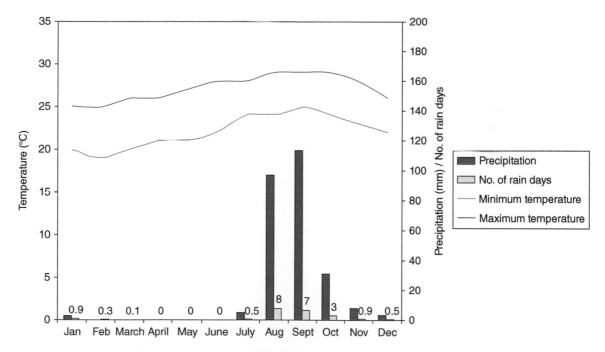

Figure 6.13 The climate of Praia, Cape Verde Islands (14.9°N, 23.5°W, 35m). This predominantly dry climate is typical of the oceanic anti-cyclones, with most rainfall late in the summer when sea temperatures are highest, allowing heavy showery rain (on average, more than 10 mm per rain day) to develop, often associated with easterly waves (described in Chapter 8). The relatively scant rainfall must be stored to last through the dry season between December and July when evaporation remains high in the north-east trade wind regime. These winds, off the Sahara Desert, frequently bring outbreaks of dust.

6.9   Questions

1. Look at Fig. 6.8. Consider the differences between these graphs in terms of temperature and rainfall. How might the significance of rainfall vary between these places?

2. Why is maximum temperature lower in July and August than in April, May, June and October in the savanna zones (e.g. Fig. 6.9)?

Notes

1   It should be noted that the data for most stations mentioned in this chapter are recorded at an open site (on an airfield), as are most across the world, so it could be argued that they do not fully reflect the temperature and rainfall of their climatic zone. In the case of tropical rainforest, the true climate is probably somewhat cooler by day, warmer by night and wetter.

2   This is expressed in terms of the annual mass of carbon fixed from the environment (i.e. the amount of carbon dioxide changed to plant material) in a given area, primarily by vegetation, which may then be eaten by animals. Total productivity includes production of carbon by animals.

3   Mesas and buttes are seen outside desert areas, in particular where there is tectonic activity or newly active erosion. The shape of this landform is dependent on the length of time it has been exposed to erosive agents (including wind abrasion), as well as the amount of rainfall. (With thanks to C.R. Twidale.)

7
Dry Environments

7.1 Background

Although in the public mind the tropics are often associated with warm moist equatorial or monsoonal climates, the characteristic climates of most of the zone are dry ones. Large swathes of the tropics north and south of the moist zone have an excess of potential[1] evapotranspiration over precipitation, and in the large tropical deserts very little rainfall occurs. As a result, vegetation is sparse or may not be present. In general, natural plant growth is dependent on water from springs or on infrequent storms.

Many types of weather characterize the world's deserts and semi-deserts. They are often windy places, as there is little vegetation to reduce the speed of the wind across the surface. Convection by day reaches considerable depth, but the atmosphere is often too dry to form convective clouds. Wild fires may sweep across areas of semi-desert. Winter may bring ground frosts and even air frosts around dawn, in particular over higher ground.

From as much as 3500 years ago, we have records of the severe weather in desert areas, the harshness of the arid zones and the plentiful harvests available where there is an adequate supply of water, as recorded in the Bible. In particular, the fresh silt-laden fertile waters of the Nile supported a long-lived and very successful society for centuries. The Tigris, Euphrates, Indus, upper Ganges, upper Niger and Colorado rivers also support relatively large populations, despite the aridity of the surrounding environment. Although less bountiful, partly due to its meagre flow, the Jordan has also allowed successful settlement over millennia.

In some cases there is a historical reason for settlement in these locally marginal zones. It is known that the Sahel has been a wetter zone than is now the case and that between about 11,000 and 5000 years ago much of the Sahara was wetter than it is now (Burroughs 2005). Population pressures, however, have increased rather than decreased settlement in these marginal areas. Indeed, the relatively fertile soils of the desert areas may be an attraction, albeit a passing one.

Elsewhere, trade and the presence of runoff waters from mountain chains have supported relatively large societies, such as those in Iran, Afghanistan and southern-central Asia.[2] Fishing and trade, as well as pearl diving (a result of sandy desert wash-off), have provided important support for coastal communities (although oil production now provides most income in the Middle East and a large proportion of income in southern California).

An Introduction to the Meteorology and Climate of the Tropics, First Edition. J F P Galvin.
© 2016 John Wiley & Sons, Ltd. Published 2016 by John Wiley & Sons, Ltd.

7.2 Wind and weather in the deserts

The lack of vegetation, locally strong thermal contrasts and a predominantly (although by no means exclusively) anti-cyclonic regime make deserts frequently windy environments.[3] There has been much study of desert winds, in particular in the Middle East and North Africa; many of the wind regimes in these areas have well-known names (Weight 2001). These named winds are associated with varying weather types and many are seasonal. Appendix 2 describes named winds in the tropical zone.

Across the Sahara, the so-called Harmattan predominates. In the northern summer these north-east trade winds sweep across a more northern zone as the monsoon sweeps north into central parts of West Africa. In the northern winter its more-southerly sweep crosses much of the desert, at times reaching the moist Guinea Coast, at the same time allowing low pressure into the desert's northern margin, along the Mediterranean coast as extra-tropical air displaces tropical air southward (Barkan & Alpert 2010). Similar winds are observed across the Thar desert of southern Pakistan and north-west India, although here the north–south progression of these winds is much less than in West Africa.

Winds are strong over the deserts due to the lack of vegetation, giving a low surface roughness, as well as a combination of steep pressure gradients and anti-cyclonic curvature. Wind speeds are further increased by day due to convective decoupling as strong insolation increases instability, thus reducing friction in the surface layer. Steep pressure gradients and anti-cyclonic curvature are particularly important over the sea, where the trade winds, associated with the subtropical anti-cyclones, are frequently strong.

The Khamsin frequently brings hot weather with very low relative humidity northwards across the Mediterranean Sea in summer. The Shamal of the Arabian Peninsula may reach gale force. The Kaus is a moist but strong south-easterly wind that blows across the Arabian Gulf during the winter. Oppressive humid weather is frequently the result in northern parts of the Gulf.

Any strong wind over a dry, clay or silt surface (or one partially formed of clay or silt, which is common in desert areas, where a lack of rainfall means that surface particles are poorly sorted) can generate dust storms, sometimes over large areas. Clay and silt are often present in deserts and satellite imagery frequently reveals these storms; dust is sometimes carried great distances away from its origin (Knippertz & Stuut 2014). Over sand, the effect is different; particles are only lifted briefly, generally bouncing along the surface (see Box 7.1). Most of the strong winds described below can lift dust or sand. De Villiers and van Heerden (2007a, 2011) discuss the local effects of raised dust in the United Arab Emirates.

Although strong winds can generate dust storms in many desert areas, some locations are favoured as they have extensive clay surfaces (usually close to rivers or in former flood plains). Much of the southern Sahara, including the Sahel region between Chad and Mauritania, the northern Sahara of central Algeria, southern Tunisia and north-western Libya, the 'empty quarter' of Saudi Arabia, the Thar desert, the Kalahari-Namib desert and the Euphrates valley of Syria and Iraq are six tropical-desert areas that often produce extensive dust storms when strong winds develop following dry weather (Washington et al. 2003). The first is by far the most significant source region and is an infrequent source of dust falls in western Europe (Burt 1991, 2014), the second is an occasional source of dust in central Europe and the Middle East (Galvin 2012; Varga et al. 2014; Tilev-Tanriover & Kahraman 2015).

Diurnal modulation of wind speed (and direction) is often evident. Winds are usually much stronger from late morning to early evening than at other times, while the air is

Box 7.1 Raising and transport of dust

Although commonly known as sandstorms, it is clay particles that are most often raised and carried long distances from desert areas as dust storms.* Sand is far too large to be lifted by convection. However, sand may be carried by strong winds by a process known as saltation, the sand bouncing along close to the ground, perhaps up to knee height (Andreotti et al. 2002). This process has a part to play in the lifting of material from the surface: as particles of varying sizes collide during saltation, the smaller ones will tend to gain kinetic energy at the expense of the larger (sand) particles, so are lifted higher. In this way silt particles may also form part of the material carried across and away from desert areas. Once away from the surface, strong winds will carry them further, especially where the boundary layer is unstable.

Weather determines whether significant amounts of dust are lifted and transported. First of all, the weather needs to have been dry for some weeks in the area from which the dust originates. Second, winds need to be strong across the surface, so that the clay can be lifted from the surface. The friction velocity is about 8 m s^{-1} and if the dust is to be lifted away from the surface the lowest layer of the atmosphere must be unstable. It is thus possible for clay particles to be lifted to heights between about 2000 and 2500 m. Occasionally, significant amounts of dust (the clay particles) may be lifted above 3000 m. If winds are strong enough throughout the lowest layer of the atmosphere, the clay particles may be kept airborne to be transported over long distances. This combination of weather is most often seen in autumn, as the air becomes cooler, but there is still strength in the sun, or in spring, when the cool air is heated strongly as the overhead sun moves poleward. Although it is rare in the summer, when winds are generally stable, significant dust may occur at almost any time of year. Clearly, wind direction determines which areas may be affected: the eastern Mediterranean and north of the Sahara have most dust when the wind is in the west. The Arabian Gulf sees significant dust during Shamals and the southwestern Sahara is affected in easterly winds.

The presence of low-level instability and, often, the presence of a capping layer of warm air, combined with an increase in wind speed with height, often means that dust is concentrated in a layer above the surface that is then able to affect areas of high ground. Where winds are lighter, near the surface, the dust tends to settle out more rapidly than where winds are stronger, above the surface.

Once lifted into the air, dust may be remarkably persistent, especially in the lower layers of the atmosphere and it often requires rainfall to wash out the vast majority of the dust. This is a benefit to agriculture, as clay from desert lands has not been subjected to much rainfall so still contains a high concentration of nutrients. (Clay is also hygroscopic and so assists in condensation from water vapour, helping clouds to form and rain to fall.)

The depth of a dust storm also has a profound effect on temperature. For instance, Slingo et al. (2006) report a 10°C drop during a dust storm in Mali.

In general, so much dust is lifted near the surface that visibility becomes very poor (below fog limits) and the effects may be seen far from the source of the dust (Galvin 2012). Less and less dust is carried to greater and

*Clay particles are very small (the longest axis <0.002 mm) and are generally flat, therefore they are aerodynamic (Williams 1983), favouring long-distance transport. Recent research has revealed that larger particles (silt) may form a significant proportion of the dust lifted from desert surfaces (Washington et al. 2003; Knippertz & Stuut 2014) and that some of this material may also be carried very long distances (Varga et al. 2014).

greater height. Nevertheless, at height the dust may be carried greater distances, often until increasing stability in the boundary layer allows it to fall out. At times, if precipitation has not scavenged the particles, the dust may travel as much as 2000 km (the distance between central Algeria and the southern UK). Dust frequently reaches Oman from central Iraq, indicating its long lifetime in suspension. Small amounts of dust may be observed for day after day over Cyprus, having been lifted over Iraq or the northern Sahara desert.

unstable to surface temperatures. Although this is evident over land across much of the world, the modulation is particularly large over deserts, where there is little or no vegetation to retard surface winds and the temperature range is large under predominantly clear skies and a dry atmosphere. As winds drop during the evening, the temperature falls increasingly rapidly, giving many inland desert areas a very large diurnal temperature range. Locally, this may be as much as 30°C and is frequently 20°C.

Hot weather is often associated with the climate of the world's tropical deserts. However, this is not always the case and cool nights are a characteristic of all deserts in winter. The Atacama desert of South America and the Sahara and Namib deserts of Africa have areas of high ground where temperatures cannot rise to the levels seen near sea level (see also sections 7.2.1 and 11.3). In winter these areas may become particularly cold and a case study of snowfall on the Hoggar massif of Algeria is described in Appendix 5.

7.2.1   The progress of frontal systems across Arabia: Kaus, Suahili and Shamal

In this chapter the named winds of Arabia have been chosen to illustrate the association of wind and weather in deserts, although similar (often unnamed) wind and weather systems may be seen elsewhere. A somewhat different progress of winds occurs on the margins of the Sahara desert; the change of wind through the year along its monsoonal southern margin is described in Chapter 8.

As presented in Chapter 4, severe weather over south-west Asia is not dependent on the development of upper-tropospheric troughs. Indeed, in winter there is a semi-permanent trough at 200 hPa over the region. It is the progress and development of mid-tropospheric troughs, best observed at 500 hPa, that determine severe weather in this region. In winter, frontal systems develop in association with these mid-tropospheric troughs. A progression of winds develops over Arabia in response. Initially, there is often a moist south-easterly Kaus, followed by a dry warm south-westerly Suahili and finally a cool dry north-westerly Shamal, marked by the passage of the cold front.

The Kaus often brings oppressively warm, moist weather to the areas around the Arabian Gulf, accompanied by layer cloud and rain. High ground, such as the Zagros mountains of south-western Iran, can experience prolonged and heavy rain, locally accompanied by thunderstorms, generated from medium-level instability. Locally, this wind may reach gale force; speeds tend to increase rapidly and shipping may be badly affected. In particular, the small wooden boats typical of the Gulf may need to remain in port. Lives have been lost on these small fishing and trading vessels, in particular in the Strait of Hormuz, where this wind can be funnelled between the Zagros mountains and the Hajar mountains of Oman and the United Arab Emirates.

The dry, and usually very hot, south-westerly Suahili follows. Winds can be strong, bringing high temperatures. As the cold front approaches, the Suahili may reach gale force. Across the south of the Arabian peninsula, however, the Suahili brings moisture during the summer monsoon; instability may be released over the Asir mountains of Saudi Arabia and the Hadramawt of Yemen.

Initially, close to the cold front that marks the boundary between the Suahili and the north-westerly airstream behind it, there is usually a medium-level cloud mass, associated with thunderstorms and, in places, heavy rain or hail (Fig. 7.1). The likelihood of heavy rain and thunder in the moist air is indicated by θ_w falling with height between 850 and 500 hPa (Potential Instability Index, $P < 0$), hail by the presence of vertical wind shear within cloud, combined with a high level of convective available potential energy. Destabilization may be assisted by the intrusion of stratospheric air into the troposphere (Browning 1997).

The worst of these frontal storms may be a significant weather hazard. Flash floods may occur, washing away the surface and inundating wadis. These storms can also be a serious aviation hazard. On 13 March 1979, a Boeing 727 was brought down in one of these storms on approach to Doha, Qatar, resulting in the loss of 49 of the 64 passengers and crew. Similar poor conditions spelled disaster for the helicopters

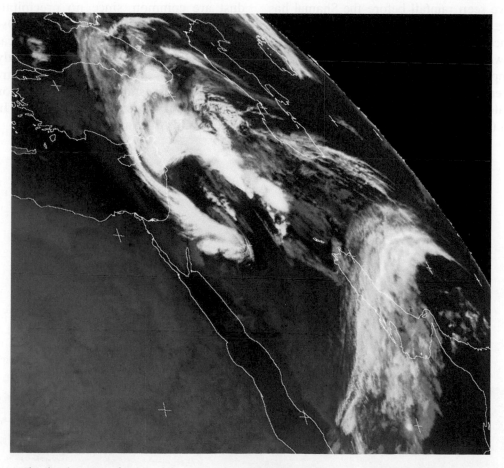

Figure 7.1 The development of areas of deep convection over the Middle East, associated with a disrupting westerly upper trough at 0100 UTC on 2 April 2006. Reddish areas indicate intrusions of dry air. Three main areas of significant convection can be seen. In the north thick (frontal) cloud with occasional cumulonimbus is affecting Syria and northern Iraq. Wrapped around the upper low is a smaller area of cloud with a few cumulonimbus tops over Jordan, Israel and northern Saudi Arabia. This system brought some very heavy rain to Baghdad, Iraq. In the south east, much of the cloud is high, although some embedded cumulonimbus clouds can be seen along the southern edge, over Saudi Arabia's Empty Quarter and parts of the United Arab Emirates – a relatively common occurrence. © EUMETSAT.

attempting to rescue American citizens imprisoned in Tehran on 25 April 1980.

The winter Shamal is the north-westerly wind that follows the cold front. Its effects are greatest in Iraq, northern Saudi Arabia and the northern Gulf states. Winds are usually strong and may reach gale force locally.

Following the potentially damaging stormy weather, the winter Shamal adds to the discomfort and danger from sand or dust.

Dust is often raised by the Shamal along the Tigris and Euphrates valleys, reducing visibility to very near zero at ground level (Fig. 7.2). Areas that have seen rainfall before the Shamal has set in are, however, spared the worst of the lifted dust, since the clay is bound together when wet.

Following the Shamal, or in shelter from it, the low temperature and humidity of the air mass have been known to allow a frost in parts of the Gulf states, as occurred on 20 January 1964 in Bahrain (Adel Deham, Bahrain Meteorological Department, personal communication) Rapid temperature falls are characteristic of the change from Suahili to Shamal. On the previous day, the largest fall in maximum air temperature from one day to the next had been recorded at Muharraq (Adel Deham, Bahrain Meteorological Department, personal communication).

Although winter Shamals are usually short-lived, lasting only a few days, early summer usually brings a long-lasting north-westerly, popularly known as the 40-day Shamal. Very high temperatures accompany the strong winds in the western Gulf. Haze and lifted dust are common, since no precipitation is associated with the summer Shamal. Over the northern Gulf states, lifted dust and occasional dust or sand storms are most frequently seen at this time of year as the sub-tropical high becomes established in more northern latitudes, but pressure falls over the Iranian plateau and Afghanistan.

Figure 7.2   Dust storm over Mali. Note the depth and density of the plume of clay particles and the presence of convective cloud above the plume. © Grant McDowall/naturepl.com.

7.2.2 The Nashi

Another wintertime wind that affects the Arabian Gulf is the Nashi. This is a cold katabatic flow channelled through gaps in the Zagros mountains of Iran, developing when the Asiatic anti-cyclone is at its peak (Meteorological Office 1994). It often reaches geostrophic speed and has many similarities to the Bora of the Adriatic (Weight 2001).

In the strong winds of the Nashi, dust may be lifted in the deserts of Iran and carried across the Strait of Hormuz into the south-eastern Arabian Gulf and the Gulf of Oman. The visibility reduction due to dust in the Nashi is often more significant than that carried by the Shamal into the Emirates of Dubai and Abu Dhabi (de Villiers & van Heerden 2011).

The duration of the Nashi is usually only 1–3 days and it often ends as a depression approaches from the west or as the anti-cyclone to the north subsides. In these cases there may be convergence with moist south-easterlies (Kaus winds) ahead of the system. Heavy rainfall is the usual result. Convergence with relatively moist Shamal air over the Arabian Gulf may also produce showers at times (de Villiers & van Heerden 2011).

7.2.3 The Scirocco

This is the most common name for hot dry southerly winds off the Sahara desert (Weight 2001). Various names are given to these winds, including Khamsin (Egypt and Malta), Ghibli (Libya) and Chili (Tunisia). Spring is the most common season for these winds.

The Scirocco develops on the eastern flank of low-pressure areas over the Sahara. Lifted dust may be carried to great altitude across the Mediterranean Sea or Red Sea to reach Europe or western Asia (Barkan & Alpert 2010).

The most notable effect is the rapid rise of temperature and fall of dew point that occurs along Africa's north coast as the Scirocco develops. On average, the temperature rises 8°C, while the dew point falls 6.5°C. Very

occasionally, associated with late springtime cold fronts, thunderstorms may develop from deepening medium-level cloud during the Scirocco.

7.2.4 The Harmattan

The common name for the north-east trade winds across the Sahara desert is Harmattan. These winds are dry, but become potentially unstable as they cross the desert and often provide the heat required to generate convection in the moist air of the summer monsoon over West Africa (section 8.6), close to the line of convergence of these air masses (Kendrew 1937).

7.2.5 Dust devils, waterspouts and hail

The great heat of the desert surface in summer often gives shade temperatures above 50°C around midday and frequently generates dust devils as a result of intense insolation. As indicated by their name, dust devils are evident from the plume of dust they lift, which may reach hundreds of metres (Fig. 7.3). The most vigorous of these dust devils can cause damage to flimsy structures and all those that develop over clay soils increase the dust load in the lower atmosphere. Occasionally, convection associated with cumulonimbus clouds may develop a tornado or large hail, resulting in damage to property and loss of life (Membery 1983a; Rodda 1983).

Odd though it may sound, hail occurs comparatively frequently in the hot deserts. The great summertime warmth of the Arabian Gulf and other water masses surrounded by hot deserts may also generate waterspouts in suitable conditions. They typically occur after hot weather, during periods of cold advection with relatively light winds and suitable vertical wind shear. They can only occur, however, where there is relatively small but conducive wind shear, changing gradually with height. The dryness of the lower troposphere means

Figure 7.3   Tall dust devil seen near the river Nile, Egypt, May 2006. © Sue Wilson.

that much of the rainfall produced from convective clouds evaporates before it reaches the ground, but large hail melts or sublimates much more slowly and is thus relatively likely to reach the ground. Large hail destroys crops, shredding the leaves and breaking the stems. It is no wonder that the people of the Middle East have found hail such a curse throughout recorded history (e.g. Exodus 9:23–25; 9:31–32).

7.2.6   Forest fires

Forest fires are relatively common in dry lands. Until recent decades these were overwhelmingly a natural consequence of the desert and, in particular, semi-desert weather. More recently, the carelessness of humanity has increased the number of fires and the destruction caused by them.

Dry thunderstorms frequently precede wet weather or occur as upper troughs move across these lands. Vegetation, where it occurs, may be exceptionally dry and combustible if struck.

Trees protruding from the landscape are natural lightning conductors and will attract ground strokes, readily catching fire. However, it is in the relatively dry savannas that the greatest consequences of fire may occur.

Where there is tree (or relatively dense scrub or crop-land) cover, fires started in a single tree may soon spread through neighbouring vegetation. The windy climate of the dry lands can readily spread these fires and as the fires become established they can generate their own mesoclimate, strengthening winds and drying vegetation along their path as a result of their great heat. Temperatures in the heart of these forest fires may reach more than 600°C, readily consuming all in their path – and hard to fight.

7.2.7   Low-level jets

An additional hazard for aviation, in particular in summer, is the wind shear associated with a strong low-level nocturnal 'jet'. In this case, the so-called jet is a strong wind close to the surface, although it is not as strong as the jet stream of the upper troposphere. The formation of a temperature inversion in the late evening decouples the surface layer from the atmosphere above. This removes the retarding effect of surface friction, allowing the air above the inversion to accelerate to super-geostrophic speeds by the end of the night.

A nocturnal jet is frequently observed between midnight and dawn along the south-eastern shore of the Arabian Gulf. Here, in conditions of light and variable surface winds (typically a north-westerly gradient with $V_g <$ 10 m s^{-1}), a low-level jet maximum in excess of 20 m s^{-1} may be observed below 300 m. This presents a serious wind-shear hazard to aircraft on take-off or landing. Figure 7.4 shows three cases where a marked temperature inversion was present and winds reached super-geostrophic speeds at low levels (Membery 1983b).

In June 1950, prior to the acknowledgement of the existence of this phenomenon, two Air France DC-3 aircraft stalled on approach to Bahrain. On 12 June, there were 46 fatalities and on the 14 June, 40 fatalities (including two investigators charged with finding out what had happened to the first aircraft). On both occasions the subsequent Courts of Inquiry concluded that the pilots had failed to keep an accurate check on their altitude and rate of descent and had flown into the sea. In reality, each pilot had encountered increasing head winds, resulting in increasing lift and, as a result, had reduced engine speed. As the head wind fell away to near zero on final approach, the aircraft were unable to maintain the glide-path altitude and had dropped into the Arabian Gulf some 5 or 6 km south-east of Bahrain International Airport.

The nocturnal temperature inversion disappears during the early morning as daytime heating quickly warms the surface and destabilizes the boundary layer.

7.2.8 Sea breezes

It would be difficult to overestimate the importance of sea breezes to coastal areas of the world's tropical deserts and semi-deserts.

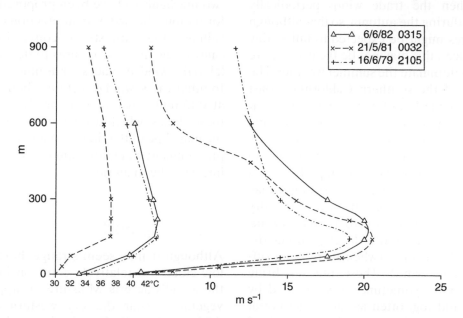

Figure 7.4 Low-level profiles at Bahrain with marked temperature inversions present. The date and time of launch of each balloon are given in the key (from Membery (1983b)).

All coastal areas in the tropics experience sea breezes, although they predominate along western and equatorward coasts. Initially there is a seaward flow, often initiated by cooling over land. As warmth increases over land, pressure falls and by late morning the airflow near the surface, up to about 1200 m, flows inland. At first, it flows at a large angle to the isobars and, where there is no barrier to block it, may reach tens or hundreds of kilometres inland. Late in the day, the flow is closer to geostrophic balance and flows almost parallel to the coast. At this time its speed often reaches its maximum, easing as temperatures fall rapidly inland around dusk.

Sea breezes are seen most often in high summer, but are relatively rare in winter. This leads to climatic features that may seem peculiar in the desert coastal zone. In general, the sea breeze stops temperatures from rising to the level seen inland and moistens the airflow, often making it feel more uncomfortable. However, as initial wind speeds need to be light (usually less than 10 m s^{-1} in the boundary layer), stronger offshore gradients may bring very warm air to the coast. This is particularly notable when the trade winds periodically strengthen during the autumn, so that although temperatures might be expected to fall as day length reduces, this may not be the case where sea breezes dominate the summer weather. The Santa Ana of the southern Californian coast may be particularly hot, with temperatures at the coast rising in excess of 45°C by day during the autumn. These winds are also warmed by descent from the Sierra Nevada. They are so dry that the fire risk increases rapidly during their onset and, of course, their strength may spread fire rapidly, especially where the dry winds reach the relatively verdant coastal zone.

Sea breezes frequently develop along the east coast of Australia (when there is an offshore gradient wind). These breezes bring moist air inland, sometimes accompanied by low cloud and fog, often as far as the Great Dividing Range and occasionally as far inland as 290 km (Mathews 1982).

There is more discussion of the sea breeze in sections 7.3 and 7.4.

7.2.9 Desert landforms

The strength of the wind in arid sandy desert areas is illustrated by their common land forms: the barchan, transverse dune, stellate dune and seif dune. The type formed is dependent on the availability of sand, the mean wind speed and its variability of direction. Where sandy sediment is in short supply, barchans grow as crescents with an apex across the prevailing wind and horns downwind (Fig. 7.5). They move downwind at between 15 and 60 m yr^{-1}, strongly dependent on barchan height, and develop when the mean wind speed is typically 10 m s^{-1} or more (Andreotti et al. 2002). The height of these dunes is typically up to 10 m. Strong winds that are variable in direction produce stellate dunes, which may grow to be as much as 250 m high (Williams 1983). If there is adequate loose sand, transverse dunes develop where there is little variability in wind direction. Most of these are only a few tens of centimetres tall. Seif dunes are less well explained; two mechanisms have been proposed for their formation. The first is an interaction of the prevailing wind with strong convection, which causes long linear structures to develop parallel to the wind, in places for as much as 200 km. In other cases, wind directions about 90° apart at different times of year are likely to form these dunes, which always have crests across the main line of the ridges (Williams 1983). The predominant wind direction in these cases is in line with the main ridgeline.

7.3 Fog and low cloud

Although it may seem strange, both fog and low cloud are relatively common in some of the world's deserts. Moisture to support scant vegetation near the dry western coasts of Africa and South America is mainly supplied by wet fogs.

Figure 7.5 Barchans in Death Valley, California on 28 October 2002. The orientation of their 'horns' indicates the prevailing wind direction – in this case, south-easterly. Note also the bare rock in the distance, emphasizing the slow rate of erosion in desert areas.

Any desert area near a source of moisture may see fog at times. It is surprisingly common along the Arabian Gulf coasts (de Villiers & van Heerden 2007b), where it may prove a very uncomfortable experience – and even occasionally a dangerous one – when the fog-point is high. Sea-breeze penetration can allow fog or low cloud to form well inland, as documented by Fisher & Membery (1988). Away from the world's oceans, especially where there is a shallow warm lake or sea, fog may form following clear nights. This is most common in autumn and winter. From personal experience, warm fog feels as if you are taking a shower whilst dressed. Its formation is dependent not only on radiative cooling, but also on sea-breeze development during the previous day, carrying moist air inland.

Low cloud and fog may also form where air is advected from relatively warm seas or lakes across rising ground, in particular during the morning (see e.g. Matthew 17:1–7, Mark 9:2–8). At times, south-easterly winds may bring low cloud and fog as far as the coastal mountains of Syria from the Arabian Gulf overnight in winter.

7.4 Severe weather in the dry tropics

A characteristic of desert areas is that what little precipitation occurs is generally of short duration and when precipitation reaches the ground it may be heavy. There is also a strong link between altitude and the likelihood of precipitation, particularly in drier areas. This provides an important resource to lowland areas, as run-off becomes rivers along which settlement is possible.

Radiosonde profiles suggest that the main anti-cyclonic temperature inversion over hot desert areas is often at about 4 km, restricting the development of isolated convective cloud (Hastenrath 1991).

Thunderstorms may develop in one of two ways over the hot deserts. Following the highest temperatures of the day, where the inversion is somewhat higher, convection may be sufficiently vigorous to form cumulus clouds. Occasionally, where there is cool air aloft – typically ahead of an upper trough – this may develop into cumulonimbus by late afternoon as the temperature inversion breaks down. Given the low dew point of air over the hot deserts, the base of these cumuliform clouds is usually above 3000 m (Fig. 7.6) and may occasionally be as high as 4000 m. Clouds may develop quickly, forming cumulonimbus in about an hour and dissolving almost as rapidly. Virga may be seen below the cloud, mamma may deepen the cloud base and in many cases the strong downdraught gusts reach the ground, the air below the cloud base having cooled to evaporate the precipitation, which does not reach the ground.

More often, there is sufficient moisture at medium levels for altocumulus to destabilize as cool air is advected over warmer air near the surface (and the semi-permanent temperature inversion at the top of the boundary layer). This occurs either ahead of an upper trough or as the upper troposphere (above cloud tops) cools during the evening. These unstable altocumulus castellanus clouds are not as deep as convective clouds in the humid tropics. Given that a cloud-top temperature of −20°C is required to cause glaciation – the process whereby water is turned readily to ice – such clouds can easily reach this level, at about 8000 m. These cumulonimbus clouds rarely have tops above 12,000 m, however, even where temperatures are falling in the upper troposphere.

In winter significant troughs may extend south across the deserts as the semi-permanent tropical upper-tropospheric troughs (TUTTs) are displaced from their common distribution (Miles 1959).

This is particularly important in the Sahara. Instability is great as the extra-tropical air spreads south. Welcome rain is often the result – at least over higher ground – and snow may fall at higher altitudes. This is described in Appendix 5.

In either case, thunderstorms may develop, with lightning bringing a risk of wildfire. Wildfire is a result of the ignition of dry vegetation by lightning and spread (where there is sufficient dry vegetation) by typically strong desert winds (Delgado Martin et al. 1997). In many cases the fire may be self-perpetuating in vegetated areas, as renewed vigorous convection from the fire itself sucks in strengthening winds. The semi-desert areas of the western USA and Australia, where most of the precipitation is from these clouds and there is dry vegetation (Hollermann 1993; Smith 2005), are particularly susceptible (although in some cases these fires might be set by discarded cigarettes or poorly managed camp fires). In the dry season savannas may also be affected by wildfires. In the savannas and semi-deserts above-average rainfall may increase the risk of wildfire. The rapid growth and thickening of vegetation increases surface cover and during the following dry season allows fires to spread readily.

Downbursts are a common consequence of precipitation from cumulonimbus clouds over the desert. Given their high altitude and the extreme dryness of the air at lower levels, most of the precipitation evaporates before it reaches the ground. As a result many desert thunderstorms are 'dry' at the surface, in particular over low ground (Knippertz 2007). The resultant cooling causes severe downdraughts with consequent hazards to aviation (Membery 1982) and, assuming no precipitation has reached the ground, a ring of dust may be seen spreading out below the cumulonimbus clouds, as shown in Fig. 7.7.

Short-wave upper troughs can also trigger the development of severe weather almost anywhere in the tropics, in particular in spring.

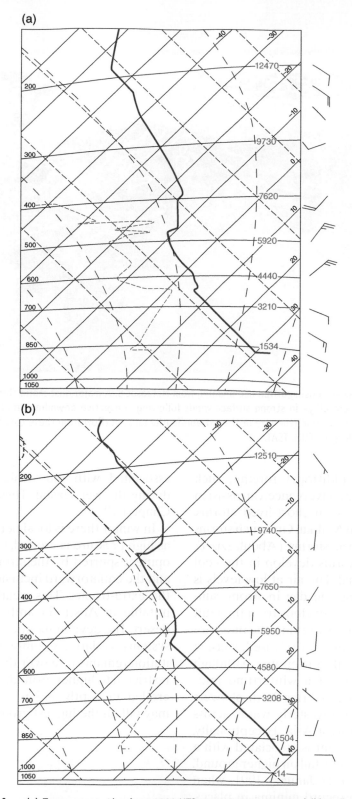

Figure 7.6 Tephigrams from (a) Tamanrasset, Algeria at 1200 UTC on 23 August 2006 and (b) Hail, Saudi Arabia at 1200 UTC on 26 August 2006. In (a) the potential for development of altocumulus castellanus between about 5200 and 6000 m can be seen. A Normand's point construction[4] suggests a base of cumulus clouds at 4500 m and free convection possible as isolated cumulonimbus towers to around 12,000 m. In (b) the layers of altocumulus, altocumulus castellanus and cirrocumulus between 5500 and 9000 m were accompanied by isolated cumulonimbus clouds with a base at about 4800 m and tops to 9700 m. Note the altitude of each of these two stations (indicated by the pressure at the surface): Tamanrasset 1364 m, Hail 1002 m.

Figure 7.7   Dust outbreak seen using difference imaging from Meteosat 8. A major dust storm (bright pink) over central-eastern Algeria and northern Libya in strong surface winds following convective downdraughts from the cloud seen to the north-east on 23 February 2006 (the size of the extraordinarily large outflow from these clouds the previous day can be seen as pale pink shades). © EUMETSAT.

Over the deserts, additional forcing, such as sea- or lake-breeze convergence, can assist in the development of such short-lived storms. Along the coast of the Arabian Gulf, such storms are known as Sarrayat storms (Abdulazziz & Essa 1994). These storms develop in the cool air of a gentle Shamal. The air at low levels is readily heated during the day in strong sunshine and a sea breeze develops from the Gulf during the afternoon. By evening, the moisture from the sea breeze may be released, causing a brief but torrential downpour. Although unusual, the passage of a wintertime mid-tropospheric trough can bring both showers and strong, cold, north-westerly winds across the deserts of south-west Asia. Exceptionally, when an active cold front is associated with a winter Shamal, snow may fall on higher ground. In a rare case on 26–27 January 1991, when temperatures fell to record minima in places, snow fell in western Iraq.

The Sahel (and other similar areas across the world) experiences torrential downpours associated with mesoscale weather systems during the wet season. These are discussed in Chapter 10.

In winter there is local periodic precipitation as high pressure is eroded, low pressure development spurred by the strong surface heating and the equatorward incursion of cool air from the extra-tropics. This weather is particularly notable in Australia and across the Sahara desert. The development of these low-pressure systems is relatively common, although precipitation remains rare over the arid lands, mostly restricted to higher ground. Tropical air may be eroded in depth and near mid-winter snow may fall on the highest ground.

7.5   The effects of desert weather

Living is hard in deserts and life is found mainly along the banks of rivers. However, it is not just the lack of water that makes deserts such a hostile environment. Tents, although relatively cool by day, can become very cold by night. Brick or concrete buildings, having a high thermal absorptivity, may become exceptionally hot in strong sunshine, the slow release of long-wave radiation making it uncomfortable indoors, in particular when the sun is almost overhead. In areas of mostly clear skies there is a need to cool buildings by day and, in winter, warm them at night. Thus there is a high demand for energy in desert areas, particularly in the developed world.

Concrete and tarmac surfaces heat up much more readily than bare-earth surfaces. With the sun overhead, concrete-surface temperatures may reach 60°C or more; tarmac considerably more due to its dark surface. Although a grass-covered surface is relatively cool, the temperature at the tips of mown grass may rise to more than 40°C under the midday sun. The effects are particularly notable in dry environments, but are also observed in the humid tropics when there is direct sunlight. Although the soles of the feet may become acclimatized to temperatures above 40°C, it is generally impossible to walk barefoot on concrete or tarmac surfaces around midday.

By day, under clear skies solar radiation may cause uncovered skin to burn and permanent damage is a common result of long-term exposure to bright sunshine.

Dehydration, however, is probably the most serious hazard in deserts, even where drinkable water is available. The hazard is increased by exposure to strong winds, even when temperatures are low. McNab's accounts (McNab 1994, 2008) of a combination of snow and strong dry winds describes the serious risks of dehydration and hypothermia on inadequately prepared soldiers. Indeed, the weather seems to have been more of a risk than gunfire from Iraqi soldiers pursuing the Special Air Service (SAS) raiders, two of the three SAS men that died in the raid having succumbed to hypothermia in the higher ground of north-western Iraq. Although uncommon, cold outbreaks occur in most years, caused by the incursion of air from central Asia, although decades separate such periods of cold weather.

Strong hot summer winds bring a different hazard. Air temperatures may reach 50°C or more and at a temperature above 37°C (blood temperature) the wind warms, rather than cools, the skin so overheating is likely. In humid air with a relative humidity greater than about 60% when wind speed is low, it becomes dangerously warm at somewhat lower temperatures and work must be limited or even cease.

Although it is dramatic enough to have your car overheat in hot, dry weather, the main danger of high air temperatures is to aviation. Gas turbine engines (as used on most commercial aircraft) produce thrust by heating and thus reducing the density of air drawn into them by a fan, thence into a compressor. The colder the air is to start with, the greater the possible reduction in density as the air is heated. Thus at high ambient air temperatures, less thrust is produced than at high temperatures. At temperatures above about 30°C, the effect becomes significant and some aircraft cannot generate enough thrust to take off, in particular those with older forms of jet engine.

The low density of warm air also reduces the lift available to aircraft on take-off and landing. This requires very long runways to be built in the tropics, often more than 3000 m in length where large passenger aircraft operate.

The problems may be further complicated by the assumption that where surface winds are light they will always allow take-off in a preferred direction. Wind shear is often a significant hazard, however, and, where available, soundings of the boundary layer (or aircraft reports) should be used by both forecasters and air-traffic control to ensure safe take-off and landing. Severe cases have resulted in the loss of aircraft, their passengers and crew (Meteorological Office 1975). In the warmest weather, operations may need to be suspended and many operators in the tropics prefer to have take-off slots during the late evening or early morning.

Although the air is cooler at altitude, the effects can be still worse in the tropics, since density falls with altitude, despite falling temperature (Meteorological Office 1994). Aircraft may need to be modified to optimize their performance from hot and high airfields.

However, the heat of the surface is a problem for all forms of transport. The surface may become unbearably hot to walk on during much of the day and the usable life of rubber tyres may be decreased.

As we have shown, lifted dust and sand are perennial hazards in many desert areas where clay soils are present. It is one of the most significant weather hazards. At the very least, it is unpleasant to be caught in a dust storm and the worst storms not only make transport (or even walking) impossible, given the very poor visibility, but may open wounds in the skin. Aircraft cannot operate in dust storms, since the dust scours and clogs engines, and aircraft in flight may even be brought down by significant suspended dust.

Even small amounts of lifted dust can make life unpleasant and this was a particular problem in Ancient Egypt. The high agricultural productivity of the Nile in near-continuous sunshine allowed large amounts of wheat to be grown and the mass production of bread. The dough was kneaded under foot in the open air in large stone vessels, but resulted in the inclusion of large amounts of dust and very gritty loaves, as has been discovered in mummified offerings in the Tombs of the Kings.

Across Australia, there is another consequence of the near-universally dry environment. Compared with stations at similar latitudes and climatic belts, lower humidity is generally observed, not least in the tropical zone. This low absolute humidity allows a greater proportion of solar radiation to reach the surface, in particular in the ultra-violet. The air is clear and in the absence of cloud, skin damage, intensified greatly by the 'hole' in the southern-hemisphere ozone layer, is a significant problem, particularly for those with pale skins (Chapter 14).

It is easy to see why weather that might be regarded as poor in the middle latitudes is a refreshing change for those living in desert areas. The popularly enjoyed monsoon fog that forms along parts of the south-east Arabian coast in summer is, perhaps, most notable.

7.6 Settlement and the over-use of scarce water supplies

In some areas, particularly those in the developed world, deserts are proving to be an attractive environment for social or personal reasons. Notably, desert areas of the western USA have been settled as areas of pleasant weather: hot, dry and sunny, but relatively comfortable. Desert metropolises, such as Las Vegas, have proved a great economic success, but at what cost to the environment? Indeed the population of the south-western USA has tripled since the Second World War (Diaz & Anderson 1995).

Furthermore, although salts, including limestone, are common in many desert areas, there are also areas of fertile clay soils, attractive for the growing of crops where there is an adequate water supply. The author has noted many crops, including rice, in the desert of southern California, requiring very large amounts of water, which readily evaporates in the dry desert environment. The building of the Hoover Dam in the 1930s across the Colorado River provided the water resources necessary for such settlement and agriculture in the US western desert, but it significantly depleted the water resource downstream, in Mexico, despite much-improved water-use efficiency (Diaz & Anderson 1995), there are also long-term consequences of such water use, including salination of the water supply due to the relatively high concentration of salts in desert soils and, over time, loss of fertility.

A similar situation applies in Israel, Palestine and Jordan: more than 70% of the river Jordan's flow is now extracted before it enters the Dead Sea. In Egypt there is concern about the use of water near the headwaters of the Nile. These rivers fertilize the deserts and are home to a number of fisheries, another important resource in many rivers of the dry tropics.

The dependence of these agricultural areas on limited water supplies brings great risks. No more so than the semi-deserts of Australia, where agriculture, including viniculture, has probably extended too far inland (Sheldrick 2005). Indeed, there are reports that the two great rivers of Australia – the Murray and the Darling – have a restricted flow into the sea following their union (Connell 2005). In turn, this lack of flow can cause salination of areas near the mouths of these rivers as seawater infiltrates the coastal plain.

7.7 Questions

1. What is a probable mechanism that brings the 40-day Shamal to an end?
2. Considering Luke Howard's cloud classification (see *Weather*, February 2003) and the subsequent explanation of the method of formation of convective clouds (Normand 1946), how would you recommend observers in dry climates and mountain zones report clouds formed by daytime heating?
3. Investigate possible limits to surface heating in direct sunshine. How does this vary with surface type?

Notes

1 The word 'potential' is important here: clearly, it is not possible to evaporate water that is not available on the surface. However, it is possible to calculate the amount of moisture that could be evaporated, if it was available, using surface wind speed, temperature and relative humidity values.

2 Southern central Asia can only be counted as tropical in summer, although it has a dry environment, in many ways similar to those described in this chapter. In summer, a weak ephemeral subtropical high lies over central and East Asia. In winter, the subtropical jet stream lies well to the south, so the area is not under the influence of tropical air.

3 In summer, pressure falls as temperatures rise over land, spawning monsoon circulations. However, prior to the monsoon rains, pressure remains high in the middle and upper troposphere. The windy conditions of the dry areas are also found around the periphery of subtropical anti-cyclones over the oceans, where seas are frequently rough.

4 A construction to reveal the probable base of convective cloud described by Normand (1946) allows the base of convective cloud to be assessed or forecast. As unsaturated air always ascends at Γ and its hmr is constant (see Appendix 4), the intersection of these two lines on a tephigram (or similar graph) shows the expected base level, although it assumes no entrainment of dry air.

8
Monsoons

8.1 Introduction

Seasonal monsoon circulations of the atmosphere are seen over all the tropical continents (Fig. 8.1). The most significant is that over Asia, although west Africa, southern Africa, northern Australia and parts of South America are also dependent on summer monsoons for the majority of their annual rainfall.

A reversal of winds near the surface twice per year and the development of large-scale deep circulation systems (Fig. 8.2) generally characterize monsoons. As a result, for most places within the area of the monsoon circulation one season is wet (summer) and the other dry (winter). There are many variations, however, and this generalization hides important local weather variations.

8.2 The summer monsoon over southern Asia

Over southern Asia in summer the distortion of the line of surface convergence of the trade winds (the ME) extends thousands of kilometres north of the equator. In theory, the convergence line forms the leading edge of the south-west monsoon, but in truth the story is much more complex.

Elementary accounts of the monsoon state that: (i) the heating of Asia causes pressure to fall, such that (ii) south-westerly winds bring a moist flow and copious rainfall across the Indian subcontinent. This by no means describes the progress of the Asian summer monsoon, however. The main complexity arises first from the distribution of mountains across the subcontinent, second, from the various water masses of the Indian Ocean and, third, from the broad-scale dynamics of the atmosphere.

Initially the northern branch of the ITCZ strengthens over the equatorial Indian Ocean as the Mascarene high-pressure centre intensifies and moves north over the southern Indian Ocean. During April, moist south-westerly winds set in at low levels south of it and the STJ moves northwards, eventually well to the north of the Himalayas (as described in Chapter 4).

In May, the surface of the Arabian Sea and Bay of Bengal becomes very warm (>29°C), allowing deep convection to spread northwards and the ITCZ to become diffuse. During the month, Thailand and Burma see wet weather as the moist south-westerlies deepen. Over much of India, however, it is only after sufficiently deep south-westerly winds in the lower troposphere can become established that the summer

An Introduction to the Meteorology and Climate of the Tropics, First Edition. J F P Galvin.
© 2016 John Wiley & Sons, Ltd. Published 2016 by John Wiley & Sons, Ltd.

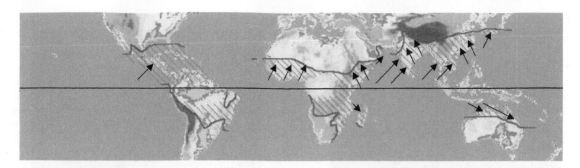

Figure 8.1 Land areas that have the majority of their rainfall in summer associated with the poleward motion of deep convection. Where appropriate, low-level wind directions that carry moist warm air are indicated. In areas where there are no arrows, winds are relatively dry, or are weak (as over South America). Shaded areas show the normal maximum extent of deep convection (where 850hPa $\theta_w \geq 22$ °C).

monsoon sets in. Before the south-west monsoon becomes established, easterlies must strengthen aloft in response to the warming and deepening of the troposphere over southern Asia (Galvin & Lakshminarayanan 2006). These reach jet-stream strength (>40 m s^{-1}) near the top of the warmed and thus deepened troposphere, forming the equatorial easterly jet stream (EEJ), centred over the southern Bay of Bengal. The EEJ may extend from the Philippines to the Gulf of Guinea at its maximum extent and usually lies between about 10°N and the equator. It forms the southern limb of an upper atmospheric circulation around a warm dome, usually centred over Tibet. During the month, humid, but predominantly dry, oppressive weather sets in across southern Asia (Fig. 8.2a). Afternoons can be very hot under almost cloudless skies, in particular over central and northern India with maximum temperatures often exceeding 45°C. By this time, the south-westerlies have an origin in the southern hemisphere as south-east trade winds (as described in section 3.3).[1] As this flow becomes established, outbreaks of deep convection develop in Nepal, Bhutan and the Indian states of Assam and Arunchal Pradesh, at times spreading across the Gangetic plains and Bangladesh.

As the cross-equatorial low-level flow becomes established to the south of India, air with a greater depth of moisture and increasing

cloud spreads north in bursts from late May or early June, accompanied by periods of rain, some of which is from deep convection. Assisted by a relatively steep pressure gradient between high pressure over the southern Indian Ocean and low pressure over northern India, the ME moves north. Surface winds become strong with speeds between 10 and 15 m s^{-1} throughout the summer, reaching gale force at times across the Arabian Sea (Membery 2001). The strength of the flow causes overturning of the ocean surface, thus cooling it (Fig. 8.2b). Where the south-westerlies are strongest, the cooling may be as much as 6°C. Along this coast and late, the south-east coast of the Arabian peninsula, the cooling is enhanced by upwelling, so occurs soon after the monsoon flow becomes established, although other parts of the northern Indian Ocean cool more gradually. Convection in the moist low-level flow is suppressed, not only by the relatively cool ocean surface, but also by coastal divergence of the south-westerly flow. Ultimately the flow becomes stable where the ocean is coolest. Over land, convection spreads north more readily and there is little rain in southern Pakistan or the Indian provinces of Gujarat and Rajasthan. Over northern India, convection may be assisted by additional (dynamical) forcing in the form of divergence in the high-level easterlies, easterly waves, orography or, locally, coastal convergence (Asnani 1993).

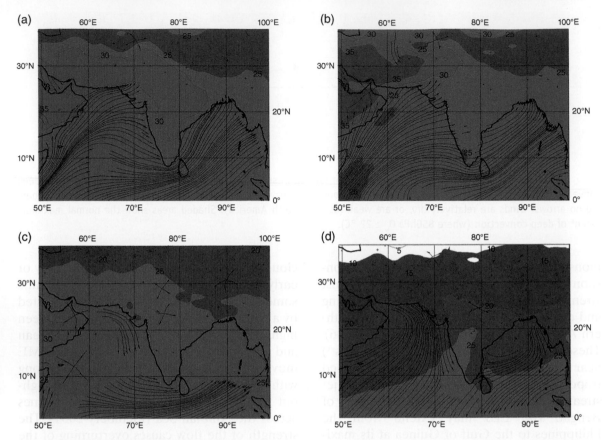

Figure 8.2 The progress of the Asian monsoon. Streamlines indicate the direction and, by their spacing, the strength of mean monthly surface wind; colours indicate mean monthly air temperature at 1000 hPa (approximately 100 m a.s.l.), ranging from white (<10°C), through dark blue (10–15°C), navy (15–20°C), purple (20–25°C), magenta (25–30°C), pink (30–35°C), to red (35–40°C). Data were obtained from ECMWF ERA-40 reanalyses (Simmons & Gibson 2000). For areas of high ground, temperatures are obtained by adjusting surface temperatures to the 1000 hPa surface using a standard lapse rate. (a) May – the 'pre-monsoon' period when surface temperatures are rising rapidly, but have not yet attained their peak over northern India, under largely cloudless skies. (b) August – the height of the summer monsoon with strong south-westerly winds across the Arabian Sea and Bay of Bengal, the focus of the monsoon rains lying across the Gangetic Plains with cool water lying off Somalia and Oman. (c) October – before the north-east monsoon becomes established winds are light across the sub-continent and there is sufficient warmth to bring rains to eastern India. (d) January – the height of the winter monsoon when cool dry north-easterly winds predominate and the weather is generally dry across southern Asia. Note the effect of sensible heating over land in all these images, as well as the comparative warmth of the Arabian Sea off the western Indian coast in January as the north-east monsoon current carries warm water from near the equator over the shallow continental shelf.

Deep convection is characteristic of the leading edge of the summer monsoon, bringing occasional thunderstorms and heavy rain. Locally, this may be accompanied by large hail. However, the flow is more stable to the south and much of the rainfall is from layer clouds, although land areas have a diurnal cycle that brings embedded convection within the extensive areas of layer cloud (Ramage 1971).

At its northward extreme, typically in mid-July, the south-west monsoon periodically reaches the Himalayas (Fig. 8.1). This mountain chain forms an area of discontinuity since the moist monsoon is relatively shallow and cannot ascend the slopes. At the same time, the ME effectively disappears, since the Himalayas protrude through the moist layer. Forcing by convergence is replaced by forcing due to ascent at this extreme northern deviation of the ME. Thus, by late summer the air with the highest potential temperature and highest total thickness is over Tibet, maintaining an easterly flow and helping to feed ascent up the slopes. Westerly winds make only occasional incursions north as far as the mountains and the ME is periodically evident over the Gangetic plains (Figs 8.3 & 8.4) during the peak of the monsoon season, associated with occasional low-pressure areas. The rainfall maximum, aided by orographic ascent, is the north, on the Himalayan slopes.

Although thunderstorms are common on the Himalayan slopes of northern Pakistan at the peak of the summer monsoon, rainfall is generally sporadic in these western areas and summer rain is generally less than that in winter, so this area does not have a 'true' monsoon climate. However, the south of the country has more rain in summer than in winter, although precipitation is poor in this part of Pakistan, much of which is desert or semi-desert. Although occasional mesoscale disturbances develop as increasing moisture spreads from the east and south into south-eastern Pakistan during the early summer monsoon, it is usually only in late summer, and then only occasionally, that rainfall reaches western Pakistan.[2]

Although the Himalayas ultimately block the moist monsoon flow, heating of the Tibetan plateau causes a diurnal cycle of winds along valleys crossing the mountain chain. Daytime anabatic winds may become particularly strong, locally reaching gale force along valleys such as that of the Sun Kosi in Nepal. These valleys are deep, being incised several thousand metres into the mountain barrier.

Despite the evident northward progress of the summer monsoon, its advance is not straightforward and rainfall usually spreads across India from the south-east.[3] Hence, we need a different mechanism to explain the progress of the monsoon than the simple motion of moist south-westerly winds. Indeed the moist flow across India is partially blocked by the Western Ghats, the mountains that mark the west coast of India (Patwardhan & Asnani 2000). There is copious summer rainfall on the western slopes of these mountains, but the area to their lee remains largely dry, in a rain shadow, the south-westerly air flow having lost some of its humidity. In order to bring monsoon rains across much of the rest of India, two mechanisms appear important.

The first is the development of the pre-monsoon trough at the surface over the Arabian Sea. The flow ahead of this trough, which may contain a tropical depression or cyclone, brings south-easterly winds and moisture from the equatorial zone across the Bay of Bengal and much of southern India, supplying a reservoir of moisture for convective development. The second important mechanism is the motion of small, low-pressure systems from east to west across Bangladesh, Bhutan, northern India and Nepal into southern Pakistan. Associated weather systems are usually carried by easterly waves. Some of these disturbances seem to have an origin in tropical revolving storms over the South China Sea, the vorticity from which is maintained at high levels and can cross the south-eastern part of the Himalayas. As the low-pressure areas cross northern India, south-easterly winds develop in their wake, bringing warm moist air from the northern Bay of Bengal. Thus rainfall is greater in the east than in the west and is periodic – in particular across northern parts – with spells typically lasting a few days, followed by several days of dry weather.

8.3 The summer monsoon over East Asia

This monsoon system must be considered separate from the South Asian monsoon, even though its driving mechanism – a fall of pressure over the central Asian land mass and the

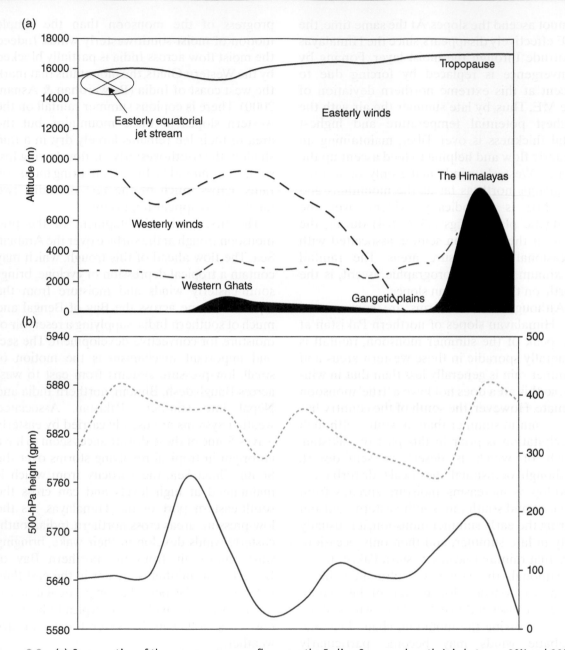

Figure 8.3   (a) Cross-section of the summer monsoon flow over the Indian Ocean and south Asia between 2°N and 32°N from radiosonde profiles at 1200 UTC on 15 July 2006. The tropopause is shown by ▬, the top of the low-level moist zone by ▬ and the level of the top of westerly winds by ▬. (b) Corresponding 500 hPa height with rainfall data for June–September, adapted from Rao (1976).

establishment of a wind system in association with it – is the same (Yihui & Chan 2005).

East of the Himalayas, the monsoon 'plum' or 'cherry-blossom' rains spread across much of

East Asia in three stages, preceding the onset of the true summer monsoon. During April, as fruit trees come into bloom, a shallow humid west- or south-westerly wind moves across

(a)

(b)

(c)

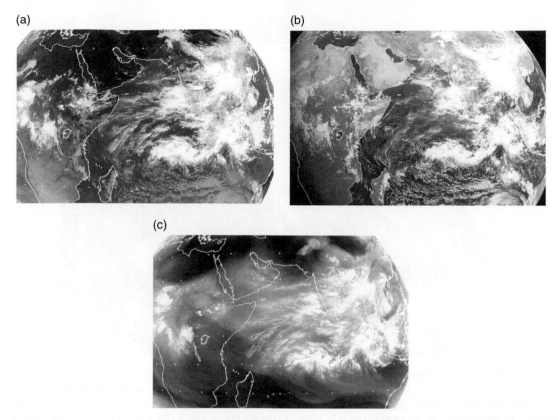

Figure 8.4   Infra-red, visible and water-vapour satellite images from Meteosat-5 of the south-west monsoon over Asia at 0600 UTC on 21 July 2007. (a) In the infra-red picture the cold cirrus clouds over the Bay of Bengal and India form elongated streaks, where the cloud tops are sheared off in the EEJ. (b) The deepest cloud, most likely to produce precipitation, is shown as bright white in the visible image. (c) The deep humid zone is indicated by light shades in the water-vapour image and can be seen to extend as far north as the Himalayan peaks. The most humid air is distorted north across India and Africa by the monsoon flows. Dry air remains over parts of south-west Asia, notably much of Pakistan, although a bulge of moist air can be see across much of the Arabian peninsula. The remnant ITCZ can be seen over the southern Indian Ocean, near 7°S. Courtesy of University of Dundee Satellite Receiving Station.

southern parts, ahead of the tropical air of the 'true' monsoon. This is followed by south to south-easterly winds that bring rainfall as a form of frontal system, known as the Mei-Yu front at the leading edge of the deeply moist tropical flow (Ding 1994). The leading edge of the humid air steps northward at intervals, centred around 23 April in southern China, 13 June around the Yangtse River valley (Fig. 8.5) and reaching some northern parts of China around 8 July (Table 8.1). As the south-easterlies reach around 45°N they bring most of this area's annual rainfall total (Asnani 1993; Ding 1994).

However, northern China usually remains north of the Mei-Yu front, the STJ having reached its northernmost excursion near 40°N. Thus, although there is a convective element in this northernmost excursion of monsoon rains, much of the precipitation is from layer clouds: altostratus and altocumulus forming a tropical front as the warm moist air interacts with the STJ. It is very rare for the tropical air mass to extend north of 40°N and temperatures rarely reach those of the tropical zone in Beijing. Ascent to the uplands of inland China aids the release of latent heat and rainfall is copious in

Figure 8.5 The Mei-Yu front over southern China at 1200 UTC on 12 June 2007 observed by MTSAT. This front, the leading edge of the East Asian summer monsoon, moves north gradually during the summer and can be seen between Burma and Kyushu, Japan. Along its northern edge, where warm moist air has advanced over cooler air near the surface, is a band of altostratus and altocumulus. At its southern edge is a line of cumulonimbus with a cloud-top temperature around −80°C, indicating an altitude of about 16 km. The deep convection formed over land where the 850 hPa θ_w was above 23°C. Courtesy of University of Dundee Satellite Receiving Station.

Table 8.1 The onset, duration and retreat of heavy monsoon rains in China (adapted from Ding (1994))

	Onset of monsoon rains	End of monsoon rains	Duration
South China (~20°N)	23 April (±5 days)	28 June (±5 days)	~75 days
Central China (~30°N)	13 June (±5 days)	18 July (±5 days)	~35 days
North China (~40°N)	8 July (±5 days)	27 August (±5 days)	~50 days

central areas, locally amounting to more than the totals in the south during the summer.[4] However, the monsoon may assist in interactions with upper-level disturbances to bring summer rainfall to northern China (discussed further in section 10.10).

Winds must back at low levels for deep humid air to spread across the region, since south-westerlies are partially blocked by high ground in Sumatra, peninsular Malaysia, Thailand, Burma and south-western China to depths between 1 and 5 km. These mountain chains have frequent rains, but an area thousands of kilometres in their lee are in rain shadow.

Rainfall is usually heavy and torrential rains are often seen. Flash flooding may occur locally.

As the rainfall peak moves north, drier conditions follow into southern China around 28 June and into the Yangtse River valley around 18 July. The peak of the heavier rain lasts about 75 days in southern China, 35 days around the Yangtse River and 50 days in northern China (Table 8.1).

The main area of rain has similarities at the surface to the South Asian monsoon. Low pressure, preceded by easterly winds, is seen in association with the precipitation area. However, development of the monsoon is aided in two ways that are not part of the monsoon in South Asia. In southern China, the development is early in the summer season, associated with the strengthening of easterly winds in the upper troposphere. This area is below the divergence at the right entrance of the EEJ. In the north of China and Korea, the STJ has a part to play in the development of the heavier rain. Even in summer the STJ is present across the north Pacific and speeds are high at times (Fig. 3.1). As the moist air moves northwards, the semi-persistent trough over East Asia again aids ascent by divergence (Lu & Lin 1982).

The northward movement (in somewhat variable, but distinct, stages related to both large-scale atmospheric dynamics and the relatively small-scale orography of East Asia) follows the poleward advance of the STJ as the north Pacific sub-tropical high-pressure area moves north (Ding 1994), allowing pressure to fall and the incursion of the warm moist air from the south-east.

8.4 Variations of rainfall in the Asian summer monsoon

The summer monsoon is highly significant in terms of rainfall across all the continents on which it develops. Most of the Indian subcontinent receives about 80% of its rainfall under its influence, mainly between June and September. The depth of south-westerly winds at the core of the Asian monsoon reaches about 5 km. However, the depth of the most humid air is no more than about 3 km (Asnani 1993). Along the southern ranges of the Himalayas very large amounts of rainfall are deposited, typically reaching more than 6000 mm (Pokharel & Hallett 2015), some

places recording amongst the highest global rainfall totals. For instance, Cherrapunji in the Assam region of north-east India, at an altitude of 1313 m, on the southern slopes of the Khāsi Hills, has an annual total of 10,824 mm of precipitation,[5] more than 5000 mm of it falling in June and July.[6] As the moist air reaches these ranges, surplus moisture is deposited as rainfall (Dhar & Nandargi 2005). However, the east–west distribution of the mountain ranges in relation to the moisture-laden winds brings considerable variation of both mean and daily rainfall in the Himalaya (Pokharel & Hallett 2015). Away from the Himalayan slopes, the zone north of 15°N also receives the highest rainfall accumulations during the summer monsoon. North of 25°N, in particular over land, heavy rainfall is somewhat more limited than further south (Asnani 1993).

Figure 8.4 shows monsoon activity across the Indian Ocean and Asia in July 2006. The main area of convection can be seen between Thailand and northern India, with more limited convective cells evident in southern India, on the Himalayan slopes and across Tibet. Air in the upper troposphere can be seen to be very moist in all these areas, whilst the Thar desert of India and Pakistan remains under the influence of subsided dry air.

Although moist south-westerlies are observed across the whole of the Arabian Sea during the Asian summer monsoon, stability and the influence of surrounding mountain chains are sufficient to keep rainfall to very low levels in coastal Yemen, Oman, Iran and Pakistan. This brings a desert or semi-desert climate, stretching from north-west India, through south-west Pakistan and Iran, across the Arabian Peninsula. Nevertheless, some parts of south-west Asia have relatively large populations, in particular in coastal areas, where trade and fisheries have long been important industries.

These industries are significantly affected by the monsoons, in particular in summer, when strong winds make sailing both difficult and hazardous. Even in these days of motorised vessels, high seas may be perilous[7] and many traders across the Indian Ocean are likely to be confined to port for at least part of the summer season. The increased numbers of fish that the monsoon-related upwelling brings, however, means that fishing tends to peak in summer off Arabia and Pakistan, despite the rough seas. Most fishing vessels of the area are small and relatively fragile. David Membery (2001) gave an excellent review of the effects of tropical cyclones and the monsoon on historical trade around the Indian Ocean.

One rather strange aspect of the summer monsoon along the south coast of the Arabian Peninsula is the effect of upwelling cool water along this coast. As well as promoting a good fishery, the upwelling water cools the overlying air. This causes condensation from the moist flow to be trapped close to the surface and, in some places, in particular the bay between Al Ghaydah, Yemen and Şalālah, Oman, there is persistent low cloud and drizzle, associated with the Khareef wind. In the relative shelter of the Şalālah Plain, which is surrounded by mountains, this weather is very popular with many an Arab holidaymaker. However, the increased moisture in the boundary layer also allows deep convection to develop from scattered high-based cumulonimbus clouds over the Yemeni and Saudi Arabian mountains.

8.5   The Asian winter monsoon

As Asia cools the subtropical high becomes re-established over the south of the continent and north-easterly winds bring dry weather across all areas north of about 10°N.

The establishment of the winter monsoon takes several months, however (Figs 8.2c,d); areas of heavy rainfall gradually retreat south in a rather erratic fashion between September and November.

8.5.1   South Asia

From late September the south-westerly winds at low levels decrease, allowing the Arabian Sea to warm. Indeed, much of the northern Indian Ocean warms, generating periodic widespread deep convection and heavy, thundery rain. Occasional tropical depressions or cyclones can be expected after the south-westerly winds become light over the Arabian Sea and Bay of Bengal between September and December (see Chapter 9).

During October, the south-east of India has another period of rainfall as north-easterly winds become established. Warm moist air is carried inland from the Bay of Bengal and a broad zone of cumulonimbus develops inland of the coast – in particular on the Eastern Ghats – over an area as much as several hundred kilometres across (Asnani 1993). This is an important period of rainfall for this region.

Whilst this rainfall is usually relatively modest and is, at best, erratic, heavy rain sometimes occurs. At the beginning of October 2009, heavy rain fell for several days in Andhra Pradesh, Karnataka, western Maharashtra and Goa. Dams feeding the Krishna River were full after plentiful rains in September (ending a comparatively dry summer monsoon season). The release of water from them contributed to serious flooding in the valley in which hundreds of people died, often as their houses were washed away. Crops, including maize, rice, sorghum and millet, were inundated and hundreds of cattle died. The extensive inundation is shown in Fig. 8.6.

The heavy rains were associated with a slow-moving depression at the trailing edge of the moist monsoon. Easterly winds on the northern flank of this low brought copious moisture inland. As the air rose towards the eastern slopes of the Western Ghats, it came under the influence of an upper-tropospheric low, centred near Pune. Dry air north and west of this upper low assisted the rapid development of deep convection, as well as its maintenance after the effects of daytime heating were

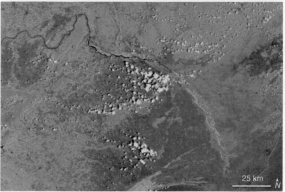

Figure 8.6 Flooding of the lower Krishna river valley, India showed by false-colour images from Terra/MODIS. The upper image shows the flooded valley on 5 October 2009 and the lower image the valley as it appeared before the flood on 10 September 2009. Courtesy of NASA Rapid Response Team.

removed. Depressions are comparatively common over the Indian Ocean at the trailing edge of the receding summer monsoon, but rarely cause widespread flooding. These are discussed in section 9.6.

Although skies are mostly clear south and west of the Himalayas during the winter monsoon, stable atmospheric conditions often lead to poor visibility. This is particularly noticeable around the industrial cities of northern India and southern Pakistan. At times fog may form in the early hours, clearing only gradually during the morning.

Unusually, moist lower-tropospheric westerly winds may return to northern Pakistan,

northern India and southern Nepal. Gradual ascent from the valleys of the Ganges and Indus may allow extensive areas of stratocumulus to form that may persist for days, while the moist feed continues. Somewhat more commonly, extensive areas of cloud may form at medium and high levels ahead of the southwest Asian trough, typically across central and eastern India. Occasional incursions of moist air will also produce cloudy skies across the Eastern Ghats and Western Ghats.

8.5.2 East Asia

The return of the winter monsoon flow is associated with the return of high pressure building from the west, quickly suppressing precipitation towards the end of September (Lu & Lin 1982).

As the moist monsoon retreats, the re-established cool north to north-easterly flow across Japan, Korea and China begins to interact with the re-established westerly flow above it, associated with the strengthening and southward motion of the STJ, close to 30°N (Fig. 3.1). Initially there is convection to middle-tropospheric levels, but maritime stratocumulus soon forms as cooling of the boundary layer continues, at times accompanied by stratus in areas near to the coast (Galvin & Walker 2007). Although there is no rainfall peak as the moist air retreats, except in southern China after 15 July, there is slight rain and drizzle at times in many areas, including Hong Kong (Guo & Wang 1981; Ding 1994). Over East Asia, once fully established around the beginning of October, the winter monsoon flow is a north-westerly, bringing very cold weather south-eastwards from central parts of Asia. Korea, Japan and much of China have very cold weather, considering their position in middle latitudes – a marked contrast to the heat and humidity of summer.

The cooling of China, away from the coast, leads to a strong interaction of airstreams and a pall of deep altostratus is common in the lee of the north–south ranges that form the eastern Himalayas from late November or early December. Widespread rain or snow falls from this cloud, in particular over higher ground. Much of central southern China endures cloudy conditions in winter, rather than summer, even though the heaviest precipitation (some of it very heavy) falls during the summer monsoon.

The thicker cloud that forms over China can be a significant hazard to aviation. Much of this cloud deck is at a temperature near 0°C, even when several thousand feet deep. Moderate, occasionally severe icing may be encountered by aircraft flying at modest height through this cloud layer. Such icing of the airframe or engine can cause loss of airspeed and stability, even, ultimately, loss of the aircraft (see section 11.2).

By late December, the ITCZ is recognizable as two bands of cloud across the Indian Ocean, close to 10°N and 10°S.

8.5.3 South-East Asia

Much of the winter season is dry across the whole of south Asia, including parts of southern China, Burma, northern Thailand, Laos, Cambodia and Vietnam. However, although areas of large-scale convection are absent, slight showers may still affect windward coasts. A north-west monsoon brings cool continental air from central China across much of the region. Drier weather also spreads briefly into the northern and central Philippines during March and early April. However, monsoon variation is less well marked over the extreme south of the region and the island nations of South-East Asia, where rain can be expected for much of the year.

The relatively cool weather of the winter monsoon can be uncomfortable for the population – in particular those who live in coastal locations, such as Hong Kong, where cool spells require warm clothing and buildings to be heated. Indeed, warnings of cold are issued to the Hong Kong fire brigade if night-time temperatures are expected to fall below 12°C,

when the fire risk becomes significant, as the rooms of otherwise-unheated flats are warmed using paraffin stoves (B-Y Lee, Hong Kong Meteorological Department, personal communication).

A case study of cloudiness associated with the South-East Asian winter monsoon is given in Box 8.1.

8.6 The West African summer monsoon

Over West Africa the ITCZ is distorted and moves north from the coast of the Gulf of Guinea as the Sahara desert becomes increasingly hot during the lengthening days. The ME moves north, a few degrees behind the position of the sun overhead at midday during April and May. The distortion is so strong that the ME becomes separated from the area of deep convection south of the ITF and associated thundery rain. In the early part of the monsoon season, convection occurs periodically over land and continues over the Gulf of Guinea. In June it becomes confined to land, north of 5°N, as the waters of the Gulf cool. By early August, the ME lies at its northward extreme, between 1500 and 2000 km from the coast of the Gulf of Guinea, over northern Africa.[8]

This monsoon circulation has many similarities to that in Asia, but also significant differences, largely due to the differences in orography and the landmass distribution of these two regions. It is also less extensive, the monsoon flow reaching no more than about 20°N of the equator and deep convection only occasionally seen north of 15°N. The monsoon flow and deep instability spread inland much more quickly here than over India, however, since there are no high mountain ranges to impede the inland flow. Nevertheless, there is significant variation across the relatively modest orography of the nations of the Guinea Coast (Owusu & Waylen 2009). Initially,

Box 8.1 Cloudy South-East Asia

Few people associate south-east Asia with cloudy, sometimes dull, weather, but these conditions are often present in parts of southern China and northern Vietnam during the winter season (Raschke et al. 2005). Extensive persistent areas of cloud develop east of mountain ranges that stretch north–south, marking the eastern edge of the Himalayas. The depth and extent of these cloud layers waxes and wanes day by day, as atmospheric conditions change. Altocumulus and altostratus layers typically affect western areas, whilst lower cloud layers are usually observed further east.

Figure 8.7 is a pair of satellite images illustrating the cloud layers. An overcast layer of altostratus, stretching several hundred kilometres, can be seen to the east of the Daxue Shan and Wuliang Shan ranges of the eastern Himalayas and a shallower broken or overcast layer of altostratus and altocumulus east of the Phou Sam Sao, which forms the border between Laos and Vietnam. Further layers of altocumulus and stratocumulus can be seen as far east as Taiwan and the southern islands of Japan.

The cloud-top temperatures of each of the areas of cloud over south-west China and northern Vietnam indicate that the tops of these layers were at around 5500 and 3500 m, respectively. This gives a clue to their method of formation.

The winter season across southern and much of south-eastern Asia is dominated by the north-east monsoon, which brings largely dry weather that is particularly cold across China. However, the monsoon flow is comparatively shallow, despite deepening as it reaches the Himalayan mountain barrier. Above the north-easterly trade winds the air is warm and comparatively humid, carried by predominantly westerly winds. The temperature contrast between the air masses yields a very stable flow.

Complications arise due to the ranges of mountains. Cooling of the high-level air from the west causes extensive condensation east of the ridges and altostratus clouds form with a base near 3500 m. The north-easterlies are blocked as they reach high ground, so the cloud does not affect areas to the west: Laos, Thailand and Burma. The height of the mountains also affects the height and depth of the cloud. Thus, the top of the cloud layers is a little above the height of the peaks of the Daxue Shan and Wuliang Shan, which reach between 3000 and 5000 m, as well as the Phou Sam Sao, which reaches around 2500 m.

The cloud depth is likely to be determined in part by the strength of the high-level westerlies. In this case, wind speeds were around 30 m s^{-1} across the crest of the Daxue Shan (below the STJ) and near 15 m s^{-1} across the crest of the Phou Sam Sao. Wind profiles, temperature lapse rates and indicated cloud depths for each of these cloud decks are shown by the radiosonde profiles from Nanning (Guangxi), Changsha (Hunan) and Huaihua (Hunan), China in Fig. 8.8.

Forecasters need to be aware of the persistence and development of these cloud sheets during the winter monsoon season. Flights at medium and low levels may encounter hazardous weather conditions within the cloud. Both layers form in the wakes of mountain ranges and moderate turbulence may be expected locally around and within the clouds, in particular to the lee of the Daxue Shan, which lies below the mean position of the STJ in winter.

Infra-red satellite imagery (Fig. 8.7b) gives cloud-top temperatures between about –13°C in the lee of the Daxue Shan and around –2°C in the lee of the Phou Sam Sao, so airframe and engine icing are also likely to be significant threats. Indeed, this persistent cloud may have been one of the main reasons for the high rate of aircraft losses on cargo flights 'over the hump' during the First World War.

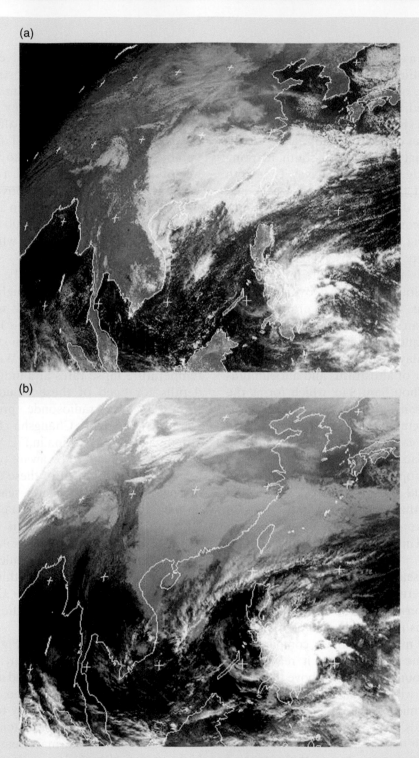

(a)

(b)

Figure 8.7　(a) Visible and (b) infra-red satellite images from GOMS showing stratiform cloud across southern China and northern Vietnam at 0600 UTC on 3 February 2006. Courtesy of University of Dundee Satellite Receiving Station.

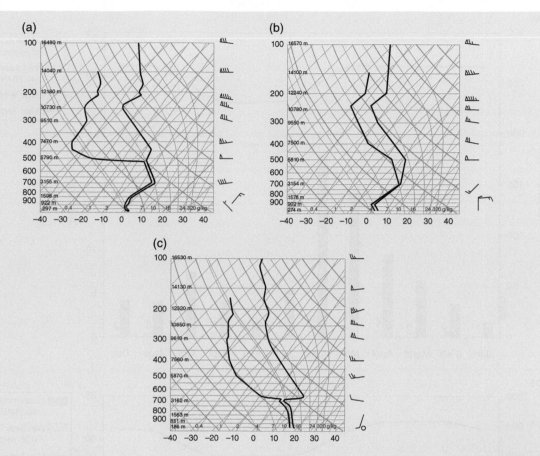

Figure 8.8 Radiosonde profiles for 3 February 2006 through the layer-cloud formations shown in Fig. 8.7: (a) Nanning, Guangxi, China (22.6°N, 108.6°E) at 0000 UTC, (b) Huaihua, Hunan, China (27.6°N, 109.9°E) at 0000 UTC (only standard-level data were available from this site) and (c) Changsha, Hunan, China (28.2°N, 113.0°E) at 1200 UTC. Courtesy of University of Wyoming, Department of Atmospheric Science.

The cloud depths indicate a broad area of overcast, locally dull, weather in central-southern China with outbreaks of slight rain or snow. Across northern Vietnam and southern coastal China, as well as the East China Sea, there are predominantly cloudy and rather cold conditions, although cloud depths are less and there are breaks in the cloud, allowing some brightness. A few slight showers associated with 'warm' cumulus and stratocumulus may be generated close to the coast by instability in the boundary layer over the warmth of the East China Sea. At other times stratus forms, bringing dull weather. This weather is frequently observed during the winter monsoon season (Atkinson 1971).

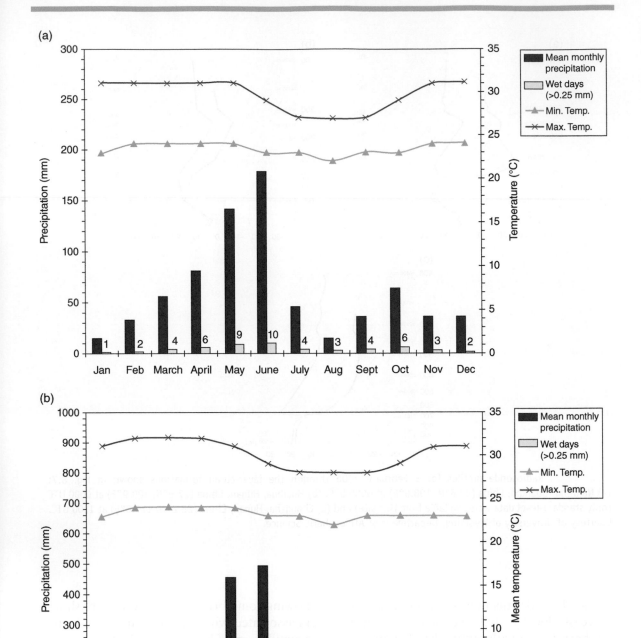

Figure 8.9   Variations in the monsoon regime across West Africa: (a) Accra, Ghana (5°36′N, 0°10′W), (b) Abidjan, Ivory Coast (5°15′N, 3°56′W),

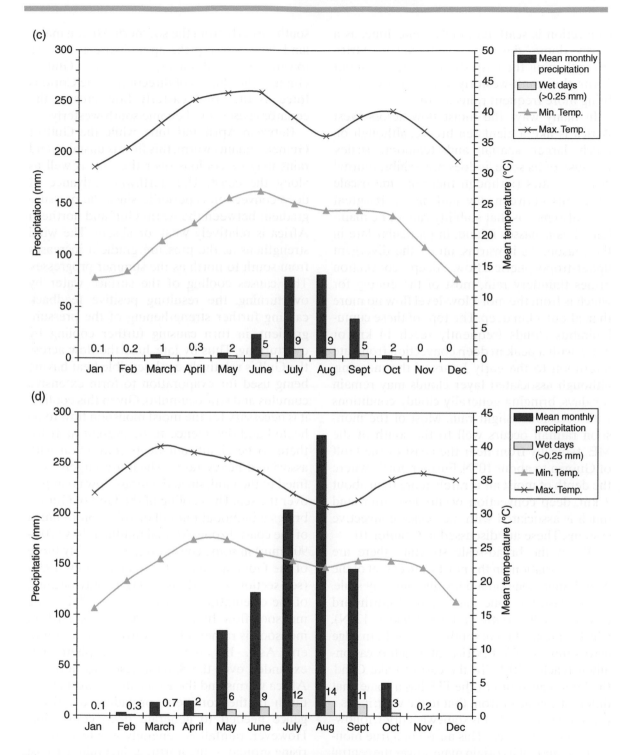

Figure 8.9 (*Continued*) (c) Tombouctou, Mali (16°46′N, 3°01′W) and (d) Ouagadougou, Burkina Faso (12°22′N, 1°31′W). Note the variation in rainfall scales.

convection is scattered, but by late June, as a warm dome develops over North Africa strengthening the easterly winds above about 3 km, the south-westerly winds at low levels bring more frequent rains north.

In many ways the moist flow across West Africa is like a giant sea breeze, although on much larger spatial and temporal scales. Because of its scale, it does not exhibit diurnal characteristics (although there are mesoscale variations between day and night). Its great scale also means that stability cannot be maintained as it passes inland, in particular late in the season. As it warms, under the divergent upper-tropospheric flow, deep convection brings thundery rain, most of the energy for which is from the moist low-level flow no more than about 4 km deep. The tops of these cumulonimbus clouds frequently reach 14 km or more, with a peak in deep convection from late afternoon to the early hours of the morning, although associated layer clouds may remain for days, bringing generally cloudy conditions with occasional slight rain. Most of the monsoon rainfall occurs well to the south of the ME, typically from near the coast of the Gulf of Guinea to about 10°N. Further north, where the depth of moist air is rarely more than about 2 km, deep convection occurs less often and much is associated with mesoscale convective systems. These are discussed in Chapter 10.

Within the broad-scale structure there are notable variations in the northward extent of the deep humid zone on a daily and monthly scale. In the east, the zone has a limited northward extent, usually south of Lake Chad (~12°N), which is about 600 km south of its northern edge across most of West Africa, although occasionally it reaches 20°N. To the east of Lake Chad, the humid air south of the ITF has a somewhat different character from that to the west: it has a greater instability, but is somewhat drier, following a long land track. This air is not a true monsoon because of its origin mainly over the central African landmass.

To the south of the ME is a moist flow that has crossed the equator. Initially this is a dry south-easterly from the southern African mainland. However, it picks up moisture as it flows towards the south-facing coast of the Gulf of Guinea. The change of direction of the Coriolis force as the south-easterly flow crosses the equator causes it to become south-westerly.

Between April and June, while the Gulf of Guinea remains warm, this flow is unstable and rains may be copious over the sea, as well as along the coast. The northward advance of deep convection is periodic, since the pressure gradient between the warm Gulf and northern Africa is relatively weak or absent. The wind strengthens as the pressure gradient increases from south to north as the summer progresses. This causes cooling of the surface water by overturning, the resulting positive feedback causing further strengthening of the pressure gradient, in turn causing further cooling by overturning. Thus by late June the flow across the Gulf is relatively cool, sensible heat having being used for evaporation to form extensive cumulus and stratocumulus. Given this cooling, it is necessary for the moist monsoon flow to be heated and divergence to be present aloft for there to be significant deep convection with associated heavy rains, which are largely confined to the land, since the moist flow is capped over the sea. The cooling of the Gulf of Guinea brings a bi-modal rainfall distribution to much of the coastal zone. Annual rainfall is low (700–900 mm) in some coastal zones – notably those of the Gold and Slave Coasts (Kendrew 1937) (see section 3.3 and Fig. 8.9a), mainly because of the orientation of this coast, parallel to the monsoon flow. In this respect the West African monsoon is rather different from that in southern Asia. By June, the troposphere has expanded over the Sahara desert as northern Africa warms and the EEJ extends west across much of the continent, providing the divergence needed to generate deep convection. However, rainfall is generally greatest on the rising ground to the north of the Guinea Coast (Fig. 8.9b). Indeed, rainfall is reduced north of areas of high ground (Owusu & Waylen 2009), although much less than is the case in the lee of the Western Ghats in India. Figure 8.9(c) illustrates the climate of the extreme northern limit of the West African summer monsoon and Fig. 8.9(d) the transitional climate, away from the coastal zone. Between the Guinea Coast and about 7°N there are two rainfall maxima, while there is a single annual peak further north. Near the northern limit, the winter is dry (Fig 8.9c). Typical of monsoon zones, high temperatures precede the monsoon rains and temperatures are lower during the wet season.

Across the north of this monsoon zone, in the Sahel, rainfall is very variable and has many of the characteristics of dry-land precipitation, although the average annual quantity usually large. The showery rainfall is heavy and may be accompanied by large hail. Flash flooding is comparatively common. (This is discussed further in Chapter 10.)

Rainfall associated with the West African monsoon is variable and has declined in recent decades. A link has been proposed to temperature and temperature differences of the North and South Atlantic Oceans, warming of the latter having displaced the mean position of the ITCZ and monsoon flow southward. However, variations occur on an annual, as well as a decadal, scale, with warm seas increasing coastal West African rainfall, while reducing it in the north (Ramage 1971). Other hemispheric-scale variations ('teleconnections') seem likely to have contributed to the changes, although the signal is weak (Atkinson 1971; Gu & Adler 2003). The recent overall decline in rainfall in the West African monsoon zone may also be due to local changes, as we will see in Chapter 12. Effects appear to be greater in the more humid southern zone, where the rainfall distribution has bimodal peaks, than is the case in the drier north. In this area, the risk of drought (relative reduction of rainfall) appears greater than in the northern savanna–Sahel zones (Gu & Adler 2003).

A conceptual model of the structure of cloud masses and air masses across the West African monsoon zones is shown in Fig. 8.10 (Leroux

2001). This shows the change southwards from (A) hot-dry (hazy or dusty) desert to (B) humid-stable transitional to (C) convective, then (D) vigorously convective, (E) deep-tropical stable and shallow tropical unstable, then (F) shallow tropical stable conditions. It can be seen that embedded convection is most likely in the centre of zone D. Layer clouds are well broken in the north of this zone (or may be absent) and, although deep and precipitating, the layer clouds in the south of the zone do not contain cumulonimbus.

8.7 The West African winter monsoon

As the year wanes and high-level easterlies decrease in strength, the ME moves slowly southward. The Sahel becomes dry in October as the moist convective zone moves south (Fig. 8.9c). Steadily more and more of the savanna lands become dry. At its furthest excursion south, the ME lies along the south-facing coast (~5°N), so that there is a dry north-easterly 'Harmattan' flow on most days from the central Sahara. The flow is generally hot by day and into the evenings, but may be rather cool following nights of clear skies over the desert, the effect seen most often across the drier areas of savanna and semi-desert.

However, the coastal zone does not become completely dry during the winter. There are occasional incursions of sea air, which, by the latter part of winter, is unstable. The reduced strength or absence of south-westerly winds weakens the overturning of the ocean and a warm Guinea current becomes established (Fig. 5.1a). Cloudy skies are relatively common throughout the year in this coastal zone, bringing occasional slight rain or showers. Deep convection occurs south of 5°N across central Africa through the winter months, although there is a pronounced peak in rainfall during the summer when temperatures fall – a characteristic seen throughout the monsoon zone (Fig. 8.11).

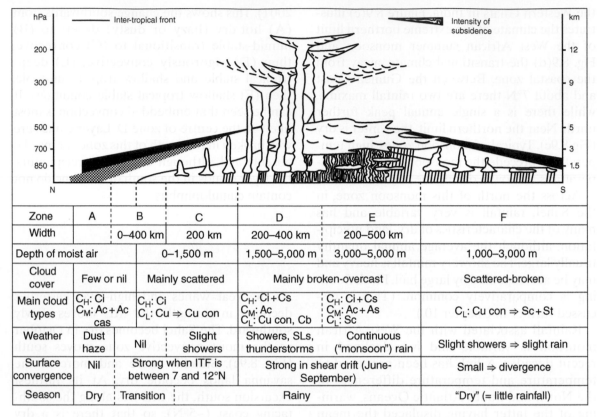

Zone	A	B	C	D	E	F
Width		0–400 km	200 km	200–400 km	200–500 km	
Depth of moist air		0–1,500 m		1,500–5,000 m	3,000–5,000 m	1,000–3,000 m
Cloud cover	Few or nil	Mainly scattered		Mainly broken-overcast		Scattered-broken
Main cloud types	C_H: Ci C_M: Ac + Ac cas	C_H: Ci C_L: Cu ⇒ Cu con		C_H: Ci + Cs C_M: Ac C_L: Cu con, Cb	C_H: Ci + Cs C_M: Ac + As C_L: Sc	C_L: Cu con ⇒ Sc + St
Weather	Dust haze	Nil	Slight showers	Showers, SLs, thunderstorms	Continuous ("monsoon") rain	Slight showers ⇒ slight rain
Surface convergence	Nil	Strong when ITF is between 7 and 12°N		Strong in shear drift (June–September)		Small ⇒ divergence
Season	Dry	Transition		Rainy		"Dry" (= little rainfall)

Figure 8.10   Meridional cross-section of the troposphere over West Africa in northern summer (adapted from Leroux (1970)). SL indicates a squall line associated with an African Easterly Wave. Zone A lies north of the ME. The (transitional) zone between the ME and the ITF is indicated by B. C is a shallow convective (capped) zone. D is the main convective region of the monsoon under the influence of upper-level diffluence. E is an area of thick layer clouds, associated with long periods of monsoon rain. F sees the return to capped shallow convection with a transition to layer clouds as surface divergence becomes re-established. This zone dominates over the Gulf of Guinea and along the Guinea coast in high summer when the total breadth of the monsoon transition and monsoon rains reaches 1000 km or more.

8.8   Rainfall and the monsoons in East Africa

It might reasonably be thought that all of East Africa across, and to north and south of the equator, would see significant periodic rainfall from the monsoons, but Somalia, Kenya and Tanzania frequently suffer droughts and have become notably dry in recent years, away from their highland interiors. Indeed, air circulations over eastern tropical Africa are distinct from those in the west, separated by the so-called Congo air boundary (see section 3.1). This separation is largely due to the very

shallow moist boundary layer (<3 km) usually seen over the western equatorial Indian Ocean (Kiangi 1989) and the relatively low temperature of the western equatorial Indian Ocean. In turn, warming of the western Indian Ocean has a profound effect on rainfall, usually bringing floods, as in Somalia in 1997 (Camberlin & Philippon 2001) and 2006.

However, there are significant complications in the rainfall pattern in eastern Africa, south of the predominantly dry Sahara desert. The first is the extensive Ethiopian plateau. Cloudy skies and rainfall are present for much of the year over this large area of high ground.

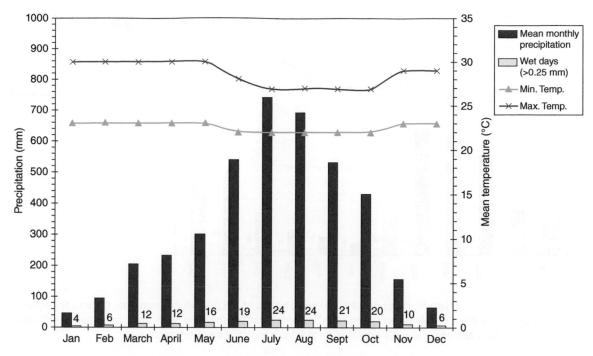

Figure 8.11 The climate of Douala, Cameroon (4°03′N, 9°42′E). Here rain occurs year-round, reaching its peak in summer, which coincides with the lowest temperatures, which are the result of increased cloudiness and evaporation.

A peak in rainfall occurs as the ME moves north across the plateau and deep convection from increasingly moist air becomes widespread. North of the plateau, however, the weather remains predominantly dry. The northern summer brings moist air from the Indian Ocean north across Somalia, although the strong south-westerly winds of the Asian monsoon are offshore and the weather is mainly dry over the lowlands of the Horn of Africa.

Secondly, there are complex interactions between air masses from the Atlantic and Indian Oceans. The one from the Indian Ocean contains more moisture and ascent up the ranges of the East African Highlands releases its latent heat. This leaves relatively warm dry air at high levels above the moist but stable Atlantic air. This stability restricts rainfall north and west of the East African Highlands, with rainfall only developing in the presence of additional forcing.

In the northern winter, north-easterly winds are generally dry, although near the equator there is a flow from the Indian Ocean and precipitation may occur over high ground. The Ethiopian plateau also generates some rainfall, although this is mainly somewhat further south than in the northern summer months.

The broad area of convection spreads south during the southern summer across much of the south of the continent, crossing the savannas. Most of the south of this continent has a long dry winter monsoon season. The ME reaches Madagascar and eastern coastal areas of Mozambique by November, but it is a very diffuse feature further west. As in north-eastern Africa, however, the southern African monsoon is not a classical form.

Over land the situation is complicated first by the presence of high ground, forming a 'ring' around most of the land mass and, second, by the convergence of air from the Indian and Atlantic Oceans. As described

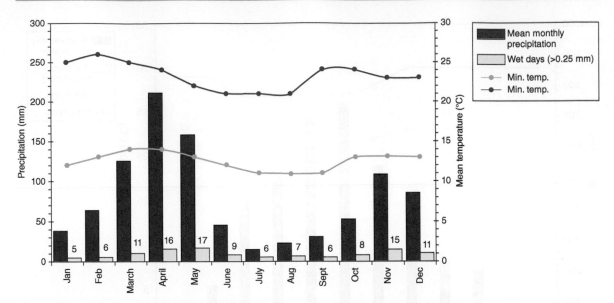

Figure 8.12   The climate of Nairobi, Kenya (1°19′S, 36°56′E), typical of the East African monsoon. Close to the equator, this monsoonal climate has a double peak of 'long rains' between February and June, with 'short rains' between October and December. The high altitude of Kenya's capital is evident in its monthly mean temperatures.

above, Indian Ocean air contains more moisture than that from the Atlantic (Leroux 2001), so air from the north-east can release more rainfall from instability than air from the west. The inland motion of moist air is restricted by orography, since the leading edge of the monsoon flow is no more than about 2 km deep. Over central Africa, it is only when the moist air can flow over the highlands that deep convection can occur. The East African Highlands, either side of the Rift Valley, absorb much of this moisture, even when the line of convergence of surface air is well to the west.

In general, the southward movement of the monsoon flow during November is accompanied by a short period of rain. By the end of the month it has reached southern Tanzania and northern Mozambique. Once again, as summer wanes, the rainfall area retreats north from mid-January to reach northern Kenya and Uganda by the end of February, bringing a period of 'long rains'. This, in turn, clears north during April and May, precipitation eventually becoming largely restricted to equatorial Africa (Fig. 8.12).

Despite the effect of orography, over south-western and southern Africa convection at the leading edge of the monsoon often forms a band extending from about 10°S on the coast of Angola, arcing south-eastwards to reach the Mozambique coast near 30°S when the monsoon reaches its peak. Off shore, the warm Mozambique current, between Madagascar and the mainland, allows the ME to retain characteristics similar to those of the ITCZ.

8.9   The South American–Caribbean 'monsoons'

Over South America there is a gradual north-ward and southward expansion of moist unstable air with only slight motion in the winter hemisphere towards the equator. Rarely does any part of South America have a double rainfall peak. The broad-scale dynamics of the atmosphere are evidently different across the Americas. There is no upper-tropospheric easterly jet stream (upper-tropospheric winds are westerlies) and the depth of the tropopause remains modest, rarely reaching more than about 16 km.

In the southern-hemisphere summer, the area of rainfall expands steadily south, crossing Bolivia, Paraguay, northern Argentina and Uruguay by late December. The area close to the equator does not experience a dry season, although the Caribbean and much of Central America is mainly dry during the northern hemisphere winter.

In the southern summer, convergence of a north-westerly wind with south-easterlies cannot generate the same level of deep convection as elsewhere, since air from the north must follow a long land track across South America, the Andes preventing air from the Pacific spreading across the continent. North-westerlies over south-eastern Brazil form a relatively dry continental flow. Here, much of the rainfall comes from mesoscale convective complexes (Laing & Fritsch 1997; Gaye et al. 2005). Nevertheless, there is considerable rainfall on coastal mountains – in particular those of Brazil's east coast and the eastern slopes of the Andes. The central plains receive somewhat less rainfall, even in the Amazon Basin and this is reflected in the areas of woody savanna that locally replace the selvas.

In the northern summer, the Caribbean experiences its wet season. Here rainfall is mainly the result of destabilization of the atmosphere and uplift provided by the mountainous orography of the islands as the ITCZ moves north. It is not a true monsoon, since there is not a marked reversal of wind direction and the flow is from the South American land mass across the Caribbean Sea, rather than from sea to land. Across this island-group, unlike most of tropical South America, there are two annual rainfall maxima, the first attributable to the northward motion of increasingly warm moist air in June and July, the second associated with the return of cooler air from higher latitudes across warm ocean in October and November. Much of the summer rainfall of the Caribbean is associated with tropical storms. In winter, the area has less rain, but most of the area does not become dry.

In the provinces of north-east Brazil, as in East Africa, the effect of varying offshore water temperature affects the climate. The cool equatorial counter-current brings periodic drought and has been a cause of a mass exodus of the population (Hastenrath 1991).

8.10 The Australian summer monsoon

The distortion of the ITCZ into a monsoon flow is perhaps less well marked over northern Australia than in most other monsoonal climates and the expansion of the troposphere is modest compared with that over Asia or north Africa. However, easterlies strengthen near the equator, over New Guinea and over Java, occasionally reaching jet-stream strength, in response to the warming of the Australian landmass. Moist oppressive air precedes convective downpours southwards into the marginal semi-desert areas of the country in summer. Almost all of northern Australia is dependent on summer rainfall and, as a result, has woody savanna vegetation, despite the proximity of parts of the north coast of Australia to the equator (the latitude of Cape York is less than 11°S). The ME reaches about 20–25°S across Australia at the height of the monsoon in late January, its greatest poleward extension usually in the east of the country, occasionally to 30°S or more. The extensive areas of high ground in central and northern Australia, and the Great Dividing Range along the east coast, enhance monsoon rains at times.

In central and southern coastal Queensland, the summer monsoon is often associated with severe weather. Areas of low pressure, some in the form of tropical depressions or tropical revolving storms (Chapter 9), form periodically off the coast, some having tracked east across the Cape York peninsula, bringing moist easterly winds and areas of heavy convective rain (Fig. 8.13). In an area that usually

(a)

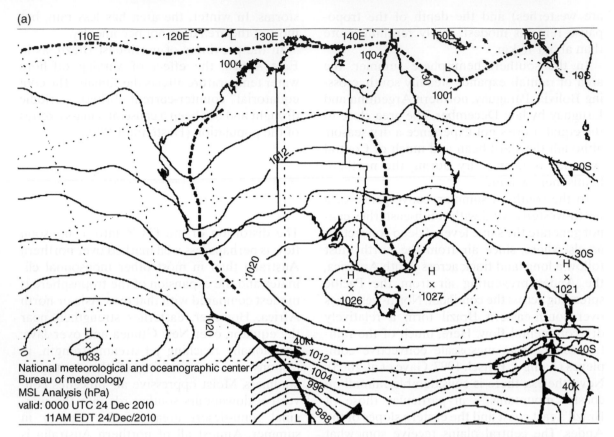

Figure 8.13 (a) Mean sea-level pressure and frontal analysis for Australia at 0000 UTC on 24 December 2010. There is a broad area of low pressure across much of the north of the country, associated with the southward advance of the summer monsoon. Although the line of lowest pressure (■ • ■ • ■ ; the monsoon trough) is indicated, the moisture that brings the monsoon weather is spread across much of the broad low-pressure zone. However, the deepest convection is associated with the tropical depression, off the coast near Cairns. These depressions bring moist air inland on their southern flank and particularly heavy rain may fall on the Great Dividing Range, around 400 km inland. The rain from this system contributed large totals to the subsequent flooding in a vast area around Rockhampton. © Bureau of Meteorology, Melbourne.

experiences dry weather, the 'wet' can be notable. Most rain falls on the Great Dividing Range and drains both west into the outback and east across the coastal plain. The rivers of this state usually have a modest flow, so heavy rainfall occasionally brings flash floods that later form a significant hazard in the relatively populace coastal zone, most towns built as ports near the mouths of these rivers. Although drought and glut are common features of this area (Fig. 8.14), with the most recent drought between 2000 and 2006, several years since 2006–2007 have experienced flooding. Notable

events occurred in Channel Country, Warrego and central west Queensland on 20 and 21 January 2007, between January and March 2008, and between January and March 2009.

Particularly severe flooding, covering an area of more than 10^6 km^2 around Rockhampton, occurred between late December 2010 and early January 2011, then in the Brisbane area in mid January 2011, the latter originating in flash floods around Toowoomba, near the crest of the Great Dividing Range. Approximately 300,000 properties were affected in Brisbane (Giles 2011).

(b)

Figure 8.13   (*Continued*) (b) False-colour image of Queensland, Australia from Aqua-MODIS at 0420 UTC on 24 December 2010. The 2 km resolution of this image allows the variety of cloud types associated with the monsoon to be seen. Modest cumulus clouds can be seen near the western edge of the image, marking the edge of the moist zone. A large shield of cirriform cloud associated with deep convection around a tropical depression covers the coastal plain inland as far as the Great Dividing Range. Occasional cumulonimbus clouds have developed over Cape York Peninsula and the Gulf of Carpentaria. Courtesy of NASA Rapid Response Team.

Perhaps more than any other continent, the strength of the Australian summer monsoon is associated with variations in tropical storm activity, in particular tropical depressions and hurricanes in Queensland. This is discussed further in section 9.7. An unusual phenomenon precedes the onset of the Australian monsoon rains in the Gulf of Carpentaria

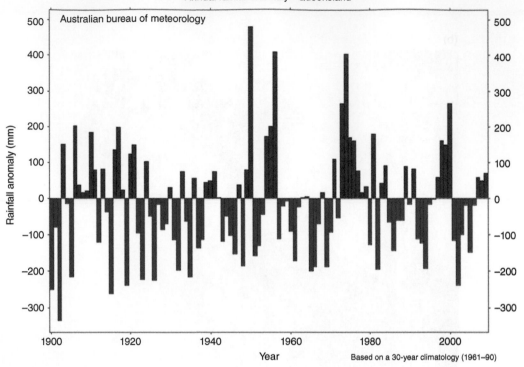

Figure 8.14 Aggregated annual rainfall anomaly for Queensland, Australia, based on the mean rainfall for 1961–1990. There is a clear periodicity in the data since the 1950s. Most recently, between 2001 and 2006 there was a long period of drought, following a very wet period between 1997 and 2000. The periodicity is linked to the El Niño–La Niña phenomenon (section 12.2). © Bureau of Meteorology, Melbourne.

Figure 8.15 The 'Morning Glory' roll-cloud near Burketown, Queensland, Australia, seen from the air. © Russell White.

between late August and early November as humidity rises. Here a complex interaction of orography, a moist shallow cool boundary layer and sea breezes sets up the 'Morning Glory' cloud roll as the ME makes a halting advance south across the Gulf (Fig. 8.15). The phenomenon is still poorly understood in detail, but development appears dependent on the shallow monsoon boundary layer being drawn over Cape York and up its mountains by a sea breeze. As this sea breeze subsides, it is overturned to form a wave trapped beneath the trade-wind inversion. This is carried across the Gulf, accompanied by a roll cloud, close to the line of surface-wind convergence. It brings a rapid wind increase to around 13 m s^{-1} and rising humidity. In some cases a line of deep convection may accompany the convergence, but in most cases the roll cloud, which may be up to 1000 km long, brings a stable flow. The overturning is due to westerly winds above the inversion, so the cloud rolls backwards and generates both lift and turbulence (Jackson et al. 2002). This can be a major hazard to light aircraft at low levels, but the lift ahead of the roll allows gliders to 'surf' at speed for several hundred kilometres (Thomas 2003).

Further south, convective rainfall may develop as a result of the interaction of tropical air with upper-level disturbances in the extra-tropics. This is not part of the monsoon system, although moisture from the monsoon flow may be entrained, assisting in the formation of these systems at the edge of the tropical air mass, as described in section 10.10.

8.11 Variable broad-scale factors affecting the monsoons

As can be seen in the descriptions above, there are important interactions between the poleward motion of warm moist air and large-scale atmospheric flows and variability in the strength of winds and development of pressure systems causes periodic or annual variations in the monsoon rains. This section describes some of the factors that influence the effects of these broad-scale flows.

There are various relationships (teleconnections) between rainfall and regional factors. La Niña frequently enhances the rainfall in countries surrounding the Indian Ocean, but research suggests that a range of factors control the rainfall of the East African, East Asian, North American and Australian monsoons. Not least of these is the quasi-biennial oscillation (QBO, discussed in section 12.4), which itself is weakly linked to the El Niño–La Niña reversals.

In East Africa, when there are westerlies in the lower stratosphere and upper troposphere, the early-season 'long' rains, which bring the largest quantity of rain and so support agricultural production in the region, are often enhanced when other factors support the development of deep convection (Ng'ongolo & Smyshlyaev 2010). There is also a possible link between drought and easterlies in the lower stratosphere and upper troposphere. However, no direct link between East African rainfall and the El Nino–Southern Oscillation is established, possibly due to the over-riding need for atmospheric conditions conducive to ascent. The most important factor is probably the strength of the Walker circulation (see section 3.1), which determines broad-scale ascent and descent over the region (and is linked to lower-stratospheric and upper-tropospheric winds).

Variability in the East Asian monsoon is most closely linked to sea-surface temperature anomalies across the north Pacific. Two modes have been found. When the east Pacific is warmer than normal, summer rainfall tends to increase in northern China, north of 27°N, but is much reduced in southern China. Higher rainfalls in the summer monsoon are linked to anomalously warm SSTs in the west Pacific, but in this case the monsoon is erratic in the north of the country. The teleconnection in these cases reduces or enhances the tropical upper westerlies, the core of the STJ moving north or south of its mean position, respectively. In the first case, this is further associated with anti-cyclonic development along the coast of southern China (Jia et al. 2010). Variations in Chinese rainfall occur approximately every 2–3 years and show some similarities to the variations in the lower-stratospheric QBO. Anomalies of similar sign are present over the north of the Caribbean.

The Madden-Julian Oscillation (MJO), described in sections 5.3.1 and 12.3, also has an important intra-seasonal effect on the monsoons of all continents, causing variations in rainfall intensity or duration as the Kelvin wave of the MJO passes (Zhang 2005). The QBO of stratospheric winds is also related to variability in the effects of the monsoon, as described in Chapter 12 (Andrews et al. 1987; Baldwin et al. 2001; Wikipedia 2014).

8.12   Questions

1. Consider the consequences on human and animal life of variations in monsoon rainfall in terms of resources and environment.
2. Does increased rainfall in northern India reduce rainfall in the south of the country, Sri Lanka and the Maldives?

Notes

1 There is a temperature inversion associated with the Mascarene high during the winter season. This provides a 'lid' to convection over the south-western tropical Indian Ocean. The Mascarene Islands include Mauritius, La Réunion and the Seychelles; they lie close to the equator in the western Indian Ocean. Relatively high pressure is seen in this area, in particular at medium and high levels during the northern winter, associated with mass descent. This high must move south before the Asian summer monsoon can become established.

2 The summer monsoon 2007 was an exception to this. In late June, a mesoscale system at the leading edge of the humid monsoon flow brought serious flooding to Karachi on 23 June. This was followed by a tropical cyclone, which moved inland across western parts of Pakistan during 25 and 26 June (Galvin 2007). The floods of 2010 also emphasize how heavy summer monsoon rainfall may be in this area. In late July and early August, disturbances maintained by rising ground and the south-eastward migration of the south-west Asian upper trough brought exceptional rainfall locally. Apart from the serious loss of thousands of lives, hundreds of thousands of hectares of crops and tens of thousands of farm animals were lost as floods spread south through the valleys of the Indus river system, across the desert areas of the country (BBC News 2010a,b; Chamberlain et al. 2010). More than 14 million people were affected by this event. The situation was worsened locally as dams were threatened with overtopping. Although built to control and smooth the supply of water to dry areas downstream, flash flooding was the result of water having to be released as reservoirs reached their limits. July precipitation in 2010 totalled 579 mm in Muzaffarabad, North-West Frontier, 528 mm in Mianwali, Punjab, and 500 mm in Islamabad (257 mm on 30 June). The July precipitation mean at this last station is 233 mm and for the whole year is 961 mm. The first 9 days of August brought large positive rainfall anomalies into the desert south of the country, worsening the swollen state of rivers, in particular the Indus. At the same time, the monsoon brought heavy rain to the Himalayan foothills of northern India.

3 It may seem strange that the south-west monsoon is often associated with south-easterly winds. Prior to the MONEX experiments of the 1970s, it was generally assumed that monsoon rains advanced on predominantly south-westerly winds, as indicated by the climatological average pressure gradient across south Asia in summer. As described in the text, however, it is necessary for winds from other directions, predominantly the south-east, to become established for sufficiently moist low-level air to advance across both India and East Asia.

4 The mean June–August rainfall at Xi-fan in central China is 750 mm, whereas that at Haikou on Hainan Island (the extreme south of the country) is 650 mm.

5 1961–1990 total; this is more than five times the annual total for most British uplands!

6 The extreme annual rainfall at Cherrapunji is more than twice this total, as given by Guhathakurta (2007).

7 Much of my father's service in the Royal Navy aboard ship was spent on the Arabian Sea and the roughest weather he encountered was during the south-west monsoon of 1942 on the destroyer *HMAS Launceston* – sufficient to make most of the crew sick.

8 A full discussion of this process is given by Leroux (2001).

9
Tropical Revolving Storms

9.1 Broad-scale convection and the development of tropical storms

In earlier chapters the formation and motion of broad-scale tropical weather systems, such as the ITCZ and monsoons, are described. These are associated with the warm humid air of the near-equatorial tropics. However, the development of areas of deep cumuliform and layer clouds, associated with low-pressure systems, is often a major cause of disruption due to flooding, landslip, wind damage and high seas. As a result, much effort is put into forecasting the development and motion of these tropical revolving storms. Such storms can be forecast ever more accurately, particularly up to about 48 hours in advance, not least by the UK Met Office (McCallum & Heming 2006) and considerable effort is put into the improvement of these forecasts (Elsberry 2006). The names of tropical revolving storms vary by location. In the Americas, they are known as hurricanes, in South-East Asia they are typhoons, and in the Indian Ocean and south-west Pacific they are known as tropical cyclones.

The forecasts of tropical revolving storms are co-ordinated by a number of designated regional specialized meteorological centres (RSMCs) of the WMO to produce a single official forecast to be used world-wide. These centres are at Honolulu (central-north Pacific),

La Réunion (southern Indian Ocean), Miami (north Atlantic and north-east Pacific), Nadi (south-west Pacific), New Delhi (northern Indian Ocean) and Tokyo (north-west Pacific). Forecasts are also prepared at various forecasting centres in Australasia, as well as Pretoria and Mauritius, although these are not officially RSMCs, but regional tropical cyclone warning centres. These centres are shown in Fig. 9.1.

Most numerical weather prediction centres running global forecast models contribute to the global tropical-storm forecasting effort. The reliability of the Met Office global model has led to the development of a web-based seasonal storm forecast for the North Atlantic based on the Hadley Centre for Climate Prediction and Research GloSea seasonal climate model (Met Office 2014a), which shows some success in the prediction of storms several years into the future (Smith et al. 2010). The Aon Benfield Hazard Research Centre of University College London also produces useful seasonal forecasts (tropicalstormrisk.com).

9.2 Tropical storm development and decline

Tropical revolving storms form initially as a cloud mass to one side of the equator, over sea temperatures greater than or equal to 27°C.

An Introduction to the Meteorology and Climate of the Tropics, First Edition. J F P Galvin.

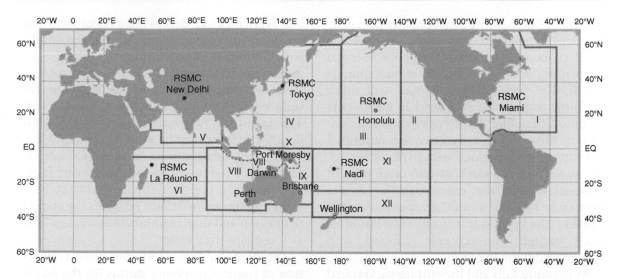

Figure 9.1   The location and areas of responsibility of designated tropical-cyclone warning centres.

Mass ascent causes pressure to fall, forming a tropical depression. However, further development is possible only where there is sufficient vorticity to develop the depression (Emanuel 2005) and then, where conditions are suitable for continued development, tropical storms (Cornish & Ives 2006). The centre of the depression is normally required to be at latitude 5° or more for such development, although very rarely deepening occurs closer to the equator. Winds that cross the equator may assist in the generation of tropical storms, particularly in the north-west Pacific (Verbickas 1998), since the convergence is from directly opposing directions. Formation is thus enhanced during the westerly phase of the MJO (section 12.3). Layer clouds usually extend to great depth in these cloud masses and certainly several thousand metres more than is generally the case in the inter-tropical convergence zone (see Chapters 3 and section 5.3).

It also appears to require air temperatures to be falling aloft for development to occur, so the area ahead of an upper trough (although not too close to it) is well suited. This typically occurs when there is a cold front at higher latitudes and at a similar longitude to that of the depression. (In this respect, the area off the Brazilian coast may seem ideal, but as explained later, this is not the case!) Dynamical ascent generated by mid-level vorticity, such as that associated with easterly waves, also aids the development of these storms (Jones & Thorncroft 1998). Indeed, few storms develop without the presence of these waves and they are a key indicator of likely development. All the factors conducive to cyclogenesis, the formation of a depression and its associated deep convection, with layer clouds throughout the troposphere, need to be present for development. Even where the sea surface is warm enough, there is sufficient vorticity and there is an upper trough present to aid development, storms may not occur (Emanuel 1988). This may explain the fact that only one hurricane has been observed off the Brazilian coast (*Catarina* off Brazil in March 2004), where wind shear in the troposphere is usually too strong to allow a tropical storm to develop, preventing storms developing in depth. (Indeed, the inter-tropical convergence zone is not usually seen to migrate south with the sun over the South Atlantic Ocean.)

As a tropical depression develops, although initially in relatively cool air, convective heating is so powerful that the depression gains a warm core as it becomes a tropical storm

(Emanuel 2005). Most convection occurs near the centre and this may allow an 'eye' to develop due to convective subsidence. This eye is surrounded by a ring of cumulonimbus clouds. The tops of these clouds will often reach the tropopause at around 16 km, in part because of cooling due to over-shooting convection through the convective 'lid' described in section 5.4. Combined with dynamical forcing, altostratus and nimbostratus often form in association with the cumulonimbus clouds, frequently forming a 'wall' around the centre. Outside the eye-wall region, convection is more limited, convergence reducing with distance from the eye-wall.

As pressure falls and the surface wind speed of a tropical depression reaches gale force (mean speed 18 m s^{-1} or more), it becomes known as a tropical revolving storm. With further development, the system becomes known by a variety of names: hurricane in the north-east Pacific, North Atlantic and Caribbean, typhoon in the western Pacific and tropical cyclone in the Indian Ocean and south-west Pacific. These storms have surface wind speeds of 33 m s^{-1} or more.

Further deepening of the low pressure at the heart of a tropical revolving storm can generate stronger winds still. Major hurricanes are defined in the Americas when wind speeds are 49 m s^{-1} or more (Emanuel 2005) and super typhoons in the north-west Pacific when mean wind speeds reach 65 m s^{-1}. Table 9.1(a) shows the Saffir–Simpson scale of tropical revolving storm intensities and their likely effects. The corresponding classification by the Japan Meteorological Agency is shown in Table 9.1(b). The significant difference between these scales and the local classification of tropical storms is evident.

Tropical depressions deepen most commonly over the north-west Pacific (mostly between March and December), south Indian Ocean (between November and April), the north-east Pacific, North Atlantic and Caribbean (between June and November), and the south-west Pacific (between December

and April). Fewer depressions occur over the north Indian Ocean (rarely during the south-west monsoon and during the change of monsoon circulation). No tropical revolving storms form in the south-east Pacific, where sea-surface temperatures (SSTs) are always too low. The need for a very warm water source is evidently crucial, as can be seen by the rapid dissolution of these storms as they run ashore, even when temperatures are high over land. The immense constant supply of latent heat from a warm ocean surface provides the energy to develop and then maintain the storm as long as it remains over water.

The heat, humidity and associated convection of tropical revolving storms lift the freezing level locally, typically to more than 5500 m in the area immediately around and above the eye. In this area, the –20°C isotherm (the level at which icing in cloud becomes less significant) rises above 9500 m (Fig. 9.2). The widespread convection warms the troposphere, which expands, creating an upper high above the storm, the tropopause often lifting above 17 km. As a result, cirrus and cirrostratus clouds rotate anti-cyclonically away from the dome at the centre of the storm – see, for example, Fig. 9.3. Occasionally, the anti-cyclonic circulation around this high may generate winds of jet-stream strength.

Inspection of Fig. 9.2 shows that cyclonic rotation occurs throughout the eye-wall region, although wind speeds are greatest up to about 4 km and decrease in the upper troposphere (Frank 1977; Anthes 1982). Studies reveal that most warming occurs above 4 km; temperature anomalies of 11°C or more have been measured (Hawkins & Imbembo 1976).

The level of the tropopause, –20°C isotherm and freezing level all decrease quickly with increasing radius from the centre of a storm. Near the outside of the circulation, the levels are typically 16 km, 8200 m and 4500 m, respectively (Asnani 1993). A typical radius for the main area of frequent cumulonimbus clouds is rarely more than 2° (Riehl 1979, Ch. 9), suggesting that the main convection area is rarely

Table 9.1(a) The Saffir–Simpson scale of tropical revolving storm damage

Category	Level	Maximum sustained wind (1-minute mean) (m s⁻¹)	Damage	Example
1	Minimal	33–42	No real damage to building structures. Damage primarily to unanchored mobile homes, shrubbery and trees. Some coastal road flooding and minor pier damage.	Hurricane Earl (1998)
2	Moderate	43–49	Some roofing material, door and window damage. Considerable damage to vegetation, mobile homes and piers. Coastal and low-lying escape routes flood 2–4 hours before arrival of centre. Small craft in unprotected anchorages break moorings.	Hurricane Georges (1998)
3	Extensive	50–57	Some structural damage to small houses and other buildings. Mobile homes are destroyed. Flooding near coast destroys smaller structures, with larger structures damaged by floating debris. Land less than 1.5 m above sea level may be flooded 13 km or more inland.	Hurricane Fran (1996)
4	Extreme	58–68	More extensive building failures with some complete roof-structure failure on small houses. Major damage to lower floors of buildings near the shore. Major erosion of beach. Land less than 3 m above sea level may be flooded, requiring mass evacuation of residential areas inland as far as 10 km.	Hurricane Andrew (1992)
5	Catastrophic	>68	Complete roof failure on many houses and industrial buildings. Some complete building failures with some small buildings blown over or away. Major damage to lower floors of all structures less than 5 m above sea level and within 500 m of the shoreline. Mass evacuation of residential areas on low ground 8–16 km from the shoreline may be required.	Hurricane Camille (1969)

Tropical cyclones not on this scale can produce extensive damage from flooding. Categories 3, 4 and 5 hurricanes are collectively referred to as major (or intense) hurricanes. These major hurricanes cause over 83% of the damage in the USA even though they account for only 21% of tropical cyclone landfalls. Sourced at aoml.noaa.gov/general/lib/laescae.html.

Table 9.1(b)   Japan Meteorological Agency classification of tropical revolving storms in the north-west Pacific Ocean

JMA category	Maximum sustained wind (10-minute mean) (m s⁻¹)	International category	Class
Tropical depression	≤17	Tropical depression (TD)	2
Typhoon	18–24	Tropical storm (TS)	3
	25–32	Severe tropical storm (STS)	4
Strong typhoon	33–43	Typhoon (TY) or hurricane	5
Very strong typhoon	44–53		
Extreme typhoon	≥54		

The definition of 'typhoon' is different between the Japanese standard and the international standard. A tropical storm with the wind speed of more than 17 m s⁻¹ is called a typhoon in Japan, while in the international standard a storm with a wind speed of 33 m s⁻¹ or more is called a typhoon.

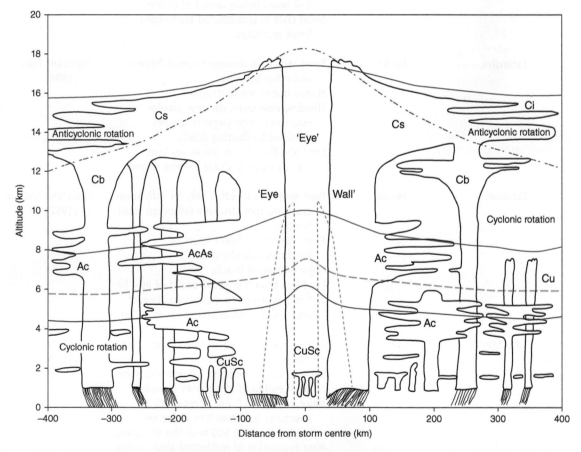

Figure 9.2   Cross-section of a mature typical tropical revolving storm. The effect of vigorous convection on the freezing level (——), 500 hPa height (– – –), –20°C isotherm (——) and tropopause height (——) can be seen easily. Cloud types: Cb, cumulonimbus; Cu, cumulus; Sc, stratocumulus; Ac, altocumulus; As, altostratus; Cs, cirrostratus; Ci, cirrus. The division between cyclonic and anticyclonic rotation is indicated by —·—. Only within the area – – – do wind strengths reach more than 17 m s⁻¹. Drawn using data from Hawkins & Imbembo (1976), Frank (1977), Anthes (1982) and the National Hurricane Research Laboratory (1970).

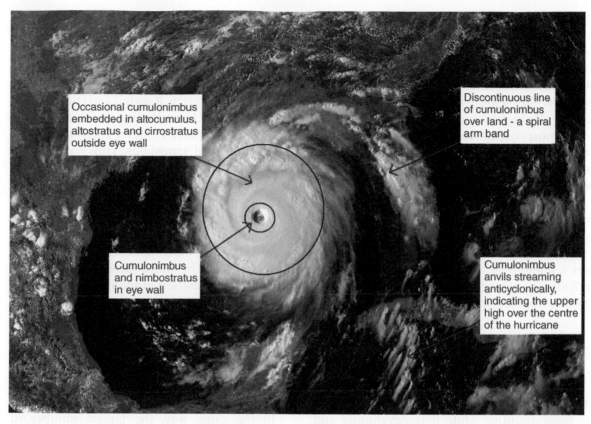

Occasional cumulonimbus embedded in altocumulus, altostratus and cirrostratus outside eye wall

Discontinuous line of cumulonimbus over land - a spiral arm band

Cumulonimbus and nimbostratus in eye wall

Cumulonimbus anvils streaming anticyclonically, indicating the upper high over the centre of the hurricane

Figure 9.3   Cloud masses associated with a mature tropical cyclone (Hurricane Katrina on 28 August 2005).

more than about 450 km across. Figure 9.3 shows the main areas of cloud in a large vigorous mature tropical revolving storm. Many tropical revolving storms are smaller than this and a few are considerably larger.

Spiral rain bands are often seen in association with tropical revolving storms (see Galvin 2005; Hole 2006). Formed in cooler air, outside the main core, spiral rain bands are often very vigorous, bringing copious rainfall and frequent thunderstorms (Asnani 1993). In many cases the rainfall of a tropical storm is heaviest within these bands (Emanuel 2005). They spiral cyclonically towards the storm centre and, although no measurements appear to have been made of their motion, are likely to move at about 8 m s⁻¹, a typical speed of the 500 hPa wind 700 km from the storm centre. These bands take approximately 6 days to complete one revolution around the core.

Tropical revolving storms decline as they reach cooler water or land. Their decline may be relatively slow in the former case, but rapid over land, the energy and wind speed of the system soon reducing. Nevertheless, where a storm runs inland over a swamp, relatively deep warm lakes, or can reach another area of warm sea before the depression has filled it may regenerate, regaining its power. Hurricane Andrew crossed Florida's Everglades swamp on 24 August 1992 and lost little of its power, allowing it to develop further as it reached the Caribbean Sea. The San Felipe hurricane on 17 September 1928 retained its energy as it moved north through Florida, sustained by the warmth of the extensive and shallow Lake Okeechobee (Emanuel 2005). Nevertheless, the potential vorticity associated with any storm may be carried long distances over land, such that westward moving tropical revolving storms over the South China Sea may re-form as depressions over the northern Bay of Bengal. Occasionally, storms that originally formed over the southern Bay of Bengal may re-generate over the Arabian Sea, having crossed southern India early in the season, while the SST remains sufficiently high (see section 5.1).

In most ocean basins, tropical revolving storms are given names. Where a storm causes major damage and destruction its given name will never be used again.

The annual average, maximum and minimum number of tropical storms in each ocean basin is shown in Table 9.2. However, there is considerable variation from year to year and there was a notable 28 storms in the North Atlantic in 2005 (the season ending very late, in January 2006) compared with 10 in 2006.

9.3 The effects of tropical revolving storms

A tropical revolving storm or depression is at its most vigorous while it is developing and during its mature stage; declining storms are relatively weak. Convection causes strong turbulent up-draughts and there is strong wind shear at high levels, where the outflow from the storm is strongly anti-cyclonic. The risk from icing in cumulus and cumulonimbus clouds is high. Rain in tropical revolving storms may be locally torrential and accompanied by hail. Tornadoes and waterspouts may also form. These factors are the most crucial for aviation and aircraft must attempt to fly round these storms, as they are deeper than the maximum altitude at which most commercial jets can fly. It is the turbulence from convection and strong horizontal wind shear, as well as the high probability of severe icing in the cumulonimbus and nimbostratus clouds near the core of the storm, that are the main risks, in particular at medium levels. The radius of the strongest winds close to the surface within a tropical revolving storm is usually small, rarely more than a few tens of kilometres (Fig. 9.2), with gale-force winds extending into the mid troposphere.

Although wind speeds are the most dramatic element of severe tropical storms, they do not usually cause the most damage associated with the system. Rainfall and storm surges are most likely to cause loss of life and livelihood (McCallum & Heming 2006) as a result of flooding and landslides. This was particularly notable in two of the most costly storms to affect the USA: Hurricanes Katrina on 29 August 2005 and Sandy (which had undergone extra-tropical transition by the time it affected the north-east of the country, having enlarged to become a superstorm) on 29 and 30 October 2012. The storm surge, combined with heavy rainfall, caused inundation of low-lying land, transport systems and, in both cases, loss of life. At least 1836 people lost their lives as Katrina moved across the Caribbean Sea and spread inland through New Orleans (see section 9.5), whilst the death toll for Sandy was 207 in Jamaica, Cuba, Haiti, the Dominican Republic and the USA.

In the same way, the largest number of deaths attributable to Super-typhoon Haiyan, which affected the central Philippines, northern Vietnam and the south coast of China between 7 and 10 November 2013, was due to the storm surge in Leyte Gulf, Philippines, rather than directly to the strength of the wind, which was likely to be amongst the strongest associated with a tropical revolving storm (Fig. 9.4). The high winds blew the storm surge into San Pedro Bay, at the north-western corner of the gulf, the water level greatly elevated by the time the surge reached the city of Tacloban, which was inundated to a level of about 6 m (van Ormondt 2013). However, a storm of similar strength had taken a similar path across Leyte Gulf, Samar and many of the other Visayas Islands of the Philippines between 15 and 17 October 1912 (Coronas 1912), so this storm was not unprecedented (Galvin 2014).

Rainfall rates often reach more than 50 mm h^{-1} from both the spiral arms and the eye-wall

Table 9.2   Annual number of tropical revolving storms for each ocean basin (using data from Emanuel 2005; Landsea & Delgado 2014; en.wikipedia.org/wiki/Tropical_cyclone; metoffice.gov.uk/weather/tropicalcyclone/observations)

Basin	Tropical revolving storm or stronger (>17 m s⁻¹ sustained winds)			Hurricane/typhoon/severe tropical cyclone (>33 m s⁻¹ sustained winds)		
	Most	Least	Average	Most	Least	Average
Atlantic	28 (2005)	4 (1983)	12	15 (2005)	2 (1982, 2013)	6
North-east Pacific	27 (1992)	5 (1976)	17	14 (1990 and 1992)	3 (2010)	9
North-west Pacific	39 (1964)	14 (2010)	26	26 (1964)	9 (1998, 2010)	17
North Indian	10 (1976)	1 (1986)	5	6 (1972)	0	2
South-west Indian	15 (1996/7)	1 (2012/13)	9	10 (1970/1)	0	5
South-east Indian	11 (1981/2)	1 (1987/8)	7	7 (many)	0	3
South-west Pacific	16 (1971/2)	2 (1981/2)	9	11 (1971/2)	1 (1979/80)	4
Globally	100 (1992/3)	69 (2010/11)	86	65 (1971/2)	34 (1977/8)	47

Data in this table are updated to 2013/14 from statistics for 1968–1989 (northern hemisphere) and 1968/69–1989/90 (southern hemisphere). Averages are rounded to the nearest whole number.

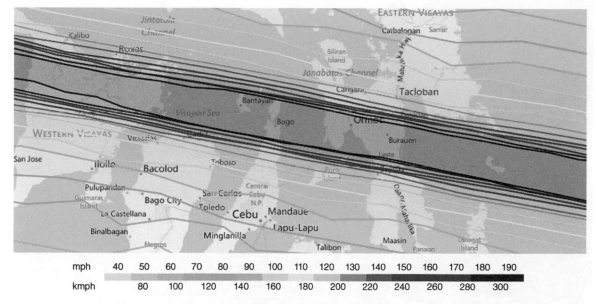

Figure 9.4   The estimated gust footprint associated with Typhoon Haiyan as it crossed the Philippines. These gusts are unusually high for a landfalling tropical revolving storm and diminished only gradually as the storm crossed the Visaya islands, probably because the warm water masses between the islands was sufficient to maintain the storm's strength. Notably, the highest gusts of the storm (over 65 m s⁻¹/240 km h⁻¹) lay to the south of the area of greatest devastation (from a storm surge) around Tacloban (compiled from observations and damage reports). © University College London.

(see Galvin 2005). Such heavy rain can overwhelm streams and rivers, causing flash floods and inundation of low ground. Settlements on either side of rivers are often flooded following rainfall from tropical storms; large populations commonly live close to these rivers. Still more damaging are the landslides that may follow this rainfall. This was very evident when volcanic mud from the Mayon volcano in the Philippines buried and destroyed many homes around Legaspi following the passage of Super-typhoon Durian on 30 November 2006. Deforestation of hillsides for agriculture or quarrying has increased the risk of landslides and many people sheltering from a tropical revolving storm have been buried by landslides, particularly in recent years. A typhoon that affected Hong Kong in 1937 killed 137,000 people, most of whom drowned.

Further inundation of low ground often occurs due to storm surge. With large populations living on coasts, floodplains, deltas or coral atolls, loss of life may be high, particularly where there is no high ground, as seen in New Orleans in 2005, Bangladesh in 1971 (Emanuel 2005) and 2007, and the Philippines in 2013 (Galvin 2014). Inundation also brings longer-term effects. Farmland may be polluted or suffer salination from a storm surge, the effects lasting years. Water may become undrinkable and, in the worst cases, may not even be suitable for crop irrigation. The effects of storm surge and flood are illustrated in Fig. 9.5, which shows large cobbles on the upper beach at Malimono, Surigao del Norte, Philippines.

Many beaches in areas where tropical storms occur indicate the level of surge from these

Figure 9.5   Stepped beach profile showing the level of storm surge (indicated by the arrow) with a slight hollow behind at Cayawan, Surigao del Norte, Philippines.

storms. There is usually a step in the beach, raised by recent surges. Figure 9.6 shows the beach at Cayawan, Surigao del Norte, Philippines, where this step is evident.

Because of the high risks associated with tropical storms, many nations in South-East Asia now have warning mechanisms in place that use telephones, radio and television to communicate information. In the Philippines, radio and television bulletins interrupt other

programming and instruct people to flee to high ground or other places of safety from areas that are expected to be affected. Although the internet, with its ability to disseminate warning e-mails and graphics, is potentially a powerful warning tool, in less-developed economies in the worst affected area its availability is restricted to relatively few.

The extreme wind speeds at low levels generate very high waves and swell that may

Figure 9.6 The effects of Typhoon Ike (locally named Nitang) recorded by poorly sorted debris at the top of Punta beach, Malimono, Surigao del Norte, Philippines. Flash flooding can be expected to have brought this layer of boulder- to sand-sized debris a few kilometres from the slopes of the Diuata Mountains, which reach around 1000 m within 5 km of the coast. The typhoon struck on 31 August 1984, crossing Surigao, Bohol, Cebu and Negros provinces. The maximum wind speed was 60 m s^{-1} at Surigao airport, just 16 km north-east of Malimono. Although initially crossing a comparatively rural part of the Philippines, its track took it across the more populous Visayas island chain and an estimated total of PP10.846 billion (US$602.489 million at the time) of damage was caused, with 1343 deaths recorded – the costliest typhoon to affect the islands to date (Alojado & Padua 2010). A storm surge also affected the coastal plain on the west of the Surigao peninsula, causing widespread flooding. This is likely to have contributed to the deposition of sand and rubble about 2 m above mean sea level.

travel thousands of miles. Occasional waves in excess of 15 m high may be generated, particularly in the poleward semicircle, where winds are strongest (Cornish & Ives 2006). Constructive interference may even produce an occasional wave more than 30 m high. Such seas can inflict severe damage and are capable of capsizing ships, even those of frigate or destroyer size (see Fig. 9.7). Indeed, it is thought that typhoons inflicted about the same amount of damage to the US Pacific Fleet during the Second World War as did the Japanese navy. Typhoon Cobra, which developed over the Philippine Sea at the end of the north-west Pacific typhoon season of 1944, was encountered by the US 3rd Fleet on 18 December. Three destroyers were capsized and lost, 146 aircraft were blown overboard or damaged beyond repair, 13 other vessels required major repairs and 9 others needed minor repairs (Calhoun 1981), with nearly 800 sailors losing their lives. Clearly, it is necessary for shipping to avoid tropical revolving storms if at all possible. Sixty-four years ago, it was extremely difficult to predict the formation and movement of storms; radar, which

can see the precipitation from these storms, had only just been added to ships' equipment (Emanuel 2005).

At low latitudes, tropical cyclones move at speeds of the order of 5 m s^{-1} or less. It therefore follows that swell will propagate well ahead of the storm. In the days before satellite imagery and numerical weather prediction, it was the swell ahead of these storms, perhaps combined with careful observation of clouds, that gave the best indication of an approaching tropical revolving storm. Clearly, ships with engines can usually outrun these storms or steer round them, but in the days of sail, extreme care was needed if a ship was to avoid the extreme hazards of these systems.

9.4   Storm tracks in the Pacific Ocean

In the Pacific there are three main genesis regions. Most important are the north-west Pacific and South China Sea areas, west of about 160°E, south of about 20°N. In this area, the MJO is most prevalent and this, combined

Figure 9.7   The aircraft carrier USS Lunga Point in rough seas generated by a western Pacific tropical storm in October 1945. Courtesy of US Navy Historical Center: Photo # NH 94876.

with a large area of very warm seas throughout the year (see Fig. 5.1), generates a large number of tropical revolving storms during a very long storm season. As they grow from a tropical depression, these tropical storms track initially west-northwest. Those over the open ocean may affect Micronesia and the Philippines, from north-east Mindanao northwards. The track usually curves north-west and, if sufficiently far away from land, north, then north-east. However, the so-called 're-curving' track takes a large number of storms ashore over or near Taiwan. Those that complete the re-curving track can affect Korea or Japan. In this area, they may transition to form a mid-latitude weather system, imparting a great deal of energy that may be carried across the North Pacific to Alaska.

Those that form over the South China Sea have little time to re-curve. The coasts of southern China and northern Vietnam have occasional encounters with these storms. Typhoon Xangsane first brought destructive hurricane-force winds to the central Philippines on 26–28 September 2006 before crossing the South China Sea to bring winds of 50 m s^{-1} to Vietnam on 30 September and 1 October. At least 200 deaths were associated with this storm.

In the north-east Pacific, under the influence of easterly waves (Berry et al. 2007) warm waters also generate many storms and this area is second in importance only to the north-west Pacific (Table 9.2). Most form near 110°W and 15°N, but soon dissipate as they reach cooler water. Despite the frequency of storms, few affect land, since most are carried away from North America, although on occasion storms may move northwards to reach the coast of Mexico. A small number of tropical revolving storms forms west of 140°W when the water temperature here reaches a peak. These relatively rare storms may affect Hawaii.

Despite very warm waters during the southern summer (see Fig. 5.1), fewer tropical revolving storms form over the south-west Pacific (Table 9.2), mainly because of the high level of wind shear commonly seen in this zone, which is associated with the formation of tropical upper-tropospheric troughs (see Chapter 4). Most tropical revolving storms in the south-west Pacific run close to the many islands of the area, typically affecting Fiji and the Solomon Islands as they track east. A few remain over the cooler waters of the Coral Sea and begin to re-curve south, then south-west. Occasionally the north-east coast of Australia, or even New Zealand's North Island, may be affected. The mountainous terrain of the latter can bring very large amounts of rainfall from these storms, which may result in flash flooding.

9.5 The formation and tracks of hurricanes in the North Atlantic–Caribbean

African easterly waves are now recognized as important in the formation of many of the tropical revolving storms of the North Atlantic–Caribbean region, as well as elsewhere (Emanuel 2005). These waves are found in the middle troposphere, usually most easily recognized at the 600 hPa level (~4500 m). The waves carry positive vorticity and associated barotropic instability across the Atlantic (Smith et al. 2010). In so doing, one of the ingredients for the development of tropical depressions is carried west. Indeed, when the forcing of these waves is absent, associated with a decrease in middle- and upper-tropospheric wind speeds, tropical revolving storms are usually absent, as was particularly noticeable in late July and for much of August 2006 in the Atlantic basin.

Over the warm water of the western Atlantic and, in particular, the Caribbean, where the water is relatively shallow and can thus warm up more easily, tropical storms may develop near the base of an easterly wave.

Most storms that form in the open Atlantic re-curve north, then north-east and may affect the east coast of the USA or occasionally Canada's Atlantic provinces, if they are carried by upper winds rapidly across the cool waters north of Cape Hatteras. Those that reach 40°N may become incorporated into a mid-latitude weather system, a process known as transition.

More rarely, storms cross into the Caribbean, where they often strengthen. Hurricane Katrina of 2005 was one such storm.

The warmth of Caribbean waters frequently generates severe tropical storms. All the Caribbean island chains and the east coast of Central America, as well as the south coast of the USA, bear the brunt of these storms. Flash flooding and inundation may affect the mountainous Central American coasts, whilst the low-lying land of the southern USA, in particular around the Mississippi delta, is at significant risk of both coastal and fluvial inundation. This does not seem to deter a growing population from moving to the sunshine of the southern US states!

Florida – a low-lying state – is particularly exposed to tropical revolving storms from both the Atlantic and Caribbean, and a large population is at risk from their effects.

9.6 Tropical cyclones in the Indian Ocean

Most tropical revolving storms of the Arabian Sea have a peculiar character, closely associated with the onset and recession of the south-west monsoon (Galvin & Lakshminarayanan 2006). In late May or early June, a depression forms close to the Laccadive Islands (~11°N, 73°E), at the leading edge of the south-west monsoon flow. Where other factors are conducive (around one year in four), a tropical storm forms from this depression and either moves north up the west coast of India or north-west across the Arabian Sea (dependent on the mid-tropospheric flow). The latter very occasionally brings copious rainfall to the Arabian Peninsula (particularly Oman), Somalia, Oman or southern Iran.[1] Crucially, once the monsoon sets in (as described in section 8.2), the strength of the wind overturns the ocean,

bringing cool water to the surface, thus preventing further storms forming. Thus these pre-monsoon storms rarely form more than once in any year (Membery 2001). However, there are notable exceptions, most recently in 2010, when an initial advance of the summer monsoon spawned Cyclone Bandu on 22 May. This storm moved north, then west, finally dying away in the Gulf of Aden as the monsoon receded south for a time. As the monsoon became properly established, another storm formed, deepening to be called Cyclone Phet on 2 June. This storm moved north and then east across south-eastern Oman to reach the Pakistani coast by 5 June.

Almost half the total number of tropical storms in the Arabian Sea form at this time. Nearly all the rest occur as the south-west monsoon recedes, over a longer season between September and November, so long as sufficient moisture remains in depth, after the strong winds of the south-west monsoon have decreased, allowing SSTs to reach 27°C. However, occasionally storms may also occur early in the summer season. Membery (1998, 2001, 2002) has written interesting and useful reviews of some notable tropical cyclones in the Arabian Sea.

Formation is less well organized in the Bay of Bengal, allowing a somewhat longer season of onset, associated with increased boundary-layer humidity and SSTs above 27°C. The presence of the south Asian upper trough allows the development of tropical storms in May and June, when the waters of the northern Bay of Bengal are warm (c. 29°C) and upper-atmospheric conditions remain conducive. Unusually, storms may continue after the monsoon flow becomes established, where the ocean remains warm and wind shear is conducive. This was the case in 2007, when a tropical revolving storm formed over the southern Bay of Bengal on 20 June and moved north-west across India's Deccan. Although surface winds soon decreased over India, sufficient vorticity remained within the system for the cyclone to re-form over the north of the Arabian Sea (where the ocean surface temperature was >28°C) on 24 June, later affecting Pakistan and Afghanistan.[2] For an account of the development and motion of this storm, see Galvin (2007); its position at 1200 UTC on 22 June is shown in Fig. 5.4. The formation of storms at the leading edge of the south-west monsoon occasionally allows storms to form simultaneously in the Bay of Bengal and Arabian Sea, as occurred in May 2010, when Cyclones Laila and Bandu developed within days of one another (Galvin 2010).

Later in the year, the season of recession of the south-west monsoon is sufficiently long over the Bay of Bengal to allow depressions or storms to form as late as December. As in the Arabian Sea, no tropical revolving storms form during the peak of the south-west monsoon season during July and August (Membery 2001). Large amounts of rain may fall over central and eastern parts of India, at times making up the majority of the annual rainfall of this area (see Chapter 8).

The Arafura Sea, north of Australia, is an important area for the formation of cyclones in the south-east Indian Ocean. The water is relatively shallow and can warm up readily during the southern summer. Cyclones can form here between December and April. The city of Darwin was destroyed on Christmas Eve 1974 by Tropical Cyclone Tracy, which had formed rapidly over the Arafura Sea.

A similar season occurs in the south-west Indian Ocean and some cyclones affect La Réunion, Mauritius, Madagascar or the East African coast. However, the development is usually confined to a relatively narrow band south of the equator (compared with the northern-hemisphere zone of formation), where waters are warmer than 27°C. However, the relatively shallow waters of the Mozambique Channel are also a good breeding ground for cyclones in the southern summer, easily warming above the required 27°C as far south as 25°S.

The tracks of the tropical revolving storms of 2006 are shown in Fig. 9.8. Many can be seen to re-curve, where the ocean track is sufficiently long, but in the southern Indian Ocean this is rare.

9.7 Tropical revolving storms in the south-west Pacific

Table 9.2 illustrates the high inter-annual variability of tropical revolving storms in the south-west Pacific (and a comparatively low mean-annual occurrence). Many years have few storms, but occasional years see more storms than would normally be expected. However, the severity of tropical revolving storms is not related to storm activity and devastating cyclones may occur in years of relatively few storms (see section 9.8).

This was illustrated by the summer of 2010–2011. Although only an average number of storms developed between November and March, including one – Anthony – that unusually reversed its direction of motion, several storms affected the Australian coast, bringing a high level of damage. Most devastation was caused by Tropical Cyclone Yasi, which struck the Queensland coast south of Cairns on 2 February 2011. A vigorous storm, categorized as a severe tropical cyclone, brought mean wind speeds in excess of 50 m s^{-1} along the coast close to its centre, an associated 7-m storm surge and consequent flooding. To the south of the storm centre, where gale-force winds extended more than 300 km from the storm centre, onshore winds and heavy rainfall caused the worst damage, adding to the problems caused by a vigorous monsoon season that had already caused widespread flooding (see section 8.10). However, although hundreds of buildings, vehicles and boats were damaged or destroyed, with swathes of trees and crops destroyed, the accurate prediction of this storm and timely evacuation warnings limited deaths to a single life (BBC News 2011a; Kelly 2011).

9.8   Variability in the development of tropical storms

A variety of factors alters the development and occurrence of tropical storms. An important factor is SST, although a warmer sea surface does not necessarily mean that more storms will develop and other broad-scale factors play their part. Storms can only develop where there is sufficient instability throughout the troposphere and the presence of cyclonic rotation in the upper troposphere is conducive to cyclone formation. This means that it is more common to see a greater occurrence of severe storms when the sea surface is warm than to see a larger number of tropical storms.

As we saw in section 8.11, there is a link between the QBO (discussed in section 12.4) in the upper troposphere and the strength of the summer monsoons. Similar links, which also relate to SST anomalies, assist or resist the development of tropical storms. However, at times and in some areas these teleconnections may be weak (especially where the ocean surface is characteristically sufficiently warm to form storms throughout the year).

However, other factors are also linked to tropical-storm activity (Wikipedia 2013). El Niño-Southern Oscillation (ENSO) years are characterized by above-average sea-surface temperatures over the equatorial eastern Pacific. This anomaly affects upper-tropospheric winds and mean sea-level pressure. Over the

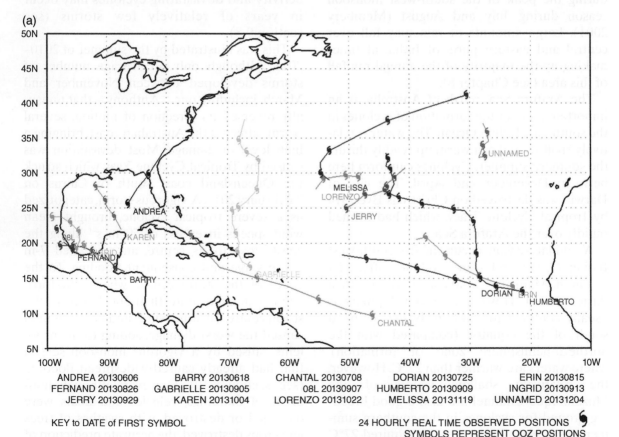

Figure 9.8 The tracks of tropical revolving storms in the 2013 (northern hemisphere) and 2013–14 (southern hemisphere) seasons: (a) North Atlantic Ocean; (b) northern Indian Ocean; (c) south-west Indian Ocean; (d) south-east Indian Ocean; (e) south-west Pacific Ocean; (f) north-west Pacific Ocean; (g) north-east Pacific Ocean. © Crown copyright (Met Office).

(b)

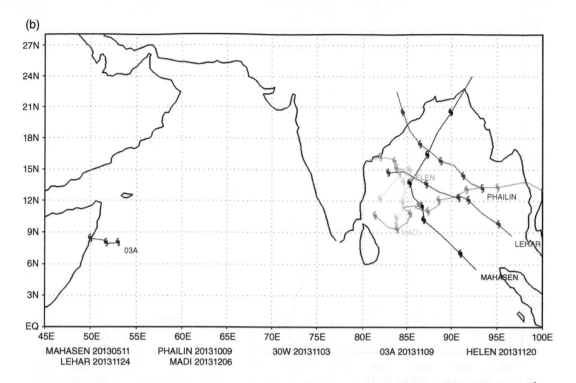

(C)

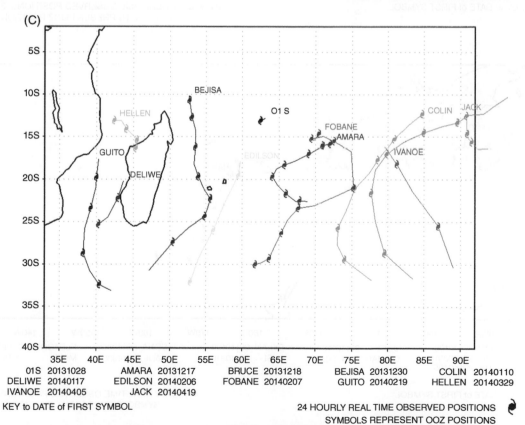

Figure 9.8 (*Continued*)

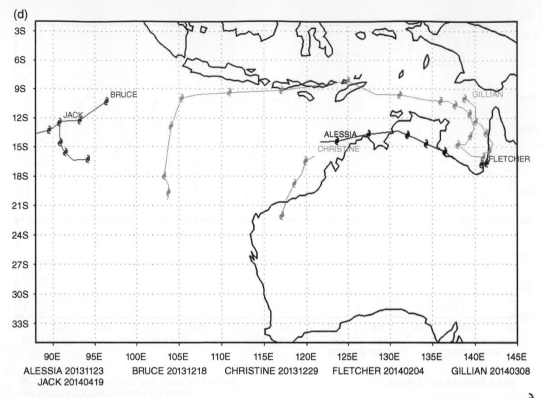

(d)

ALESSIA 20131123   BRUCE 20131218   CHRISTINE 20131229   FLETCHER 20140204   GILLIAN 20140308
JACK 20140419

KEY to DATE of FIRST SYMBOL   24 HOURLY REAL TIME OBSERVED POSITIONS
                                                         SYMBOLS REPRESENT OOZ POSITIONS

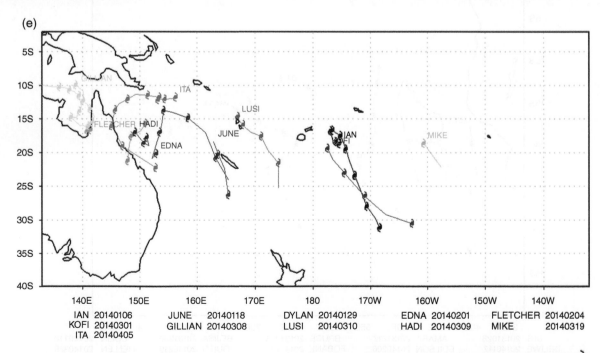

(e)

IAN 20140106   JUNE 20140118   DYLAN 20140129   EDNA 20140201   FLETCHER 20140204
KOFI 20140301   GILLIAN 20140308   LUSI 20140310   HADI 20140309   MIKE 20140319
ITA 20140405

KEY to DATE of FIRST SYMBOL   24 HOURLY REAL TIME OBSERVED POSITIONS
                                                         SYMBOLS REPRESENT OOZ POSITIONS

Figure 9.8   (*Continued*)

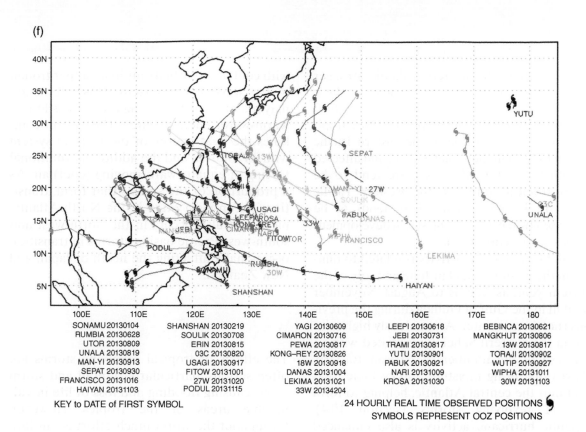

(f)

SONAMU 20130104	SHANSHAN 20130219	YAGI 20130609	LEEPI 20130618	BEBINCA 20130621
RUMBIA 20130628	SOULIK 20130708	CIMARON 20130716	JEBI 20130731	MANGKHUT 20130806
UTOR 20130809	ERIN 20130815	PEWA 20130817	TRAMI 20130817	13W 20130817
UNALA 20130819	03C 20130820	KONG–REY 20130826	YUTU 20130901	TORAJI 20130902
MAN-YI 20130913	USAGI 20130917	18W 20130918	PABUK 20130921	WUTIP 20130927
SEPAT 20130930	FITOW 20131001	DANAS 20131004	NARI 20131009	WIPHA 20131011
FRANCISCO 20131016	27W 20131020	LEKIMA 20131021	KROSA 20131030	30W 20131103
HAIYAN 20131103	PODUL 20131115	33W 20134204		

KEY to DATE of FIRST SYMBOL

24 HOURLY REAL TIME OBSERVED POSITIONS
SYMBOLS REPRESENT OOZ POSITIONS

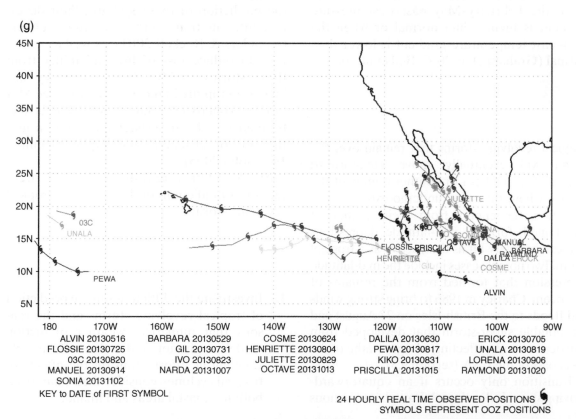

(g)

ALVIN 20130516	BARBARA 20130529	COSME 20130624	DALILA 20130630	ERICK 20130705
FLOSSIE 20130725	GIL 20130731	HENRIETTE 20130804	PEWA 20130817	UNALA 20130819
03C 20130820	IVO 20130823	JULIETTE 20130829	KIKO 20130831	LORENA 20130906
MANUEL 20130914	NARDA 20131007	OCTAVE 20131013	PRISCILLA 20131015	RAYMOND 20131020
SONIA 20131102				

KEY to DATE of FIRST SYMBOL

24 HOURLY REAL TIME OBSERVED POSITIONS
SYMBOLS REPRESENT OOZ POSITIONS

Figure 9.8 (Continued)

Caribbean and the western Atlantic, stronger upper-tropospheric westerlies increase shear and increase mean sea-level pressure, which suppresses hurricane development. More tropical storms may develop in the eastern Pacific region (although the signal is weak and affected by other factors), whereas there tend to be fewer tropical storms in the western Pacific basin. In La Niña years there tends to be increased tropical revolving storm development worldwide (section 12.2).

Atlantic hurricane activity is related to rainfall in the western Sahel in June and July of the current year, as well as to August–November rainfall in the Gulf of Guinea during the previous (northern) winter. Anomalously high rainfall in the western Sahel is associated with an increase in the number of African Easterly Waves that bring moisture and cyclonic rotation into the central Atlantic, helping to form tropical revolving storms (section 10.4). Atlantic hurricane activity is also enhanced when the February–May east–west pressure gradient is higher than normal or when the west–east temperature gradient is lower than normal (Graham et al. 2009; Bell et al. 2000).

9.9 Extra-tropical transition

Tropical revolving storms that remain over the ocean will eventually move equatorward towards the northern or southern limits of the tropics, where they may undergo extra-tropical transition, the energy from their rotation, if maintained, forming mid-latitude depressions that can be every bit as destructive as a tropical storm. Two notable cases include the mid-latitude depression that formed from the remains of Hurricane Charley in 1986 to bring strong winds and floods to the British Isles on 27 August and the notable super-storm that developed from Hurricane Sandy, affecting a large swathe of the east coast of the USA (section 9.3).

Transition only occurs if an equatorward-moving storm is developed through interactions with cool high-vorticity air in an upper trough to the west of the system. This interaction may maintain a storm long after it has left its warm-water energy source. The system combines the heat energy of the tropical system with its rotation and the additional rotational forcing of equatorward-moving (cool) air. As a proportion of the storms that form, transition is seen most often in the North Atlantic basin, but also occurs in the north-west and south-west Pacific basins, storms in transition occasionally affecting the north of Japan and New Zealand.

9.10 Conclusion

The effects of tropical revolving storms are often serious, particularly rainfall and storm surge, causing flooding and loss of life in vulnerable areas of the tropics. However, throughout the world much effort is put into the prediction of these systems: their development, motion and expected effects (Saunders & Rockett 2001), as well as action to help reduce loss of life and injury from these storms (WMO 2012). Accuracy has increased rapidly in recent years (e.g. see Met Office 2014a,b), especially where models may be improved by forecaster intervention and the use of ensemble techniques (Heming 1999; John 2006).

The Met Office (2014h) verifies the accuracy of its predictions of the motion of tropical revolving storms.

9.11 Questions

1. Investigate the difficulties in the prediction of tropical revolving storms and the reasons that numerical weather prediction models might not predict their effects well.
2. Looking at information on the effects of tropical cyclones, consider their main risks, both for coastal areas and inland.

Notes

1 Tropical Cyclone Gonu brought very heavy rain and flooding to Muscat and Oman on 5–6 June 2007. A total of 70 mm of rain fell at Seeb, Muscat, where the mean monthly rainfall in June is 1 mm. Twelve people were reported killed and 20,000 were evacuated. The severe-gale-force winds uprooted trees and damaged buildings. Electricity and water supplies were interrupted. This was the most damaging tropical revolving storm in the area for decades.

2 This storm was exceptional. It is only about once in 100 years that a storm affects the coast of Pakistan and runs north into southern Iran and Afghanistan.

10
Mesoscale Weather Systems

10.1 Introduction

The large-scale convective systems of the tropics are discussed in Chapter 5 and tropical revolving storms in Chapter 9, but significant severe weather frequently occurs outside or near the edge of the large-scale convection of the ITCZ. Mesoscale convective systems (MCSs) bring heavy rain, hail and thunderstorms to both the middle latitudes and the tropics (Zipser 1981).

Mesoscale systems are formed by a variety of mechanisms, including sea- and land-breeze circulations, barotropic instability and mesoscale changes in high-level vorticity. All are associated with warm moist air at low levels.[1]

10.2 Mesoscale convective complexes

Amongst the most important of the MCSs are mesoscale convective complexes (MCCs). MCCs are defined as having a cold-cloud shield area of at least 15×10^4 km^2 (Maddox 1980, 1986; Maddox et al. 1981), although most grow to be much larger. They tend to occur near the poleward edge of the equatorial humid zone, usually during the summer. More than 90% of MCCs develop over land, so their distribution favours the northern hemisphere, where two-thirds of the total occur.

An important pre-requisite for their formation is low humidity at high levels above warm moist air below about 3000 m, assisted by upper-tropospheric forcing. The areas within which they are likely to form can usually be identified using water-vapour satellite imagery as 'dark' dry zones above low-level air with a wet-bulb potential temperature (θ_w) greater than or equal to 22°C (Roca et al. 2005), typical of the humid zone. The fall of θ_w with height indicated by this superposition of air masses increases instability. Thus these systems are different from the broad convective zone of the ITCZ, within which the air is generally humid in depth (and instability is limited). The formation of MCCs cannot be forecast by assuming that similar areas will experience similar weather over a matter of a few hours to a few days, since they do not form in areas of pre-existing medium- and high-level clouds.

Laing and Fritsch (1997) noted that MCCs tend to form preferentially in areas to the lee of high ground, where there is increased vorticity as a result. Thus most are observed in savanna and semi-desert regions in the South American Pampas, Central America, the southern USA, the African Sahel (see section 10.4), Mozambique, northern India, southern China and northern Australia. An example of a series of water-vapour satellite images, showing the development of an MCC over Mozambique, is shown in Fig. 10.1.

Although of comparatively small area, these systems are amongst the most important in the tropics because they bring copious rainfall, often to relatively dry areas. They are a source of severe thunderstorms, flash flooding and torrential downpours, sometimes including large hail. Despite the research into their distribution, they remain difficult to forecast with reliability in all but the short range (up to about 12 hours). Associated with strong winds, resulting from downdraughts and severe turbulence, these systems should be avoided at all costs by aircraft, not only on take-off and landing, but also in flight. Although a minimum cloud-top height of about 12 km is specified for an MCC, in the tropics most have cloud tops above 14 km by the time they are mature. This is close to the maximum height at which airliners can operate.

Tropical convection may be notably enhanced ahead of upper-tropospheric troughs (evident at the 200 hPa level). These troughs were introduced in Chapter 4 and there is further discussion in Chapter 11. Uplift, associated with divergence ahead of the system, engages tropical moist air and bands of convective cloud with frontal characteristics develop. TUTTs often have a role to play in the development of tropical depressions and the subsequent formation of tropical revolving storms (Chapter 9). Deep convection may also develop near the base of the upper trough (in sub-tropical air), particularly where this over-rides moist air near the surface. Indeed, hail often develops in this type of system. Significant cloud development can be expected where easterly waves engage with the base of an upper trough, particularly over land.

10.3 Sea- and land-breeze convergence zones

Within the group of MCSs, an important formation mechanism is that of land- and sea-breeze circulations. These develop in many parts of the humid tropics.

In areas where winds are generally light, notably where the trade winds are weakest, either seasonally or as a result of the shelter of mountain ranges, sea-breeze/land-breeze circulations develop readily. As a sea breeze (or enhancement of a gentle onshore breeze) develops during the day, moist (though often somewhat cooler) air is drawn inland, supplying moisture for the development of cloud. By night – in particular where there are near-shore mountains – cooler air is brought into coastal areas as the land breeze sets in (Galvin 2004) (locally in the form of mountain winds (Oke 1987)).

The development and decay of cloud masses is particularly notable over South-East Asia, especially over and to either side of the Strait of Malacca, fed by sea-breeze/land-breeze circulations. By day, cloud develops over the Malay Peninsula and Sumatra, aided by the rising ground of these land masses. The air is very warm and moist over the strait, so the sea breeze does not cause significant cooling; uplift on the slopes allows the release of a great store of latent heat. At night, however, cooling over the land causes a land breeze, aided by katabatic drainage flows that develop on either side of the strait. These converge over the warm water and deep convection is a frequent result. Notable thunderstorms and squalls occur over the strait, posing a hazard to small vessels. The change in convection pattern from late in the day to late at night can be seen in satellite imagery (Fig. 10.2).

Figure 10.1 A sequence of water-vapour satellite images from Meteosat on 13–14 March 2006. An MCC can be seen developing and decaying over a period of 18 hours at 6-hourly intervals between 0900 UTC and 0300 UTC (~1115 and 0515 zenith time) over southern Africa, in a 'dry slot' over southern Mozambique seen most clearly in (a). Initial development is as occasional cumulonimbus before 1500 UTC (b),

(c)

(d)

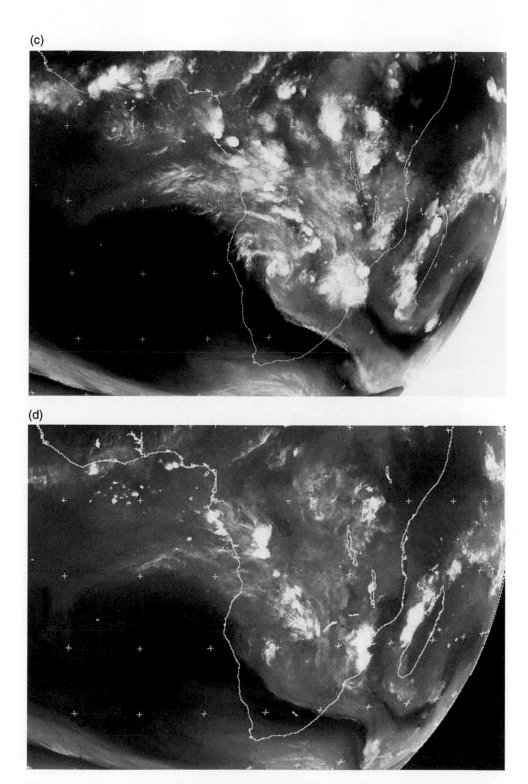

Figure 10.1   (*Continued*) resulting in a large embedded cloud mass by 2100 UTC (c). Most convection has died away by 0300 UTC, although an area of layer cloud remains (d). This had disappeared by 0600 UTC, but its signature was evident as a humid troposphere. Also of interest in these images is the migration of deep convection from central Madagascar by day to the Mozambique Channel by night and a similar diurnal change over and around Lakes Victoria and Malawi. A diurnal cycle of convection can be seen over land in many areas © EUMETSAT/Crown copyright (Met Office).

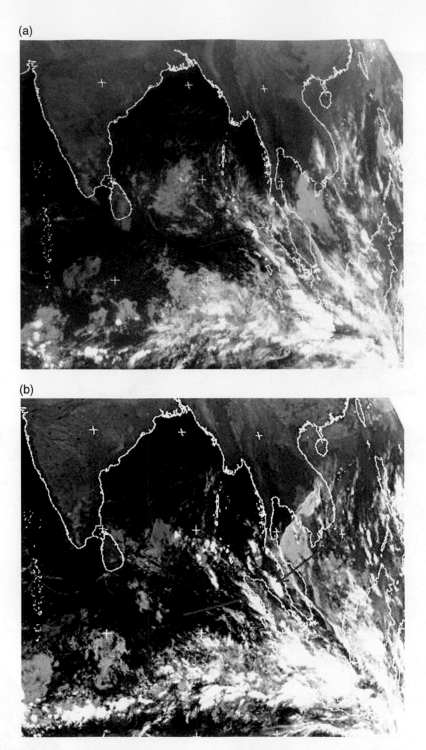

Figure 10.2 Infra-red image from Indian Ocean Data Coverage (IODC) of the north-east Indian Ocean on 30 December 2005. A change in preferential development of cumulonimbus clouds (much of it embedded in altocumulus) can be seen over Sumatra, the Malay Peninsula and the surrounding seas. (a) 1500 UTC (late evening, local time), (b) 2100 UTC (the early hours of 31 December, local time). Most cumulonimbus development can be seen over Sumatra in (a), the development over the Indian Ocean comparatively modest (even in the ITCZ, west of Sumatra and Java). In (b) cumulonimbus clouds are preferentially forming over the Strait of Malacca and Andaman Sea. Although southern Sumatra has vigorous cumulonimbus development, more cumulonimbus clouds are now forming in the ITCZ to the west, and northern Sumatra shows little sign of convective cloud. Courtesy of University of Dundee Satellite Receiving Station.

A similar diurnal change is also seen frequently during the southern summer over the Mozambique Channel and Madagascar (Fig. 10.1).

Large MCSs develop over eastern India as sea breezes develop and combine with airmass ascent up the Eastern Ghats between Vishakpatnam and Calcutta. Development begins at around 1200 local time (0630 UTC) and convergence with the south-westerly flow inland allows rapid development, peaking at around 2100 local time (1530 UTC). By this time a large area of cumulonimbus, often embedded in layer cloud, may reach 15 km, bringing an area of heavy rain as much as 400 km inland. The development of these systems is an important component of the monsoon system, since little rain-bearing cloud reaches north-east India from the south west (as discussed in section 8.2). By day, convection over the Bay of Bengal is suppressed, but it develops overnight as the flow from the land is re-established and the sea-surface temperature is once again higher than that over the land. Although most convective rain occurs in the summer monsoon, in particular as the monsoon advances and retreats, relatively high sea-surface temperatures bring occasional rain to eastern coastal India in the winter monsoon season. When there is an offshore (northerly) component in the regional surface wind, sea breezes bring warm moist air at low levels from the Bay of Bengal which may be released in the presence of suitable atmospheric forcing.

In Vietnam and southern China, sea-breeze development also promotes deep convection. In these cases, formation occurs when low-level winds are in the west, although in the latter case, winds in the boundary layer need to be veered from the usual south-westerly flow of the summer monsoon. Deep embedded convection develops in the presence of both the sea breeze and increased medium to high-level vorticity. Cloud tops may reach 16 km at the height of the summer monsoon. Over Vietnam, the line of deep convection may sometimes be seen to spread inland onto the Chaîne Annamitique, along the Laos–Vietnam border (approximately 100 km inland).

Large lakes can also cause reversals in day- and night-time wind systems. Most notable is Lake Victoria in East Africa. The formation and decay of cumulonimbus clouds over the lake from late evening to mid-morning and over the surrounding land, especially the mountains to the east, from late morning to late evening was studied by Lumb (1970) (this is discussed further in section 11.2).

10.4 Easterly waves and squall lines

Easterly waves with a period of about 3–5 days are observed throughout the ITCZ, but are generally weak and of little significance (Asnani 1993, Ch. 8). Mid-tropospheric waves have been measured in the eastern Pacific, where they are associated with the individual areas of cloud along the ITCZ (Fig. 10.3) that move west with a speed of about 5–10 m s^{-1} (Zipser 1981), breaking the near-continuous line of the ITCZ present at other times. Many of these waves are associated with so-called 'squall lines' of gusty convective downdraughts. Although minimal in the central Pacific, their amplitude grows as the waves move west, often bringing severe weather. They are also associated with the development of tropical revolving storms as outbreaks of westerly wind develop just to the north or south of the equator (Meteorological Office 1994).

The reason for the variation in effect of easterly waves appears to be variations in their driving mechanism, most importantly barotropic instability, caused by the north–south wind-speed component of the waves and the associated variation in vorticity. However, a detailed study of fast-moving West African squall lines in July and August 1992 by Roca et al. (2005) indicates that development is dependent on the intrusion of very dry air (relative humidity < 5%) from the extra-tropics

Figure 10.3   Infra-red satellite image from GOES-W showing cloud clusters associated with easterly waves over the equatorial Pacific at 1200 UTC on 14 July 2007. These clusters generally move west at a speed of around 8 m s^{-1}. A tropical depression (04E) was forming in one of these waves, close to 15°N, 115°W. Courtesy of University of Dundee Satellite Receiving Station.

into the middle and upper troposphere over the Sahel, increasing the upper-tropospheric potential instability. This is combined with mid-tropospheric easterly waves and moist air (relative humidity > 60%) below 600 hPa. The dry intrusion is present for at least 24 hours prior to the development of the squall line. These squall lines move at more than 13 m s^{-1} and last for 12–24 hours, sometimes longer.

In the case studied by Zipser (1969), the lower-tropospheric moist easterlies were overlain by westerlies with a low relative humidity above about 4 km. Propagation was associated with the incursion of air with a low value of θ_w to low levels (c. 600 m), causing rapid convection to medium levels ahead of the downdraught.

Over central and West Africa, African easterly waves (AEWs) bring the most significant weather, where they may form MCCs, often including squall lines. These are associated with local circulations that bring convective rainfall to the African Sahel. Up to about 50% of all the rainfall in the Sahel occurs from squall lines (Gaye et al. 2005) and nearly all

West African rainfall north of 12°N (Leroux 2001, Ch. 17).

The importance of these MCCs is also reflected in the distribution of rainfall. A recent study in Nigeria (Ologunorisa & Chinago 2007) showed that the southern coastal zone receives the highest annual rainfall, with a drier area lying to the north, over the centre of the country. In the north of the country there is a second zone of high rainfall (decreasing northwards) and this zone also experiences the highest number of thunderstorms. It is this area that receives around half of its rainfall total from MCCs. Nonetheless, this rainfall is scattered, varying considerably from place to place. In this way, it has some of the characteristics of dry-zone precipitation.

Despite their importance, these MCCs can have devastating results, in particular in drier areas, where there is little vegetation. In heavy rain soil may be washed away by flash flooding; hail may strip trees and crops of their leaves or fruit.

In Africa, the formation of easterly waves is dependent first on the enhanced heating of the Sahara desert during the summer, which brings moist air at lower levels northwards across West Africa to form the ITF – a monsoon flow (Leroux 2001, Ch. 10). However, this moist air is cool and thus stable. It therefore needs further forcing to develop convectively. The second requirement is the development of westward propagating mid-tropospheric waves in association with the shallow moist monsoon zone (see Fig. 10.4a). Barotropic instability results from the vorticity generated within the waves and the summer increase of easterly winds in the upper troposphere (Emanuel 2005). Indeed, wind speeds in the middle and upper troposphere are a limiting factor. It is only when the wind speed in the mid-tropospheric waves (at around 600 hPa) exceeds about 15 m s^{-1} and upper-tropospheric easterlies exceed about 40 m s^{-1} at about 120 hPa that squall lines can form. Divergence in the upper-tropospheric easterlies brings dynamical forcing for the systems and appears essential for their development. A typical AEW and squall line, with associated meteorological fields, are shown in Figs 10.4 and 10.5.

In many cases, but not always, the convective development is just behind (east of) the trough axis. Before the squall line arrives, the weather is dry and sunny, there is intense rainfall, sometimes with large hail, associated with the squall and scattered showers follow the passage of the squall line. As the trough passes, deep convection quickly ceases, although precipitation generally continues from deep layer clouds for some hours. Thus the main area of embedded convection occurs in the western portion of the cloud mass, the eastern half generally consisting only of layer clouds (Atkinson 1971, Ch. 11). AEWs are evident mainly from their signature between 700 and 500 hPa, and may be poorly indicated at other levels (Emanuel 2005).

Because of the need for dry air aloft, squall lines usually develop close to the boundary of moist air (850 hPa, $\theta_w \geq 22°C$) at low levels. Easterly surface winds usually follow the squall line and may assist in the formation of the system. These winds allow interaction of warm unstable desert air from the Sahara

with the relatively stable, but moist monsoon flow (Thorncroft & Blackburn 1999; Sow et al. 2005).

However, the interaction between the squall lines and AEWs is peculiar, since the squall lines travel west somewhat faster than the waves, at about 10–12° longitude per day (Cornforth et al. 2005). Although most form towards the rear of the trough in an AEW, they grow to a maximum as the trough axis is approached, then decline (often turning left) as they move ahead of it. This agrees well with the typical lifetime of these systems. Slower-moving systems have been observed to last up to 3 days (see Gaye et al. 2005). In this way, African waves differ from those in other parts of the world, since their form is the imposition of an easterly wind in the middle and upper troposphere on a prevailing surface south-westerly to provide the additional dynamics for uplift.

During the northern summer season, once moist air has advanced across West Africa, convection occurs in each trough and may form an MCS (Todd & Washington 2005) when there is sufficient upper forcing (visible as a 'dry slot' in water-vapour imagery). The very significant event between 29 August and 3 September 2009, associated with the westward progression of a mid-tropospheric trough from eastern Chad, across the Sahel to Burkina Faso, Mali, Guinea, Senegal, Guinea Bissau and The Gambia, is discussed in Appendix 7.

Mesoscale organization and propagation are also notable across central Africa (Todd & Washington 2005), usually in the northern summer, when air is more likely to be dry aloft. The effects of high ground are likely to be another important source of forcing in the formation of these mesoscale systems.

The relatively short lifetime and great sensitivity of these systems to atmospheric conditions makes them very difficult to forecast, even at short lead times up to a day ahead. However, if a mid-tropospheric wave is expected and there is upper-atmospheric forcing in the

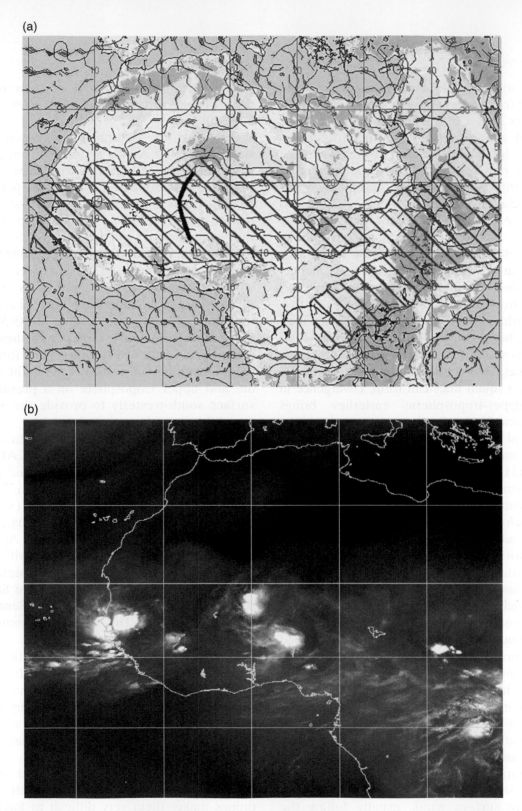

Figure 10.4 The association of moisture and easterly waves over west Africa at 0600 UTC on 31 July 2007. (a) The analysed 600 hPa flow from the Met Office Unified Model is shown as fleches and the shaded area shows where the 850 hPa θ_w is greater than 22°C. The axis of the 600 hPa trough is also marked. (b) Water-vapour imagery at 6.2 µm shows upper-atmospheric forcing as a 'dry slot' to the north of the mesoscale convective system at 0° longitude. © Crown copyright (Met Office).

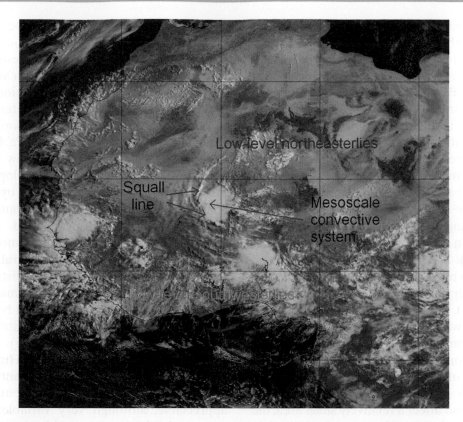

Figure 10.5 False-colour image from Meteosat-8 at 0800 UTC on 31 July 2007, showing the typical components of a mesoscale convective complex and squall line within an AEW. The squall line and relative low-level air flow are indicated. This can be compared with the 600 hPa flow 2 hours earlier, shown in Fig. 10.4. © Crown copyright (Met Office).

presence of sufficient low-level moisture, forecasts of development have some reliability, although estimating the position of a squall line 24 hours ahead remains difficult.

AEWs often propagate west across the Atlantic Ocean, where they are important in the development of Atlantic tropical depressions and revolving storms, bringing vorticity in the middle and upper atmosphere, as well as additional moisture to seed their development (Emanuel 2005, Ch. 16). There is a known association of the waves with tropical storms in the Caribbean too, although the correlation is weak. In this case, African waves, which usually become small as they pass over the Atlantic, are again amplified as they reach the Caribbean, helping to spawn depressions and tropical storms.

Where easterly waves develop elsewhere, there is a superposition of high-level westerlies over easterly winds in the middle and lower troposphere within which the waves develop.

10.5 Mesoscale convective systems in northern India

A phenomenon frequently seen over northeast India during the pre-monsoon season (March–May) is the development of so-called 'nor'westers'. These are MCSs that develop over Bihar or Jharkhand states and move south east to affect the Gangetic plains around Calcutta (West Bengal). Prior to the development of the rains associated with the south-west monsoon, pressure falls over northern India, but mid-tropospheric winds remain north-westerly. The development of the nor'westers appears to be triggered by boundary-layer convergence, which brings sea air from the Bay of Bengal, feeding ascent over heat lows inland. Strong winds and local downpours, including hail, result as cloud tops develop that may locally reach the stratosphere, above 16 km, embedded in layer cloud. Development begins at around 1630 local time (1100 UTC) and well-developed systems reach Calcutta by around 2230 local time (1700 UTC), fading away to leave little cloud or precipitation soon afterwards (Mukhopadhyay et al. 2005). Where trough lines are present, these MCSs may transform into linear features. From imagery in the paper by Mukhopadhyay et al. (2005), it seems that these systems usually fade away before reaching Bangladesh, although climatological data for Cherrapunji show that some winter disturbances reach the Khasi Hills of north-east India.

Although MCSs are small (diameter c. 150 km) and affect a relatively small area, the resultant weather may be a serious hazard to aviation and agriculture. A key sign of their formation is their south-eastward drift (an unusual direction for much tropical weather). Signs for their likely development are dry air at high levels over northern India and northern Pakistan, combined with low surface pressure and a moist boundary-layer flow from the Bay of Bengal or Arabian Sea. Radiosonde profiles were a poor indicator of the development of the storms described in Mukhopadhyay et al. (2005), although there was the potential for the release of very high levels of CAPE (~3000 J kg^{-1}).

10.6 Depressions in north-west India, north Pakistan and Afghanistan

Small-scale low-pressure systems that bring cloud and occasional rain or snow are a feature in north-west India and northern Pakistan between December and May. Occasionally, although rarely, this precipitation is from MCSs. These have a somewhat different origin and character from those of the nor'westers.

In a similar way to the depressions of the Arabian Gulf discussed in Chapter 7, pressure falls in response to short-wave upper troughs typically evident at 500 hPa. Moisture is drawn into the developing low pressure from the Arabian Sea. The development of these low-pressure systems is aided by vorticity generated by flow over the mountains of the border of Afghanistan and Pakistan. Nevertheless, rainfall is scant and most of this area receives less than 125 mm of rain in winter. Although the season of these disturbances lasts from November to May, in many years there is little or no precipitation in November or December, most occurring later in the season (Mukhopadhyay et al. 2005).

Only the slopes of the Himalayan–Karakoram ranges, where significant condensation is the result of mass ascent, receive appreciable, although very variable, rain or snow, and Srinagar (at an altitude of 1586 m) receives more than 65% of its annual rainfall total between December and May. Here the summer monsoon brings only a modest secondary rainfall peak, contributing less than 30% of the annual total.

10.7 Cross-equatorial flows

In Chapter 8 we saw that cross-equatorial flows are important in summer monsoon circulations, the long sea track bringing copious moisture in the lower troposphere. However, cross-equatorial flows also occur away from the continents, in particular when the ITCZ is displaced north or south of the equator. In every case the flow reverses direction as it crosses the equator due to the reversal of the Coriolis force. Large volumes of rain can result from these weather systems, especially when there is marked convergence in the boundary layer.

Often this marked convergence is periodic: a phenomenon known as the MJO, which is discussed further in sections 12.3 and 10.8).

Where the displacement of the meteorological equator is great, there is the greatest potential for heavy rain, not least when the moist flow encounters a mountainous island, allowing the release of its potential instability.

Locally, the varying depth of the moist troposphere can have a profound effect on the weather. As noted in Chapter 1, the easterlies of the equatorial zone are usually moist only to mid-tropospheric levels and even the summer monsoon westerly flow may only be moist to about 5 km. However, where there is little wind shear with height, as happens in association with the Asian monsoon flows and at times in the western Pacific, a long sea track can lead to moisture to a depth greater than 10 km. This was the case over the Philippines between 17 and 21 July 1972. This depth of moisture (associated with warm advection) was near stationary over one location, allowing large accumulations of rainfall and localised flooding. As Gordon (1973) reported, deep warm air brought four days of rain, some of it heavy. Commenting on the synoptic situation, Riehl (1974, 1979, Ch. 8) noted that the advection closely matched the rate of radiative cooling in the area, and the low wind shear in the south-westerly flow to about 230 hPa over the Philippines created a near-stationary system.

10.8 Mesoscale convective systems in the Gulf of Guinea

Anomalously high sea-surface temperatures over the Gulf of Guinea in winter may combine with trade-wind flows across the equator to produce unusually heavy rain associated with convergence of winds in the boundary layer (Fig. 10.6).

The MCSs that produce this heavy rain may commence over Cameroon or Equatorial Guinea and extend some way inland across Nigeria, Benin, Togo, Ghana, Ivory Coast and Liberia at times as they travel west. Their progress is similar to that of the squall lines of the West African monsoon (section 10.4), between 5 and 10 m s^{-1}, their speed determined by the mid-tropospheric flow. Each MCS tends to wax and wane, but may be identified for 48 hours or more.

These systems are comparatively rare, requiring an unusual combination of factors, including the propagation of the MJO (section 12.3) across the Atlantic just north of the equator.

10.9 Local convection

Within the ITCZ cloud systems often develop as mesoscale clusters. Studies in the late 1960s and early 1970s suggest that MCSs must have an area greater than about 25×10^4 km^2 to bring significant tropical weather (Hamilton 1975; Global Atmospheric Research Programme 1969). This is convenient for aviation weather forecasters since a likely limit of accuracy and size on aviation weather charts is weather systems no more than 1° (~100 km) across[1].

Tropical MCSs fall into four basic types: amorphous, elongated, quasicircular and vortical (Barrett 1971). Most frequently we see the first type, as discussed in the satellite-imagery analysis of Hamilton (1975). The last type is the least frequent and is always associated with tropical depressions or storms (see Chapter 9).

Given the requirements for deep instability in the tropics discussed in Chapter 3, revisited in Chapter 9, the Red Sea and Arabian Gulf are notable areas that can trigger the development of small but intense clusters of cumulonimbus clouds, especially during the summer. Convection is only inhibited by the subsidence inversion and the dry air above it. When the air at higher levels is moistened somewhat (in particular near the base of an upper trough) and when the inversion is weak, deep convection is often the result and may be seen across the width of either sea. Jackson (1988) reviewed

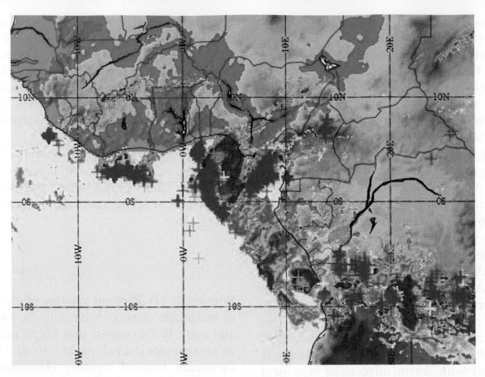

Figure 10.6 MCS development over the Gulf of Guinea at 1200 UTC on 15 March 2013. Surface winds (not shown) indicated that there was an incursion of south-westerly winds originating in a small area of high pressure near 4°S, 3°E. Development was within a shallow mid-tropospheric depression and cloud tops reached more than 13 km (pink tones) over sea-surface temperatures above 29°C. Returns in near-real time from the Met Office sferics ATD lightning-detection system are shown as red (within the last 10 minutes) and yellow (older returns, up to 30 minutes previous) crosses. An area of stable altocumulus can be seen east and south-east of the main convection area. In addition, there are smaller areas of deep convection: from the surface south of Ghana, Ivory Coast and Sierra Leone. Unstable altocumulus and cumulonimbus clouds developing from them can be seen in the top left-hand corner (a common feature in winter in north-west Africa). © Crown copyright (Met Office).

the typical temperature changes in the Arabian Gulf through the year. Whilst only the Gulf of Oman is likely to be warm enough occasionally to generate deep convection in February, temperatures are high enough between April and November to allow deep convection over the Gulf. Maxima are reached in late August, when the sea-surface temperature lies between 31 and 35°C throughout the Gulf and is above 27°C in the Gulf of Oman.

High ground often initiates local convection and this will be discussed further in Chapter 11. Over extensive mountain ranges mesoscale lines of cumulonimbus, embedded in layer cloud, are frequently seen. This is most often the case in summer, when insolation is greatest

and there is a larger area of ground that is not snow covered. Nevertheless, even in winter the area affected can be large. The area where these effects are most notable is the Tibetan plateau (Fig. 10.7). Comparison of infra-red satellite imagery through the day shows that the southern slopes of these mountains and much of the Tibetan plateau heat and cool around 15°C. Thus convection is most pronounced during the afternoon and evening.

In winter, these showers bring snow at higher levels and this generally settles above a height of about 2500–3000 m where the lower troposphere has cooled sufficiently. Very occasionally, lower levels may see lying snow for a time, as was the case on 14 February 2007 in

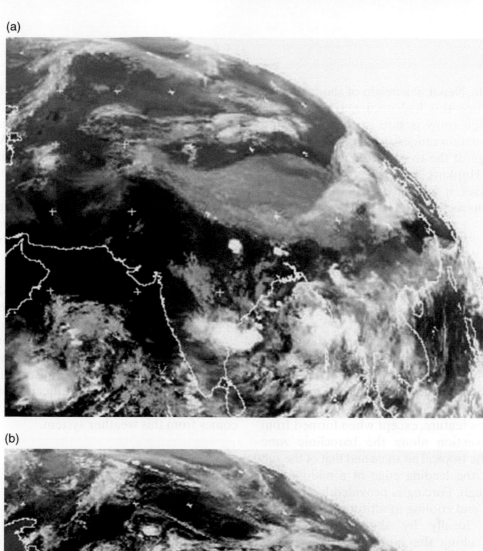

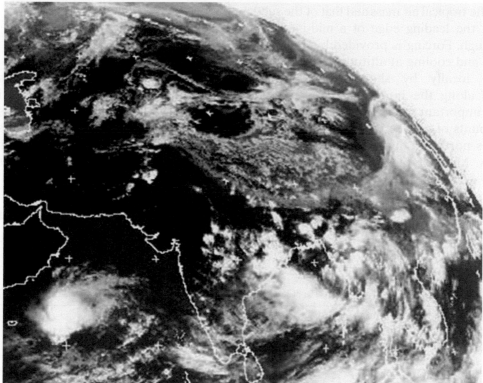

Figure 10.7 The heating and cooling of the Tibetan plateau. A temperature range around 15°C is evident from these infra-red satellite images: (a) 0000 UTC (~0600 local time) and (b) 1200 UTC (~1800 local time) on 24 May 2006. Courtesy of University of Dundee Satellite Receiving Station.

Kathmandu, Nepal, at a height of about 1300 m; the first time this had occurred in 62 years. So, although snow is rare in the tropics, the higher mountain ranges at the poleward continental edge of the tropical zone regularly see snow (D. Hopkins, personal communication). In rare cases, it falls at even lower levels, as discussed in section 7.4.

10.10 Extra-tropical interaction with moist tropical air masses

Over Australia and China the monsoon flows (Chapter 8) may interact with extra-tropical weather systems. These weather systems form along the sub-tropical jet stream (Chapter 4), but are only weakly frontal. Layer clouds are not a major feature, except when formed from deep convection along the baroclinic zone between the tropical air mass and that of the sub tropics, at the leading edge of a mid-latitude upper trough. Forcing is provided by cyclonic curvature and cooling at altitude, and is often enhanced locally by short-wave features. Although along the borders of the tropics, these are important weather systems bringing large amounts of convective rain hundreds of kilometres north or south of the moist monsoon flow. Most of the rain falls on higher ground, suggesting that forced uplift over mountains is required for deep convection (Chapter 11). Much of the summer rainfall of northern China, New South Wales and Victoria in Australia is of this form. Flash flooding may be the result in areas that usually have little rainfall.

The 'pineapple express' was introduced in Chapter 4. In the northern hemisphere the development of a mesoscale cloud band in the right-entrance development area of the equatorward filament of the STJ brings significant rainfall to the west coast of North America in winter. Under the influence of upper-tropospheric cooling (see section 3.4), this band of cloud becomes unstable and may bring heavy rain as the south-westerlies in the middle and upper troposphere carry it north-eastwards. Much of the significant rainfall of the coasts of California and northern Mexico comes from this weather system.

10.11 Conclusion

Although much of the precipitation in the tropics occurs from large-scale cloud systems, some of the most significant weather is often associated with mesoscale systems, such as tropical revolving storms (discussed in Chapter 9) and those resulting from local circulations or mesoscale dynamics, such as

Table 10.1   Temperature (T), wet-bulb potential temperature (θ_w), geopotential height (H) and relative humidity (RH) data from Bamako, Mali at 1200 UTC

Level (hPa)	1 September				2 September				3 September			
	T (°C)	θ_w (°C)	H (dam)	RH (%)	T (°C)	θ_w (°C)	H (dam)	RH (%)	T (°C)	θ_w (°C)	H (dam)	RH (%)
925	23.6	23.3	81	77	21.4	22.5	80	88	21.2	22.4	81	89
850	18.8	21.7	155	77	17.6	22.9	153	95	18.4	22.0	154	81
700	11.2	20.0	319	54	9.0	20.6	317	82	13.2	20.4	320	47
600	1.4	18.0	446	50	2.8	20.6	445	80	2.8	20.3	445	71
500	−5.9	19.0	591	38	−5.5	21.0	589	88	−5.7	19.1	592	49
400	−16.1	19.8	762	11	−15.1	21.6	761	78	−15.7	20.3	764	38
300	−31.5	22.1	972	8	−29.3	22.0	972	56	−31.3	22.5	974	19

easterly waves. These systems also bring the greatest likelihood of hail and thunder in the tropics, as well as very heavy rain in small areas. Thus the weather from these systems is the most likely to be destructive.

The ability to forecast these systems is critically dependent on surface temperature and atmospheric stability, as discussed in Chapters 5 and 11. Where development is over the sea, mesoscale systems may persist overnight, as long as the sea-surface temperature remains above about 25°C. It is important that the forecaster has access to accurate surface temperature analyses and at the Met Office the introduction of Operational SST and Sea Ice Analysis (OSTIA) (National Centre for Ocean Forecasting 2009), a global dataset useable at model resolution, assists the production of timely and accurate forecasts. Over land, where there is the marked diurnal cycle described above, it is important that atmospheric numerical models can represent the atmospheric conditions likely to form these systems and, above that, that forecasters can recognize their development using model products. The introduction of mesoscale models allows additional accuracy in these forecasts and a mesoscale model with 13 km grid length for north-east Africa, south Asia and the northern Indian Ocean is routinely used by forecasters at the Met Office.

Increasingly, forecasters also use poor-man's ensemble techniques (as well as ensemble models (John 2006)) to select the most likely outcome. Not only can atmospheric models be compared, but those of ocean temperature too. Given the critical dependence of many mesoscale systems on both the temperature and moisture source, it is important that areas of warm ocean can be predicted.

10.12   Questions

1. Why are mesoscale weather systems important in our consideration of tropical meteorology and climatology?
2. What problems does the forecaster have to consider in predicting the effects of mesoscale weather systems?

Note

1 For large-area high-level aviation forecast charts (see Appendix 4), a minimum width of about 300 km is the realistic limit to the area of a significant cloud mass, close to the required accuracy for such charts.

11
Forecasting Clouds and Weather

11.1 Background

Most of the significant areas of cloud in the tropics are slow-moving and are not subject to the same form of development and decline as systems at high latitudes, so forecasting them with the assistance of computer-model products is usually straightforward. Nevertheless, forecasters need to ensure accuracy, adding value to products and making sure that the development and motion of faster-moving systems are predicted accurately. A good knowledge of the development and motion of tropical cloud systems is needed by forecasters.

Although much of the cloud in the tropics is formed by convection, layer clouds formed by mass ascent are also important in certain areas and both bring significant weather. The ITCZ was introduced in Chapter 5 and it will not be discussed in detail here, save to say that much of the significant convective cloud over the tropical oceans occurs within the area in which the trade winds meet.

The development of deep convective cloud is favoured in certain areas where dynamical processes combine, as described in Box 11.1.

11.2 Distribution of significant cloud

Until very recently it has been necessary for forecasters to use empirical rules alongside numerical weather-prediction models to forecast zones of significant convection in the tropics. In recent years the development of high-resolution cloud-resolving models has begun to improve the ability to predict significant convection in the tropics (Grosvenor et al. 2005; Shutts 2005; Willett & Milton 2005). Nonetheless, the empirical rules continue to be applied as a check of high-resolution model output, which still has significant difficulty resolving convection. Even with the assistance of models, it is necessary for the forecaster to know well the conditions required for the formation of mesoscale weather systems. In this, forecasters are aided by the slow rate of change of weather patterns in the tropical zone. Thus, we can use a combination of relative humidity in depth and satellite imagery (as discussed in Box 11.2) to indicate where the most likely areas that both localized convection and convection embedded in layer cloud can be expected.

An Introduction to the Meteorology and Climate of the Tropics, First Edition. J F P Galvin.
© 2016 John Wiley & Sons, Ltd. Published 2016 by John Wiley & Sons, Ltd.

Box 11.1   *The formation of clouds and the importance of 'forcing mechanisms' in their development*

Convective cloud is commonly seen in most parts of the world, including deserts and ice caps, although over vast areas it is seen only rarely. In the humid tropics it is the prevalent cloud (often accompanied by areas of layered cloud). Over the open oceans there is little diurnal variation, but the change of temperature through the day promotes convective cloud formation as temperatures rise above a required temperature, the convection temperature, whilst the fall of temperature from late in the day tends to restrict (shallow) convection.

However, most convective cloud is comparatively modest, developing until it is no more than 1000 or 2000 m deep. The deepest of these clouds can produce occasional showers of rain by a process known as agglomeration,

when cloud drops combine to become sufficiently heavy to descend, thus collecting more cloud drops and growing to fall as rain. However, deep convection (with cloud-top temperature below 0°C) requires additional 'forcing mechanisms' to reach the upper troposphere and produce the characteristic heavy rain (sometimes hail or, over high mountains, snow) of the humid tropics. Both large cumulus and, in particular, cumulonimbus clouds are also characteristic clouds of the humid tropics, but these only form where conditions are particularly favourable. A list of favourable conditions is given in Table 11.1.

The development of convective cloud to reach the upper troposphere allows precipitation to form by a different process, the Wegener–Bergeron–Findeisen mechanism

Table 11.1   Conditions favourable to the development of deep convection

Convergence at low levels	This is probably the most important factor in the formation of deep convective clouds. Convergence is produced as the components of wind, ahead, say, of a trough, or within a depression are in opposing directions (e.g. as generated in association with the MJO), thus forcing air to rise.
Warm moist air at low levels	Where the moisture and temperature (combined) of air at low levels exceeds that in air above, convection may be expected, provided other factors are present).
Diffluence in the upper troposphere	If air in the upper troposphere decelerates and diverges, upward motion occurs, assisting the development of deep convective cloud.
Large values of CAPE at medium levels	Convective available potential energy – indicated by a difference between the temperature of an ascending convective 'bubble' of air and that of the environment – allows saturated air to rise as condensation occurs. As long as a rising parcel of air remains warmer than the surrounding environment, it will continue to rise. However, the temperature difference generally needs to be large if convection is to reach the upper troposphere. An ideal combination for MCC development is a 'warm nose' that is only overcome as surface temperature reaches a maximum, over which the temperature difference between the environment and the rising air parcel is 5°C or more.
Upper-tropospheric vorticity	Potential vorticity in the upper troposphere – usually indicated by a so-called 'dry slot' in water-vapour imagery – is associated with mass ascent. This favours convective ascent through the comparatively dry air of the tropical upper troposphere.
High ground	As air passes over high ground, convergence may be expected. A change of surface cover is often also present, including an increase in tree cover or lying snow. Evapotranspiration from these trees or the sublimation of snow helps increase moisture too.

(Wallace & Hobbs 2006; Wikipedia 2014b), which requires cloud-top temperatures below −4°C (Wallace & Hobbs 2006). In this process the cloud top is formed from a combination of supercooled water drops and ice crystals, the former predominating at higher sub-freezing temperatures and the latter prevalent as the temperature falls below −20°C. Within the cloud, the supercooled water drops fall relative to the ice crystals (although both may be ascending in a convective up-draught. The larger these coalescing drops grow, the greater their fall speed, relative to ice and water drops forming the cloud, so that the large drops grow at the expense of small ones. Very large drops will break to form two or more smaller ones, which may then go on to produce more large drops. The upward motion within a cumulonimbus cloud may allow drops to grow for an hour or more if descending drops can be swept upwards by new thermals. In this way hail is formed, following several – sometimes many – cycles of ascent and descent within cloud. Ultimately, snow (at higher levels), rain or hail will be sufficiently large and up-draughts insufficient to keep them within the cloud and precipitation occurs. In most cases this reaches the ground, but where the cloud base is high, some or all may evaporate. The low fall speed of snow allows sublimation (forming water vapour) within only a few hundred metres, although large rain drops may fall as much as 3000 m. Nevertheless, precipitation formed by the Wegener–Bergeron–Findeisen mechanism is usually heavier than that formed by agglomeration, having grown over a comparatively long period in deep cloud.

Forcing mechanisms

The development of convective cloud is not purely random. The tropical atmosphere, above about 3000 m, is usually dry (relative humidity < 70%), as described in Chapter 5. For deep convection, 'forcing mechanisms' are needed. Table 11.1 shows these main factors.

Among the most important of these is convergence at low levels. This is readily available (in the presence of sufficient moisture) along ranges of mountains, as well as close to trough lines or associated with the passage of the MJO (section 12.3). Potential vorticity (PV) in the upper troposphere is also a significant factor, although the dry air usually associated with PV is, at least at first, an inhibitor of convective cloud development. Once cloud develops through this dry air, its energy is vigorous, and the cloud is well able to produce hail and thunder.

In the tropics, most convection forms scatter cumulus and cumulonimbus clouds from around local midday until late evening; at other times, no more than well-separated 'isolated' cumulonimbus clouds are likely to develop. Development is preferred in areas where there is potential vorticity in the upper troposphere (revealed by 'dry slots' visible in water-vapour satellite imagery, e.g. Grahame et al. (2015)), upper-level divergence and low-level convergence (e.g. Koech (2015)).

Over land, the diurnal change is large. Because of the diurnal modulation of convection, cumuliform clouds are generally absent late at night and in the early morning over all but areas of high ground. However, some deep convection develops from altocumulus castellanus that forms in response to the cooling of the atmosphere, often after the surface temperature has fallen below the level at which cumulus clouds can form.

Figure 11.1 shows that the character of cloud systems tends to vary with location across the world, even though most are MCSs.[1] Over South America, MCSs tend to be much smaller than over the Pacific or Atlantic Oceans. Over Africa, some MCSs are associated with so-called African easterly waves (section 3.1).

Box 11.2   The use of satellite imagery and products derived from scans at varying wavelengths

Meteorological satellites scan at a variety of wavelengths, covering emission or reflection, between the visible (~0.6 μm) and the far infra-red (~13 μm). The usefulness of a particular channel varies both by time of day (visible wavelengths cannot be used at night) and by application; different satellites use various wavelength combinations.

Geostationary satellites, which orbit over a single point above the equator, allow meteorological features to be tracked and their development observed.

The highest resolution is from visible imagery from dedicated meteorological satellites, which is currently ~1 km at the sub-satellite point from Meteosat. Deep thick clouds (large cumulus and cumulonimbus in particular) show up well in visible imagery, whilst thin high clouds can hardly be seen at all.

Infra-red imagery is available at several wavelengths from Meteosat. The most important of these are 6.2 μm, which allows upper-tropospheric water vapour concentrations to be observed, and 10.8 μm, the standard infra-red channel.

At 10.8 μm the coldest cloud shows up brightest, so that brightness is inversely proportional to cloud-top temperature. This is very useful in the tropics, since temperatures vary little at a particular height and, within an error band of a few hundred metres, the cloud-top temperature gives the cloud-top height, as described in section 5.4.1. However, low cloud shows up poorly at this wavelength, especially when there are only small amounts, and may not be distinguishable from the surface. These infra-red images are available day and night at 15-minute intervals and can be merged from the five geostationary satellites to form a global cloudscape.

Water-vapour imagery at 6.2 μm identifies areas that are sufficiently moist for cloud development, but, more importantly, also reveals areas of high potential vorticity, which show up as dry (black). These areas, where the upper troposphere is dry, may see rapid cyclogenesis and mesoscale convective system development. A wavelength of 7.3 μm is also available from Meteosat and a greater depth of tropospheric humidity is evident at this wavelength, which is likely to be useful in the tropics.

Combinations of wavelength are increasingly used to derive a number of useful products. The subtraction of the radiance at one wavelength from the radiance at another is used to produce a colour (red, green, blue) image. Ice-crystal cloud can be differentiated from water cloud; blowing dust, volcanic ash or fog/low stratus can stand out from the surface below (Millington 2006).

A review of the uses of satellites in meteorology can be found in Brugge and Stuttart (2003).

Chapter 15 presents a range of remote-sensing data developed both to aid the forecaster and to improve numerical weather prediction.

Over the Himalayas, convection is localized, developing by day on the southern slopes. Over the tropical North Atlantic, many cloud systems occur as curls of cloud within easterly waves (Atkinson 1971). A discussion of important mesoscale weather systems in the tropics is given in Chapter 10.

Other convective features are those associated with large lakes. A notable (near-permanent) example studied in some detail by Lumb (1970) is the cycle of convective development over and around Lake Victoria, the centre of which is at 1°S and 33°E. The lake is about 200 km from west to east and 250 km from north to south – the

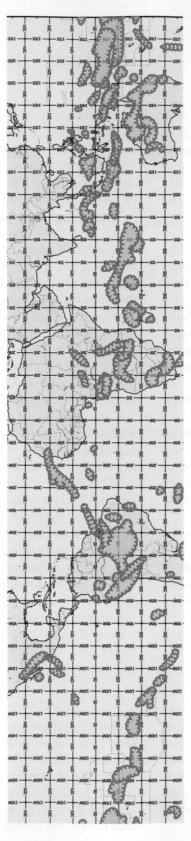

Figure 11.1 Areas of significant deep tropical (and some sub-tropical) convection at 1200 UTC on 26 January 2015. Lines of cloud are in typical positions ahead of upper-tropospheric troughs across southern Africa, the central-southern Pacific Ocean, Mexico, the south-western Atlantic Ocean, as well as between Brazil and Senegal. The southern-summer position of the ITCZ, about 5°S of the equator, is very clearly shown, although the quiescent phase of the MJO (section 12.3) has brought largely clear skies to the eastern Tropical Pacific. The Indian Ocean also typically shows two filaments in the ITCZ, one near 10°S and the other near 5°N, and some deep convection can be seen north of the equator in the Atlantic Ocean. The Australian summer monsoon is evident as an area of significant cloud over the Northern Territory and Queensland, as is some pre-monsoon convection over inland Australia.

largest lake in Africa. The diurnal cycle of convection is due to the development of a lake breeze–land breeze cycle. Lake breezes enhance convection over high ground around the lake by day. The storage of heat by the lake, combined with gentle land breezes, generates convection over the lake by night. By early afternoon (1200 UTC), as temperatures rise over land, a northwesterly lake breeze develops, typically converging with the prevailing wind at the crest of the East African highlands just to the east of the lake. Vigorous convection is the result, often becoming embedded in layer cloud spreading out from the cumulonimbus towers. During the evening and night, land-based convection ceases and now the accumulated warmth of the lake becomes the focus of convection, so that embedded cumulonimbus may be seen centred on the lake. These clouds are usually vigorous, generating thunderstorms. To quote Leroux (2001, Ch. 4): 'The fringes of Lake Victoria rank alongside mountain masses as sites of greatest thunderstorm activity. At Bukoba, the average number of days when thunderstorms occur is 226. At Kampala, the figure is 222, at Entebbe, 211, and at Kisumu, 202…' Similar diurnal cycles occur over and around oceanic islands, such as Puerto Rico (Atkinson 1971), although they are not usually seen to be semi-permanent features.

The main source of variability in the development of deep convection over and around Lake Victoria is the MJO, which is described in section 12.3.

Layer clouds formed from convection in the tropics are typically dense, having formed from air with a high humidity mixing ratio and so have a high water or ice content. Layer cloud development and maximum depth normally occur late in the day, after convection has fed moisture from below into layer cloud in the middle troposphere. The density of tropical clouds, as well as the great depth of clouds formed from supercooled water, increases the risk of airframe and engine icing, which can be significant, particularly below about 5000 m altitude. However, medium-level clouds are often limited in extent compared with the cirrus and cirrostratus that spread out from the tops of cumulonimbus clouds in the ITCZ. Much of

the ITCZ is occupied by around 4/8 cover of altocumulus and stratocumulus, as suggested by Fig. 11.2, with overcast skies observed only beneath larger areas of cumulonimbus locally on windward mountain slopes or extensive cirriform anvils (although see Chapter 9). This is supported by the change in mean relative humidity with height in the tropics, which reduces to low levels at altitude, in particular above about 5500–6000 m. The reason for the reduction in mean relative humidity in the upper troposphere is the disengagement of the upper troposphere from the lower troposphere. This may be manifest in a rapid change of wind direction at the top of the humid zone. Above the discontinuity, the air is much drier than below it.

On the western edge of the Amazon Basin layer cloud can be seen as an almost permanent feature from satellite imagery on the slopes of the Andes west of Iquitos, Peru and observations of layer cloud suggest near-permanent fog about 2500 m in this area, associated with a narrow line of 'cloud forest' (Wikipedia 2014a). This suggests a typical level for altocumulus clouds over land in the humid tropics, where they occupy a depth between about 2500 and 3000 m.

A marked discontinuity is seen over the hot deserts. Taking Kuwait as a typical example, using a daytime screen temperature of 40°C and a dew point of –15°C, the air is highly unstable, often to the 500 hPa (5500 m) level. However, with marked dryness and subsidence prevailing, there is normally little or no low- or medium-level cloud. There is, however, relatively poor visibility right up to the 500 hPa level, with clean air and good visibility above.

11.3 The effect of high ground as an elevated heat source

Throughout the tropics, high ground is an important trigger for deep convection. The lapse of temperature over rising ground is typically around –8 K km^{-1} (Γ + 2 K km^{-1}). This enhances convection over mountains, so a

Figure 11.2 Typical convection and resultant layer clouds within the ITCZ. (Another good example appears as the front cover of the journal *Weather*, July 2006.)

good knowledge of the main areas of high ground across the world (including plateau lands) is necessary for the production of good forecasts of convection.

In the tropics, the effect on convection of areas of high ground is particularly notable. The mountain ranges of the tropics that frequently enhance convective development are shown in Fig. 11.3. Convection may continue or be enhanced by additional dynamical forcing as an air mass crosses high ground. The mountain chains (or their foothills, where the mountains protrude above about 3000 m) that most often generate enhanced deep convection include:

- the Atlas Mountains (D. Hopkins, personal communication), the East African highlands (Koech 2015) and the Madagascan highlands in Africa
- the Arabian highlands (Al-Maskari & Gadian 2005; Ghulam & Dorling 2005), the

Western Ghats (Asnani 1993), the Himalayas, Arakan Yoma, Chaîne Annamitique, and the mountains of Malaysia, Indonesia and the Philippines (Galvin 2004) in Asia
- Australia's Great Dividing Range
- the Central American sierras
- the Andes of South America
- the east Brazilian highlands.

Orographic effects are particularly important at the height where relative humidity begins to fall, even in the deep humid zone (Fig. 1.3b). In most of the tropics the air is dry above 1500–3000 m, although in some places moist air extends to around 5000 m. Through the moist layer, the air is very humid, the humidity reducing the diurnal range of temperature, in particular close to the ocean. Where humidity is lower, there is a greater diurnal range of temperature. The result is that potential temperature is higher over significant high ground in the equatorial humid zone from mid-morning to late evening, promoting

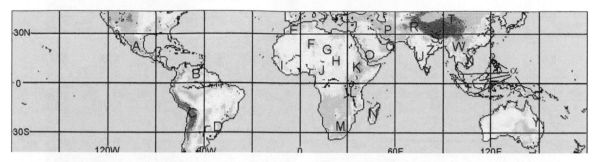

Figure 11.3 The main areas of high ground in the tropics: A, the Central American sierras; B, Venezuelan highlands; C, Andes; D, Serra Geral; E, Atlas mountains; F, Hoggar; G, Tibesti; H, Ennedi/Marra plateau; I, Fouta Djallon; J, Dorsale Camerounaise; K, Ethiopian highlands; L, East African highlands; M, Drakensburg mountains; N, Madagascan highlands; O, Hijaz-AsBr; P, Zagros and Alborz mountains; Q, Jabal Akhdar; R, Ku-e Baba-Hindu Kush-Karakoram-Baluchistan highlands; S, Himalayas; T, Kunlun Shan; U, Western Ghats; V, Sri Lankan highlands; W, Daxue–Hengduan–Wuliang ranges; X, Chaîne Annamitique; Y, Great Dividing Range; Z, Eastern Ghats; á, the mountains of the Philippines, Indonesia and Malaysia. These mountains, or their foothills, enhance rainfall, particularly in drier parts of the tropics, when sufficient moisture is available and conditional instability can be released as air ascends across them.

convection that is only limited by areas of extensive layer cloud that may form over the high ground, thus reducing instability (Atkinson 1971, Ch. 3). However, the development of deep convective cloud is also promoted by dynamical uplift as the air rises over the mountains, so that deep cloud and precipitation may remain through the night.

11.4 Tropical upper-tropospheric troughs

To a large extent, fronts are inactive features at their equatorward limit, where pressure is generally high. However, their incursion into the tropical zone can be an important trigger for severe weather. As at higher latitudes, fronts can be identified as a change of air mass (i.e. a baroclinic zone between differing air masses is present) and the STJ often provides high-level forcing for their development. The characteristic feature of these disturbances is a large-scale trough in the upper troposphere. Upper troughs are slow-moving, semi-permanent features in the STJ, as discussed in Chapter 4.

Although large-scale layer clouds on these fronts may be well broken or thin, the development of areas of embedded cumulonimbus on their warm side can be important, ahead of the

upper trough, particularly in maritime areas. These fronts are generally slow moving and are almost always warm fronts at low latitudes, within the trade-wind (easterly) zone at low levels. With westerly winds aloft and easterlies at low levels, these fronts are not associated with significant advection of heat or moisture.

The incursion of cold air associated with the upper troughs, in particular when they sharpen (as shown in Fig. 11.4), is a major trigger for cumulonimbus development. Where the water (or land) is warm, for example across much of the tropics, the ingredients are present to form large areas of embedded cumulonimbus. This process is more important in winter than in summer and is more important in the southern hemisphere than the northern.

It can be a surprise to many visitors to Brazil's central east coast that there is often cloudy weather with outbreaks of heavy showery rain. It is particularly wet on the mountain slopes above Santos and Porto Allegre, where an average of more than 1250 mm of rain falls each year. Rio de Janeiro lies close to this local rainfall maximum and has rain throughout the year, amounting to more than 1000 mm. This is associated with one of the semi-permanent upper troughs in the STJ, which is a semi-permanent feature over the western South Atlantic Ocean. In extreme

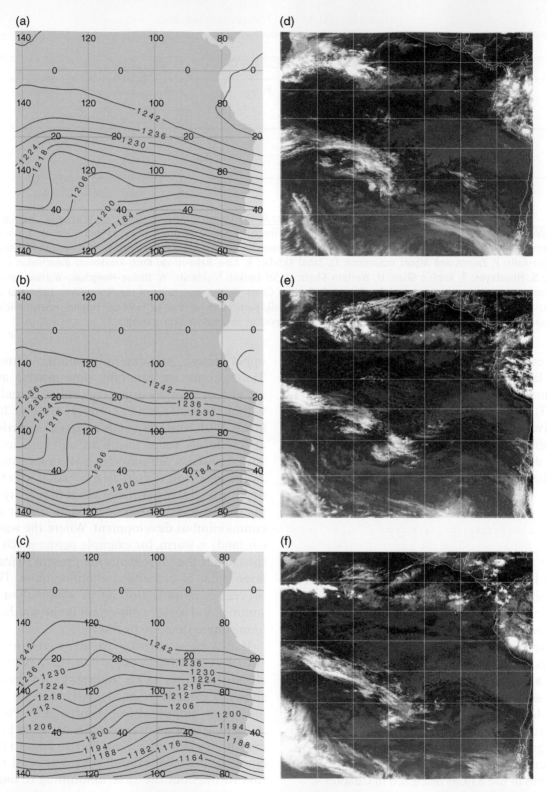

Figure 11.4   The sharpening and relaxation of an upper trough. (a)–(c) Contours of 200 hPa height (decametres) over the South Pacific in mid-December 2006. (d)–(f) Note the increase in cyclonic curvature at the base of the trough, which provides dynamical forcing to aid the development and later the wane of deep convective cloud, shown by infrared satellite imagery.

cases, the associated semi-permanent cloud mass on the eastern side of South America has been noted to cross the equator and one such event brought severe weather to Venezuela (Atkinson 1971). Periodic floods occur when mesoscale convective storms bring rainfall associated with this trough to the Sierra da Mantiquierra, north-west of Rio de Janeiro and Sao Paulo. One of these slow-moving storms, which caused serious flooding and the loss of over 500 lives (BBC News 2011b) is shown in Fig. 11.5.

Notable storms also develop in the northern hemisphere at times. One of these brought very heavy rain to Morocco between 30 March and 1 April 2002, as reported by Fink and Knippertz (2003), and in winter they may be an important local source of precipitation in the deserts (Miles 1959). The presence of high ground was particularly important in the development of these storms over the arid highlands of the Sahara desert.

Other typical areas of formation include the south-west Pacific, in particular around the Solomon Islands and Fiji, or across eastern Polynesia and the central Pacific. The global pattern of tropical upper-tropospheric troughs can be seen in Fig. 11.6.

Although these troughs move slowly, not all can be seen clearly in Fig. 3.1. This is because they are not stationary and there is variability in the cloud cover within the troughs as they migrate. Only the troughs of the South Atlantic and south-west Pacific can be seen throughout the year, their effects on the radiation budget spread over a relatively wide range of longitude. Nevertheless, troughs are associated, in part, with the cloudiness of south-eastern China, south-eastern Brazil, the south-eastern USA and the islands of the south Pacific.

11.5   Effects of severe convection on aviation

Icing and turbulence are the main hazards from deep convection for aircraft, although in suitable atmospheric conditions (slight wind shear with height), large hail can also be significant.

Airframes have been dented and windscreens smashed by ice. Either of these may cause cabin decompression and so is a danger to the passengers, pilot and crew. Hail may be found throughout much of the depth of a cumulonimbus cloud and, at times, may reach the surface in the tropics. However, it is most often found near the peripheries of the tropics, where wind direction changes only slowly with height – a necessary condition for its development.

Electrical storms (many of them severe) always accompany MCSs. These pose a significant hazard. Instruments and navigation systems may be affected, posing a threat to passengers and crew. At the surface, lightning strikes may cause loss of life by direct strike or their effect on trees or buildings. In dry areas, serious fires may be started by these storms.

11.5.1   Icing of aircraft

Ice accretion occurs in supercooled-water cloud, which may be present down to temperatures as low as $-40°C$, although the risk quickly reduces in all but cumulonimbus clouds as the temperature falls below $-20°C$. The supercooled water is induced to freeze as it comes into contact with engine or airframe surfaces, where these are below freezing. Most aircraft have either a heating system or hydraulic 'boots' that expand and contract to clear any ice accretion on the leading edges of the wings. Nevertheless, severe icing can overcome these systems and pilots need to be aware of the likelihood of significant ice accretion.

Where ice builds up on wings or helicopter rotor blades, it causes irregularities in the airflow across them, reducing or distorting lift. If the icing affects an engine, it may reduce airflow through the turbines or reduce the efficiency of propeller blades. Indeed, some older turbojet or turboprop engines may suffer icing at temperatures marginally above $0°C$, if there is a need to reduce the speed of the airflow entering the compressor fans. Piston-engined aircraft may suffer carburettor icing and this too can occur in cloud a little above freezing due to the expansion of the air as it enters the carburettor.

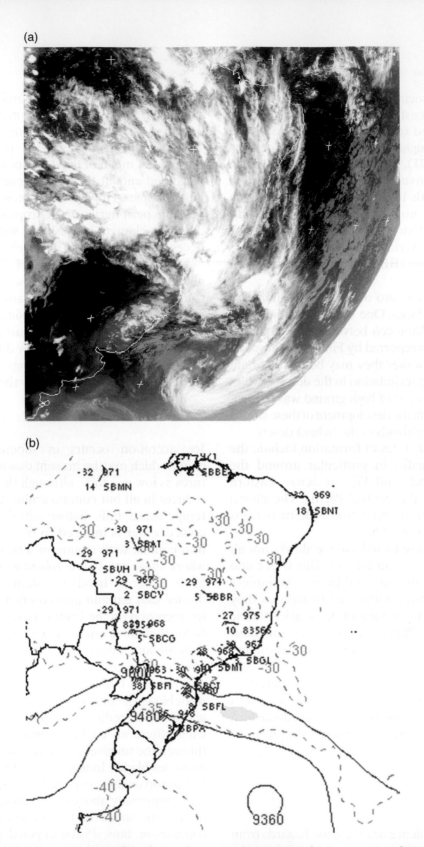

Figure 11.5 (a) Infra-red satellite image of Brazil at 0000 UTC on 13 January 2011, showing an MCS northwest of Rio de Janeiro. (b) This storm formed near the base of a semi-permanent upper trough, which has formed a sub-tropical depression near 40°S, 45°W, as shown on the 300 hPa analysis. (Shading shows an area of wind speed ≥40 m s⁻¹; heights are in geopotential metres.)

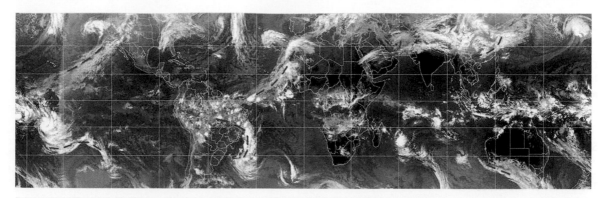

Figure 11.6 Infra-red satellite image showing a typical global cloud pattern at 0900 UTC on 14 February 2005. Cloud lines (marked with dashed lines) associated with active tropical upper-tropospheric troughs can be seen east of Brazil, over north-west Africa, in the southern Indian Ocean (associated with a tropical revolving storm), over Iran and the southern Arabian Gulf, in eastern China, over the Coral Sea and over the South Pacific (associated with a tropical revolving storm) © EUMETSAT/Crown copyright (Met Office).

In the tropics, since the liquid-water content of clouds is high, icing is often a greater risk than at higher latitudes. In deep layer-cloud, in particular that formed from convective towers, icing may be severe at 0°C. It has been noted that the greatest icing risk is in cloud producing, or about to produce, precipitation, where there is a high concentration of large supercooled water droplets. This is most often seen in extensive tropical clouds at 4500–5500 m – the height range through which supercooled layer clouds are most prevalent in the tropics – especially where there is an extensive sheet of layer cloud. Typically, the separation of the 0°C isotherm and the –20°C isotherm is about 4000 m between altitudes of 4500–5000 m and 8500–9000 m in tropical layer clouds, bringing a risk of significant ice accretion. However, most aircraft cruise above 9000 m so icing is mainly a problem for short-haul flights and aircraft climbing or descending gradually.

Icing is also an increased risk in the humid tropics due to the high frequency of cumulus congestus and cumulonimbus clouds. These clouds bring the highest likelihood of icing, as the depth through which significant icing can occur is great. However, aircraft may be able to fly around these clouds, especially if they are not embedded in extensive layer-cloud, since cumulus clouds never occupy more than half the sky and even cumulonimbus clouds rarely do (from the perspective of a pilot). This is emphasized in the *Handbook of Aviation Meteorology* (Meteorological Office 1994, p. 214). However, the severe icing associated with large cumulus feeding the layer clouds cannot be ignored.

Most icing is no more than moderate[2] in layer clouds, although well-developed MCSs may have sufficiently dense cloud associated with them to cause severe icing (especially at lower levels). Favoured areas for moderate icing include the periphery of large MCSs and tropical revolving storms (which can include nimbostratus, within which severe icing occurs). In particular, however, it will be frontal zones intruding into the tropics that most often have the greatest extent of layer cloud likely to cause icing, often through much of the middle troposphere.

Well-developed altocumulus castellanus that has not deepened to form cumulonimbus may also be associated with severe icing.

The effect of orography on icing can also be significant (Meteorological Office 1994, p. 130). Severe icing has been observed in cloud forced to rise over mountains at temperatures as low as –25°C.

The effects of the various levels of icing are described in Table 11.2.

Table 11.2   The effects of the various levels of icing and turbulence on aircraft

Airframe icing criteria

Category	Description
Light	No change of course or altitude is necessary and no loss of speed is caused.
Moderate	Change of altitude and/or heading considered desirable. Ice accretion increases, but not at a rate sufficiently serious to affect the safety of the flight unless it continues for an extended period; air speed may be lost.
Severe	Change of altitude and/or heading considered essential. Ice accretion builds up and begins to seriously affect the performance and manoeuvrability of the aircraft.

The effects of turbulence

Category	Description
Moderate	Difficulty walking through the aircraft, rather severe and frequent rolling of the aircraft and general passenger discomfort. This is the degree of turbulence is that usually associated with towering cumulus, passing through typical frontal conditions and in the vicinity of isolated cumulonimbus clouds.
Severe	Loose objects in the interior of the aircraft become dislodged, the aircraft tosses, it is difficult to retain flight altitude and passenger discomfort is marked. This degree of turbulence is that usually encountered in vigorous cumulonimbus clouds.

11.5.2 Turbulence in cloud

Away from the surface layer, most turbulence in the tropics is associated with convection. Convective up-draughts are frequently a source of moderate to severe turbulence. However, outside convective clouds and the near-surface layer there is little turbulence through much of the tropical troposphere as winds generally decrease above the boundary layer to a minimum at middle-tropospheric levels in all but the central equatorial zone. However, wherever there is significant wind shear (a rapid change of wind speed or direction) in layer clouds, turbulence is likely. At the tops of deep layer cloud, the risk may be somewhat less important than at lower levels due to the reducing density of air (at 5500 m air density is about 55% of its value at sea-level). However, density changes relatively slowly with height (about 10% for every 1500 m of ascent in the middle troposphere) and moderate turbulence can occur in cloud composed predominantly of ice crystals. After all, layer cloud forms beneath atmospheric discontinuities

that may be associated with wind shear. The most likely areas are thus (i) the equatorward limits of disturbances associated with upper troughs, (ii) any layer clouds that have formed over significant mountain ranges, (iii) monsoon circulations and (iv) within the circulation of tropical revolving storms. Away from these areas, turbulence is more likely to occur in altocumulus than in altostratus clouds.

Nevertheless, severe turbulence is extremely rare in layer cloud in the tropics, although significant turbulence may be encountered in cirriform cloud associated with cumulonimbus clouds (Zobel & Cornford 1966) above 12,000 m.

The effects of moderate and severe turbulence are described in Table 11.2.

11.6   Questions

1. What do we mean by 'significant' in terms of the cloud formations discussed here?
2. Why do we need to forecast significant weather (in the terms described above)?

Notes

1 MCSs are clusters of convective clouds (often including altocumulus castellanus), usually associated with thunderstorms and often containing layer cloud formed by spreading out of the convective cloud. Their longest axis is 100 km or more in length.

2 The rate of icing is expressed in terms of mass per unit area per unit time ($kg\ m^{-2}\ s^{-1}$).

However, drop density is proportional to icing rate and is available from numerical models so is more useful to the forecaster. The Australian Bureau of Meteorology uses these definitions: light icing: less than $0.5\ g\ m^{-3}$ supercooled water in cloud; moderate icing: 0.5–$1.0\ g\ m^{-3}$ supercooled water in cloud; severe icing: more than $1.0\ g\ m^{-3}$ supercooled water in cloud.

12

The Variability of Weather and Climate Change in the Tropics

12.1 Introduction

Tropical climates are affected by three significant forms of large-scale variability (as well as smaller-scale variability, not covered in this chapter): El Niño–La Niña, the MJO and the QBO. The first has effects on an approximately annual time scale, the other two affect the weather on a regional time scale of about 10 days. In addition, a longer-term but slow change occurs due to anthropogenic influences.

12.2 El Niño–La Niña

El Niño and La Niña, introduced in Chapter 6, are the names given to phenomena associated with changing SSTs in the equatorial Pacific Ocean. El Niño is indicated by SSTs above the normal in the eastern tropical Pacific – an area in which the water is normally relatively cold. It is often, although not always, accompanied by a modest depression in SSTs in the western Pacific. La Niña is a warming of the western tropical Pacific – an area where the water is usually warm. The change occurs over an approximately 5-year time scale, although with considerable variability, as indicated by Fig. 12.1. However, the change is not bi-modal and many years, between those of El Niño and La Niña, are neutral, with SSTs near to their normal (University of Illinois, Department of Atmospheric Sciences 2010).

Both El Niño and La Niña are associated with a change in mean sea-level pressure across the Pacific, known as the Southern Oscillation,[1] that causes a relative change in wind flow, which is normally from west to east across the basin. A significant change in rainfall occurs in the areas where the water is anomalously warm, whilst the relatively cooler-water areas see a rainfall deficit.

El Niño was first identified by fishermen working the normally productive seas along the west coast of South America. The warming caused a collapse in the fishery, since the plankton on which the pelagic fish of this area feed depend on the high levels of dissolved oxygen only found in cool water. El Niño usually reaches a peak at Christmastide and was named after the Christ Child. Increased rainfall in the eastern tropical Pacific associated with El Niño affects relatively few people, although flooding may occur in normally dry coastal areas of South and Central America. However, there is a significant effect on the flora and fauna of the Galapagos Islands. These semi-desert islands close to the equator have a unique collection of plants and animals, generally dependent on the maintenance of dry sunny weather (Oña & DiCarlo 2011). In addition, rainfall tends to be enhanced over East Africa

An Introduction to the Meteorology and Climate of the Tropics, First Edition. J F P Galvin.
© 2016 John Wiley & Sons, Ltd. Published 2016 by John Wiley & Sons, Ltd.

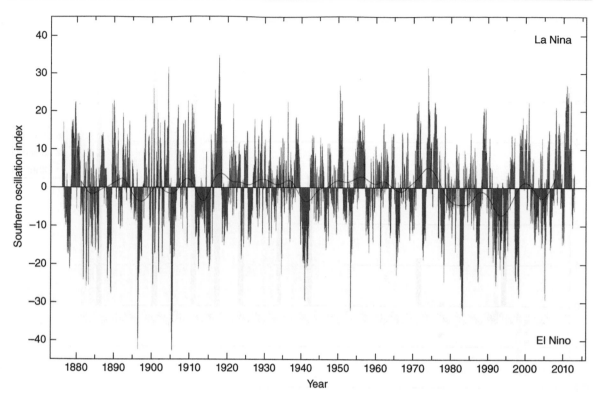

Figure 12.1 The values of the Southern Oscillation Index 1876–2011 are a measure of the variation between El Niño and La Niña.

(section 8.8) due to a weakening of the Walker circulation, as large-scale descent of air over East Africa (as well as the eastern Pacific) reduces. The effects of the Southern Oscillation extend beyond the tropics. Most notably, ENSO tends to change circulation systems across the world through large-scale teleconnections (Bjerknes 1969; Burgess & Klingaman 2015). The global mean temperature tends to fall in La Niña years, whilst El Niño years tend to be warm (Fig. 12.2).

La Niña potentially affects a greater number of people than does El Niño, its effects on rainfall largest in densely populated areas of South-East Asia, as well as along the north-eastern coast of Australia, although the droughts associated with El Niño also have dramatic consequences. During the marked La Niña of 2010–2011, strong links were noted to the development of low-pressure systems over Thailand and associated severe flooding

(Gale & Saunders 2013) and there may have been some link to the pre-monsoon rainfall that brought flooding to much of Pakistan. In addition, there were more tropical cyclones than normal along the Queensland coast in summer 2010–2011 (Met Office 2011h).

There is some evidence that El Niño and La Niña are becoming more common and there may be a link to anthropogenic warming of the atmosphere. However, given the relatively short length of the observational record in the tropical Pacific (especially at sea), this is difficult to prove conclusively (Marshall 2011). In addition, the changes in El Niño may be more complex than a simple change in strength or frequency.

There may be several years between El Niño and La Niña events and, indeed, there may not be a change from one to the other, neutral conditions sometimes occurring between them (marked by small values of the SOI).

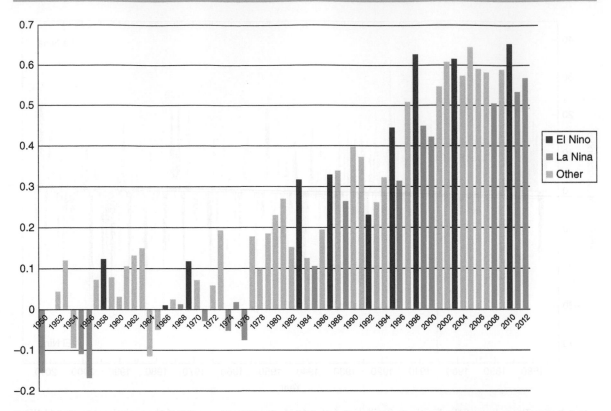

Figure 12.2   A comparison of El Niño and La Niña years with the rise of global mean temperature since 1950 (relative to pre-1950 climatology). Years with an anomalously high difference in SST between the western and eastern tropical Pacific are marked blue, whilst those with an anomalously low difference in SST are marked red (NOAA 2013).

12.3 The Madden–Julian Oscillation

The MJO, first revealed by Madden and Julian (1971, 1972, 1994) is on a smaller time scale than El Niño–La Niña, but is very significant in its effects on tropical rainfall between about 0°E and 160°E. It can be seen at times in all parts of the tropical zone. Occurring at intervals between 30 and 90 days, with a typical interval of about 50 days, the MJO is manifest as an increase and decrease in tropical convection, associated with variations in the easterly (zonal) component of near-surface and upper-tropospheric winds (Fig. 12.3). In turn, the MJO interacts with El Niño–La Niña, the Walker Circulation (section 3.1) and smaller-scale convectively generated phenomena, so that not all rainfall variability in the western Pacific and

Indian Oceans is predictable. Indeed, the MJO remains a very difficult phenomenon to forecast given these interactions and the variability of the time between successive MJO developments (Zhang 2005).

The MJO (at least in its more active form) is highly dependent on SST and does not develop when SSTs are comparatively low in the eastern Pacific and across the Atlantic. It links a fall in mean sea-level pressure to the development of westerly lower-tropospheric winds and associated convergence, accompanied by enhanced easterly winds at upper levels (as well as increased easterly surface winds just ahead of the system), followed by a period of rising mean sea-level pressure and suppressed convection that travels eastward near the equator from the eastern coastal regions of Africa to the central Pacific at a speed around 5 m s⁻¹

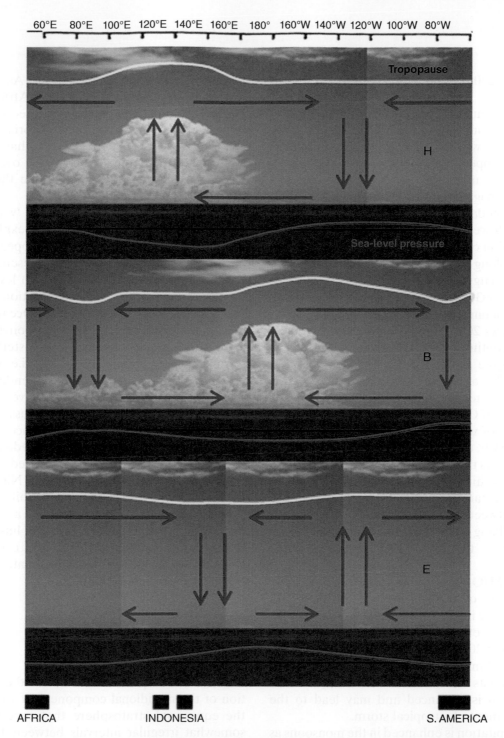

Figure 12.3 Longitude–height schematic diagram along the equator illustrating the fundamental large-scale features of the MJO through its life cycle (from top to bottom). The cloud symbol represents the convective centre, arrows indicate the zonal (east–west) circulation, and curves above and below the circulation represent perturbations in the upper tropospheric and sea-level pressure. At point H convection is initiated near the East African coast, but is generally suppressed elsewhere, particularly over South-East Asia. At point B rainfall is enhanced over South-East Asia, but it is drier than normal in most other areas. At point E convection is generally suppressed, except in the south-west Pacific. Note the relatively rapid eastward movement of the tropospheric Kelvin wave (expanding the troposphere, compared with the area of convection, which lies over the area of lowest pressure, but ceases when there is a dry feed of air in the trade winds (adapted from Madden & Julian 1971, 1972).

(Zhang 2005). The convergence causes an increase in deep convection and associated rainfall, mainly within the ITCZ (as shown in Fig. 10.6), but also within the monsoon zones and areas where other phenomena, such as tropical upper troughs, are seeding the atmosphere with moisture and ascent. The speed of MJO propagation gives it a lifetime of approximately 30 days within its area of most significance. However, there is some evidence for an association of the MJO with precipitation variability along the west coast of North America (Mo & Higgins 1998; Jones 2000; Bond & Vecchi 2003), in South America (Paegle et al. 2000; Liebmann et al. 2004) and in Africa (Matthews 2004), suggesting that this tropical 'wave' continues east, taking about 90 days to encircle the globe. This may also suggest that the MJO is a more-or-less continuous wave, as also suggested by satellite and wind data for the Gulf of Guinea (Fig. 10.6). However, using the area of westerly lower-tropospheric winds as a marker, the MJO can be seen to accelerate as it crosses the eastern Pacific and the Atlantic basin, reaching speeds around 35 m s^{-1}. Only one wave can be positively identified over the area between East Africa and Polynesia at a time (Zhang 2005) and any MJO signal is very weak over the eastern Pacific and Atlantic basins.

The MJO has an important role in the generation of tropical cyclones (Chapter 9), particularly in the Atlantic and western Pacific Oceans (Nieto Ferreira et al. 1996; Maloney & Hartmann 2000; Hall et al. 2001; Higgins & Shi 2001; Leibmann et al. 2004). As the westerly wind bursts of the MJO develop, cyclone formation is enhanced and may lead to the development of a tropical storm.

Precipitation is enhanced in the monsoons as the MJO passes and the increase in rainfall is sufficient to cause a seasonal variation in the apparent effects of the MJO: there is a northern hemisphere peak in July and August associated with the south Asian monsoon and a corresponding peak in February associated with the Australian monsoon (Fig. 12.4) (Zhang 2005).

In some areas the effects are subtle. As a new MJO develops near the east coast of Africa, the diurnal cycle of lake-breeze/land-breeze convection around and over Lake Victoria weakens (discussed in section 11.2), so that much less (night-time) rainfall is expected over this large lake, although it may continue through the night on surrounding mountains.

The development of each MJO 'wave' is highly complex and has not yet been explained satisfactorily. However, forcing appears to depend largely on SST, the presence of enhanced humidity to relatively high levels in the troposphere and cycles of convection. Over the warmest parts of the Indian Ocean, the MJO commences as an area of deep convection close to the area of convergence of easterly and westerly winds within the ITCZ (i.e. either north or south of the equator, dependent on season). However, as the MJO wave moves east, the area of convection is associated increasingly with the westerly surface winds, the convergence line tending to precede it.

Considerable work is being carried out to study the link between El Niño–La Niña, the MJO and extra-tropical climates. Whilst there seem to be such teleconnections, as might be expected from a system strongly linked to SSTs, they are relatively weak and can only contribute to the likely seasonal climate of any region, even within the tropics.

12.4 The quasi-biennial oscillation

The QBO is a periodic change in the direction of the meridional component of winds in the equatorial stratosphere that reverses at somewhat irregular intervals between 15 and 35 months (mean 28–29 months), the reversal causing warming (Andrews et al. 1987; Baldwin et al. 2001). The wind regime alternates between the 'normal' easterlies and winds with a rather irregular westerly component. The wind reversal propagates downwards at somewhat less than 1 km per month until it dissipates at the

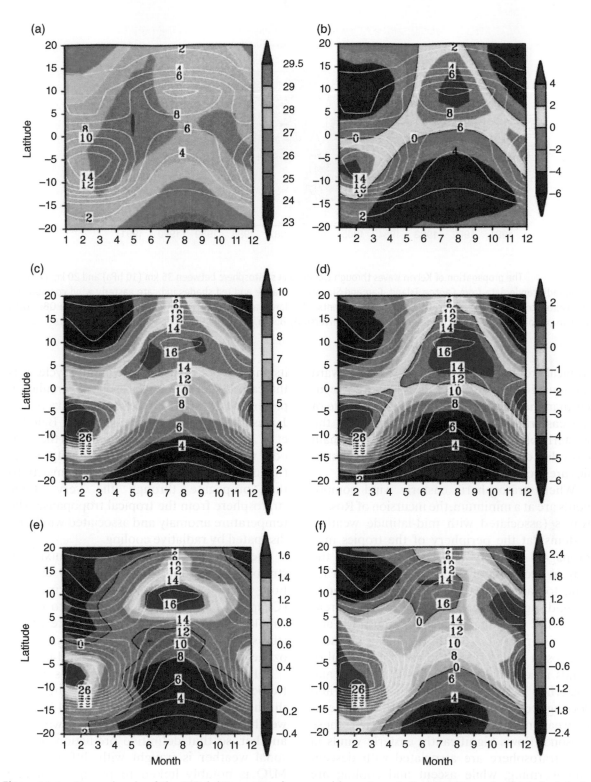

Figure 12.4   Seasonal cycle of the MJO (white contours) measured by variance in its 850 hPa zonal wind (m² s⁻²; top) and precipitation (mm² d⁻²; middle and bottom) averaged over 60°E–180°E and each month for 1979–1998. The background colours (with zero indicated by black contours) are mean (a) SST (°C), (b) zonal wind at 850 hPa (m s⁻¹), (c) precipitation (mm d⁻¹), (d) surface zonal wind (m s⁻¹), (e) 850 hPa moisture convergence $\nabla\bullet(q\boldsymbol{V})_{850}$ (g kg⁻¹s⁻¹), and (f) 925 hPa moisture convergence $\nabla\bullet(q\boldsymbol{V})_{925}$ (g kg⁻¹ s⁻¹). From Zhang & Dong (2004).

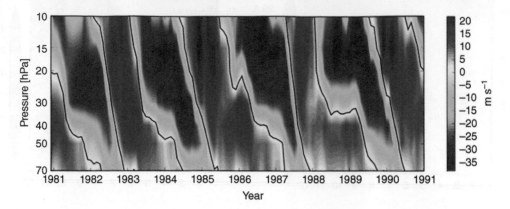

Figure 12.5   The propagation of Kelvin waves through the tropical stratosphere between 35 km (10 hPa) and 20 km (70 hPa) using radiosonde data from Canton Island, Gan and Singapore. Orange and red shades indicate easterly wind components and yellow to blue shades indicate westerly components. The downward-propagating reversal is known as the QBO, with a mean period of 28–29 months. Courtesy of Freie Universität, Berlin, originally published in Naujokat (1986).

tropical tropopause (Fig. 12.5). The downward motion of the easterlies is usually more irregular than that of the westerlies. The amplitude of the easterly phase is about twice that of the westerly phase. At the top of the QBO domain easterlies dominate, while at the bottom westerlies are more likely to be found.

When stratospheric easterly wind components are at a minimum, the incursion of Rossby waves (associated with mid-latitude weather systems) at the periphery of the tropics may propagate westerly wind components all the way to the equator at these levels. Easterlies tend to carry mass equatorward as motion is around high pressure over the poles, whereas westerlies aid the poleward transport and replenishment of ozone in the higher latitudes (Heaps et al. 1999). As the general reversal of winds is seasonal – westerlies in summer and easterlies in winter, driven by heating and cooling around the poles – the QBO represents a weakening and strengthening of these seasonal flows. Significantly, easterly winds in the stratosphere are associated with descent and warming, while ascent and cooling are associated with the anomalous westerly phase (Reynolds 2000).

The QBO was discovered in the 1950s, but its cause remained unclear for some time, although its phase is not related to the annual cycle, as is the case for all other stratospheric circulation patterns. By the 1970s Richard Lindzen and James Holton (Andrews et al. 1987; Baldwin et al. 2001) postulated an explanation: the periodic wind reversal is driven by gravity waves (mainly the result of overshooting convection) that propagate through the very stable stratosphere from the tropical tropopause. The temperature anomaly and associated winds are dissipated by radiative cooling.

The effects of the QBO include mixing of stratospheric ozone by its secondary circulation, modification of monsoon precipitation (Chapter 8) and an influence on stratospheric circulation in the northern hemisphere winter (the sudden stratospheric warming related to ozone depletion described in section 14.2.1).

All of these large-scale processes in the tropical atmosphere have a modifying effect through teleconnections to mid-latitude weather. In addition, they interact with one another, so that definitive prediction of seasonal weather is fraught with difficulty. The MJO is notably linked to precipitation and seasonal temperature cycles in North America with onward effects on North Atlantic pressure patterns (Burgess & Klingaman 2015). Similarly, El Niño–La Niña has modifying effects on the progression of the mid-latitude Rossby waves that carry frontal weather systems of the north Atlantic and north Pacific Oceans. (Effects are hard to recognize in the poorly observed and generally much more progressive weather systems of the Southern Ocean.)

12.5 A discussion of anthropogenic climate change

The characteristics of the climates of the tropics have been discussed in Chapters 6, 7 and 8 based on climatological normals, which are mean values for periods of 30 years, updated at 10-year intervals. This chapter discusses the effects mankind has on the climate.

As presented in Chapter 2, the radiation balance is affected by the composition of the atmosphere, so there can be no doubt that climate is changing as a result of the activities of mankind that alter atmospheric composition (Collins & Senior 2002). A variety of greenhouse gases (GHGs) absorb and re-emit long-wave radiation emitted by the earth in response to insolation. Although, in turn, there is a change in the emission of radiation to space (through a complex process of absorption and re-emission of many cycles), changes in the concentrations of GHGs in itself alters the amount of radiation lost to space. This is due to changes in the temperature of the atmospheric column (Pidwirny 2013). As GHG concentrations increase, some of the re-emitted long-wave radiation is retained and the temperature rises. Although there is some doubt about the total increase of these gases in the atmosphere (in particular carbon dioxide, CO_2), which have existed at very small concentrations since the beginning of the Industrial Revolution, the greatest uncertainty concerns feedback mechanisms and the probable effects on the climate of the increase. Perhaps the largest area of uncertainty (and, to some extent, inconsistency) is in the effects of changes in water-vapour concentrations. Gaseous water is a more powerful GHG than CO_2 and can exist in much higher concentrations than the other GHGs, especially near the surface in the tropics. It is the effects of warming by anthropogenic gases on the concentrations of water vapour that produce much of the overall atmospheric warming predicted by climate models. It is therefore important that we understand the feedback processes associated with water vapour, including any resultant changes in cloudiness and its transport into the upper atmosphere, if we are to understand fully the effects of industrialization and development on the climate (Solomon et al. 2010; Pidwirny 2013).

Although industrialization is considered largely to blame for the increase in GHGs, the effects of climate are not restricted to the industrialized areas of the world since, as we saw in Chapters 2, 3 and 4, the winds of the world carry air to all of its corners. It is also necessary to include the effects of agriculture and construction in calculations of the change in GHGs. These are discussed in Chapter 13.

Possibly the main complication in the climate-change discussion is the effects of feedback. The knock-on effects of changes in temperature (and precipitation) are likely to bring about a change in the 'steady state' of the climate, such that a new global series of climates will become established, once the GHG concentration and temperature rise beyond a point that will allow the current climatic conditions to continue. It would be very difficult to reverse changes once a new global climatic regime becomes established, although it is not yet clear what level of GHGs would cause the change.

There are clearly big questions to be asked about how to control these effects, but the world needs to work together to ensure that development and prosperity can come to the tropical zone, whilst at the same time ensuring that the damage from climate change is minimized. This topic is explored in more detail in Strangeways' (2011) article in *Weather*.

12.6 How is climate likely to change in the tropical zone?

The attempt to find the likely effects of changing climate is the main challenge facing climatologists. Many centres run computer models in an attempt to discover the likely change of temperature, rainfall and other meteorological factors. There is considerable variation between these models, dependent on their physical parameters (formulation, the effects of the surface, physical interactions etc.), grid length and, to some extent, stability. Errors grow in much the same way as in numerical weather predictions.

The fact of increased GHG concentrations in the atmosphere and resultant warming is now well established (LeTreut 2007). The Intergovernmental Panel on Climate Change (IPCC) Fifth Assessment Report (IPCC 2014) gives average warming across the globe between about 1.5 and 4.8°C by the end of the 21st century, not least in the tropics, where there is the greatest warming over land, mainly, although not exclusively, in areas of dry climate (in line with the predictions shown in Fig. 12.6. However, Fig. 12.6 clearly shows that the greatest uncertainty exists in areas where the greatest warming is expected). Over the sea, the likelihood is that the greatest warming

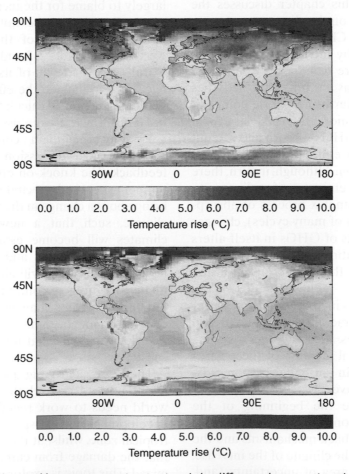

Figure 12.6   The global ensemble average warming (top) and the difference between the 5th and 95th percentiles of the predicted temperature rise (bottom) for a doubling of CO_2 concentrations since the pre-industrial era from a 130-member ensemble run of the Met Office Hadley Centre PRECIS model. From Met Office (2004b).

will be close to the equator (although warming will be comparatively modest) (Fig. 12.6).

Any change of climate can take one of two forms as temperatures rise. Either the new climatic state has a similar variability to the current climate or the variability increases. In the first case the number of extreme weather events will change little, but in the latter case there would probably be more extreme events, posing a risk of greater danger – more torrential rain, more droughts, more tropical revolving storms, more forest fires – even above the increase to be expected as temperatures rise (Saunders & Lea 2008). Clearly, these scenarios could have very different outcomes and much research now goes into discovering which is more likely.

There also remain some uncertainties about regional effects, in the tropics as elsewhere. However, studies increasingly focus on the regional scale and, as knowledge increases and model grid-length decreases, studies will assess likely effects more locally. The Met Office Hadley Centre, using versions of the Unified Model, plays an important part in this work. Despite the variations in climate-model predictions, the effects of rapid climate change are serious and may affect the people of the tropical zone more seriously than the inhabitants of the extra-tropics (e.g. Brilliant 2007).

Deforestation and forest fires, as well as increases in GHG concentrations in the atmosphere, will change the climate of the tropics (Andreae et al. 2004), continuing the pattern observed since the 19th century. Overall, climate models predict that temperatures will rise, although some parts of the tropics may see only small increases[2]. However, the changes are likely to affect both the Hadley Circulation (which brings warm air poleward) and the Walker Circulation, changing areas of broad-scale ascent and descent in the tropical zone. Increasing variability is likely and the climate is likely to be less stable, with GHG concentrations increasing (Cox et al. 2000).

One source of uncertainty is the relatively short observational record of much of the tropical zone (already mentioned in section 1.1). Unlike Europe and much of North America,

there is not a historical record dating back to the early 19th century; indeed, few observatories have more than 60 years of data. Many long-term data that do exist are not truly representative of all the climatic zones, often representing only capital cities or coastal trading ports. The network now reflects the development of aviation, records dating back only perhaps 20 or 30 years. Although longer-term records exist for much of India, the records are poor in most other tropical countries, particularly in Africa. Nonetheless, there is evidence of warming from tropical climate records, supporting the consensus that global temperature is rising, amounting to about 0.9°C in Darwin (Bureau of Meteorology 2014) and about 0.7°C in India (Attri & Tyagi 2010) as a whole over the past 100 years, for instance.

However, there is consistency in some areas and uncertainty in others, not least in the tropics. The models agree that El Niño is likely to be more common (or, at least, that the eastern tropical Pacific will be warmer). However, in other areas there is much less consistency (Fig. 12.6). The variation in the models appears greatest in terms of precipitation in the monsoon zones, some areas seeing a decrease in precipitation despite an expected regional increase, matched to warming, particularly in drier areas (Fig. 12.7). Given the expectation that El Niño will be more prevalent, it might be expected that monsoon rainfall would decrease (Met Office 2004a). This does seem to be the effect of El Niño in Australia and in much of the Indian Ocean region, but the relationship is weak and, for instance, Indian rainfall appears to have many sources of variability (Mooley & Parasarathy 1983). However, a warmer atmosphere has a much greater capacity to carry water vapour and form deeper clouds that could produce more precipitation, provided local instability is maintained and broad-scale dynamics support the development of deep convection (in the tropics). This instability would need to be maintained over land, producing rainfall where it is most needed. In turn, more precipitation might be expected to reduce surface temperature (by evaporation),

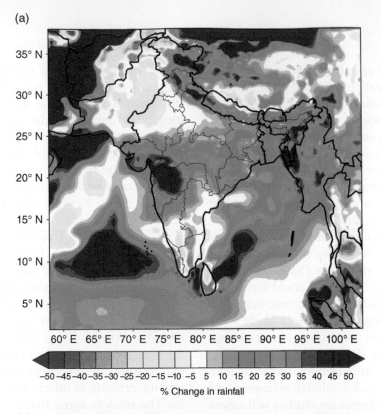

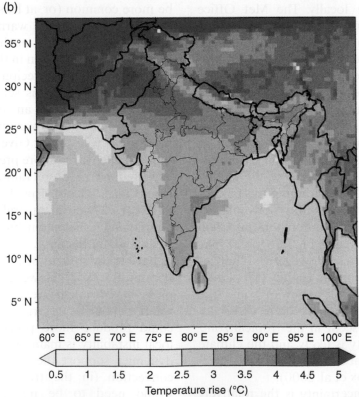

Figure 12.7 The predicted percentage change in June–September (a) (monsoon) precipitation and (b) temperature in south Asia between the early 21st century and the 2080s for the SRES A2 scenario from the PRECIS climate model.[3] From Met Office (2004b).

but could also help to form more convective clouds. Clearly, there remains great uncertainty not about the clear science of a warming atmosphere, but the likely regional, not to mention local, effects of global warming. Overall, climate models suggest, with a comparatively high degree of certainty, that monsoon rainfall will increase in south Asia as the climate warms, partly at the expense of rainfall south of the equator in the northern summer (Turner & Annamalai 2012; Dai 2013). However, Dai (2013) also revealed that soil moisture deficits are expected to increase in equatorial South America, the Caribbean, much of Africa and Australia, although soils may become a little more moist in parts of the Sahara, continuing recent trends.

Currently, there is no clear signal as regards the shift of tropical plant and animal species in response to climate change. Over the past 60 years in Australia there has been a mean equatorward movement of species in the tropical savanna zone (away from the desert), but a mainly poleward move of species from the desert (albeit with considerable variation), so it is not clear that increasing temperatures will move species away from the equator (van der Wal 2013), suggesting that there is a risk that their movement with the migration of climatically suitable zones may be insufficiently rapid.

With such great uncertainty in much of the tropical zone it will be difficult for adaptation measures to be undertaken, even though there is a clear need for preparedness (Wilby & Dessai 2010). The only way forward for most tropical nations is to work towards a reduction in GHG concentrations. This may be very difficult both financially and culturally.

The effects of tropical agriculture on the climate are discussed in Chapter 13.

12.7 Modelling climate change

Since the early models of the atmosphere including the effects of changing concentrations of GHGs were first developed in the late 1980s there has been a tremendous advance in the capability of climate models. Atmospheric and oceanic components are now linked, a most important factor in the representation of tropical climate. The resolution of the climate models is now about 150 km globally, with up to about 40 atmospheric levels in the vertical,[4] extending into the lower stratosphere, and 40 oceanic levels on a 1 km grid-length, thus bringing improvements in their ability to model temperature, precipitation and other changes (Met Office 2014c). Increasingly, models include parameters that represent the changes in atmospheric constitution more realistically and have feedback loops, allowing the effects of changes to be modelled more realistically.

However, the models remain imperfect. One of the major factors preventing numerical models forecasting with accuracy beyond a week or so ahead (apart from the small, but significant, inaccuracies in measurement) is their relatively coarse resolution. They cannot model convection explicitly, nor the effects of mountains, since convective clouds and individual mountain peaks have details about an order of magnitude smaller than the model resolution. They must be parametrized, that is, represented by a formula duplicating their broad-scale effects. In particular for climate models it seems that mesoscale representation of ocean-surface temperature and currents is important, matching with the broad-scale flow to depth within the oceans (Met Office 2011a). It is only when we are able to model small-scale gyres and their effects on the overlying atmosphere that we can have full confidence in the detail of models of climate change.[5] Nonetheless, the models can give realistic outcomes at their current resolution, using current components, within a range of probable solutions. In order to obtain these, small differences are made to model initial conditions, producing an ensemble of results (discussed also in Chapters 9, 10 and 13). Climate models do not present forecasts of day-to-day weather in the form 'there will be increased rainfall in Nairobi on 2 October 2050' nor are they forecasts of wind direction and speed for a specific day (or even a month). Rather, they indicate a likelihood of increased rainfall and magnitude of temperature change over a long period years or decades ahead. Many of these are explicitly related to broad-scale connections between weather and mean flow, such as ENSO, the strength of the Hadley Circulation and changes in the Walker Circulation, which are discussed in section 3.1. Confidence in the reliability of climate predictions is increased by a comparison between the models of different centres. This also allows areas of uncertainty in the predictions to be revealed so that work to improve them may proceed.

This uncertainty is also reflected in projections of the likely influence of a warmer world on the development of tropical revolving storms in the North Atlantic. Current projections show a broad spread of results. Although a small increase (of about two storms per century) seems the most likely outcome in the period 2006–2050, in the period 2051–2100 there is a range of about +4 to –5 storms per century, with a mean of –0.4 storms per century, from the models used to assess the effects of climate change (Villarini & Vecchi 2012).

12.8 Conclusion

It is clear that all tropical nations need to be aware of the variability within climate zones due to inter-annual and intra-seasonal effects, as well as the broader-scale influence of anthropogenic climate change. Adaptation will be one of the measures all nations will need to use in a warmer world and, not least for reasons of cost, these adaptations must be proportionate, running alongside the global approach to reduce the effects of GHGs (Hall et al. 2012). Nations also need to prepare for the variations of rainfall and other weather elements due to the MJO, QBO, El Niño and La Niña. Nations should work together to improve society's

resilience, sharing the growing knowledge of these effects (Hewitt et al. 2012).

12.9 Questions

1. Consider an approach to tackling climate change in the tropics and how it might affect inter-governmental relationships, as well as the economy.
2. How can we deal with scepticism about climate change, remembering that numerical model errors increase with time and with reduced resolution?
3. What allowances must climate change scientists make for the various factors that vary climate and that may have opposing signs?

Notes

1   The Southern Oscillation is measured by the mean sea-level pressure difference between Tahiti (149°W) and Darwin (131°E), an indicator of expected rainfall anomalies across the tropical Pacific Ocean. (The Southern Oscillation is often used alongside El Niño in the acronym ENSO.)

2   These assumptions about changing climate and its effects on agriculture assume that GHG concentrations will increase throughout the 21st century (IPCC 2014).

3   Scenario A2 of the *Special Report on Emissions Scenarios* (IPCC 2000) assumes a world of independently operating, self-reliant nations, continuously increasing population and regionally oriented economic development.

4   This is approximately half the horizontal resolution and half the vertical resolution of most global weather models, so uses only about an eighth of the number of grid points.

5   This does not mean that output in the familiar form of the surface pressure and weather analysis could ever be produced; rather, the modelling of smaller-scale features allows a more realistic representation of the mean state of the atmosphere and the mechanisms of heat and momentum redistribution (Collins et al. 2011).

13

Tropical Agriculture

13.1 Agricultural productivity and tropical environments

High mean temperatures and generally large amounts of solar radiation due to the high elevation of the sun at midday and relatively small amounts of layer cloud (at least away from mountains) make agriculture highly productive in the tropics where there is sufficient rainfall or irrigation water. This has allowed agriculture to be a large contributor to the economy of tropical nations, in some cases bringing significant political influence. A very large range of produce can be grown in the tropics with sufficient value for inter-continental export. However, in turn this has led to the exploitation of the people of many of these nations.

As has been shown in earlier chapters, the climates of the tropics are very variable on a range of scales, so we cannot consider climate and weather without reference to their effects. Most notably, the great variety of crops of the tropical zone is generally well adapted to the varying climates of the tropical zone. Changes in climate and severe weather are each likely to affect agricultural productivity, which is particularly important in the humid zone, where plentiful rainfall combines with high solar elevation to bring great productivity, limited only by altitude and availability of nutrients.

This chapter contains a brief overview of agriculture in the tropics, as well as its effects, noting the increasing human population in all tropical climate zones. In many places humans have altered the distribution of plant and animal species, in almost all areas using plants and animals for food or labour. These changes in turn change the climate as significant changes are made to the appearance of the environment.

The sensitivity and effects of the tropics on global climate are important. Small changes in rainfall patterns, when experienced over a long period of years, have a great effect on the vegetation. This in turn affects the populations dependent on it. Small changes in temperature have, within the past 2 million years, changed rainfall patterns in the tropics, making some areas drier and others wetter (Burroughs 2005). The global climate is regulated by conditions in the tropics: the source of energy for agriculture in the middle latitudes is partly supplied from the tropics, tropical forests absorb a large proportion of global carbon dioxide production and protective ozone is created in the tropical stratosphere. It is thus increasingly important for us to study not only the weather and climate of the tropics, but also the

An Introduction to the Meteorology and Climate of the Tropics, First Edition. J F P Galvin.
© 2016 John Wiley & Sons, Ltd. Published 2016 by John Wiley & Sons, Ltd.

inter-relationship between tropical climates, agriculture and humankind.

13.2   Agriculture in the humid tropics and the effects of forest clearance

Agriculture has resulted in the clearance of forest in many parts of the humid tropics, although it is the semi-deciduous forest that has been subject to most clearance historically. In part this is due to the relatively rich, but leached brown (iron-rich) or grey (kaolin-rich) soils of this zone, and also to patterns of settlement and the slightly favourable monsoon climate.[1] Forests are ideally suited to the humid tropics with roots able to tap dissolved minerals below the surface soil layer (Ellis & Mellor 1995).

As the forest is cleared, in particular near major rivers, to open areas for agriculture and as a source of wood for construction and paper making or to grow crops, there is a profound

effect on the climate. Clearly, growing populations require more arable land, but the planting of cash crops remains a contentious issue. In the tropical rain forests of South-East Asia many areas have been either cleared or the forest changed to a secondary form, often of palm and fruit trees (Fig. 13.1). Initially the clearance was to provide wood for housing, but latterly it has been to grow rice – the characteristic subsistence crop. In semi-deciduous ('monsoon') forest, crops such as tea (originally brought from China to India and East Africa), coffee and chocolate, grown largely for export, have often replaced the woodland. Elsewhere, oil crops, tropical fruit or groundnuts are the principal crops where forest has been cleared. The first of these is becoming increasingly important in the rainforest zone, at the expense of the natural forest (e.g. Carlson et al. 2013). A short-term secondary financial benefit of oil-palm production is wood for export.

Forest clearance has an unfortunate and significant effect on climate and the effects of weather (Fig. 13.2). As crops or grassland

Figure 13.1   Much of the humid tropical zone has now lost of its primary forest (darker shades), in particular in coastal populated areas, such as here at Benoni, Camiguin, Philippines, where coconut palms form a secondary forest (lighter shades) with relatively easy access and much-altered local radiation balance.

Figure 13.2   In mountainous areas, the removal of many trees (often precariously clinging to mountainsides) has unfortunate effects in heavy rain. Here, in the valley of the Bued River, near Baguio, Benguet, Philippines, soil erosion is evident on the distant, east-facing sides of the valley above the maize fields of the valley bottom.

replace woodland, the daily temperature range rises, transpiration decreases and lower surface humidity reduces cloud formation, most of which is a result of convection in the tropics. Thus crops may require additional watering in the drier areas. In addition, CO_2 may be released, with the potential to cause an increase in global warming (Carlson et al. 2013). However, the main feedback is a reduction in the hydrological cycle. Under the influence of unstable airstreams, an unusual feedback is the increase in cloud-base height of cumuliform clouds, reducing the amount of rainfall that may reach the ground before evaporating.

In West Africa, where research has concentrated on the dry-climate (Sahel) zone, recent studies reveal that there has been a greater reduction of rainfall in the humid coastal zone of the Guinea Coast, associated with Sahel drought. Deforestation is likely to have contributed to this reduction, as well as broad-scale atmospheric changes. Reductions of about 20% have occurred in the south of Ghana, for instance, compared with about 10% (with considerable local variability) in the north (Kiangi 1989). As well as jeopardizing valuable agriculture where rainfall was considered sufficient, the drier coastal zones, where there is insufficient rainfall for farming, are still less likely to be able to support agriculture (see Chapter 6).

As trees are cleared, there is a serious effect on the tropical latosols.[1] Without trees, leaf-fall and fallen trees cannot contribute to the recycling of minerals, so the soils are more easily leached of essential nutrients. The lack of a leaf canopy increases the intensity of rainfall at ground level, so the thin humus layer can be washed away more easily. Where crops are grown commercially, it is often necessary for large amounts of fertilizer to be added, even where relatively nutrient-rich river waters from mountains such as the Himalayas are used to water them. Although the potential for productivity is very high, especially where there is a continuous growing season, the costs can also be high.

Clearance and degradation of forests also emit carbon to the atmosphere, further affecting global climate. Net deforestation has contributed around 30% of the historical rise of CO_2, mainly from deforestation in the tropics. Tropical forest cover declined by 250 million hectares between 1980 and 1990. The forestry sector is currently the third largest contributor of global GHG emissions and is a larger emitter than transport.

Tropical forests are also vulnerable to climate change, with some models predicting widespread loss of the Amazon rainforest due to climate change (Betts et al. 2004; Cox et al. 2004), although any such impacts are highly uncertain.

Severe impacts of climate change on tropical forests may be more likely if the forest is already affected by forestry activities (Betts et al. 2007). Forest degradation may therefore increase the likelihood of climate-carbon cycle feedbacks, accelerating the rise of CO_2. Protecting tropical forests and the carbon they store is now being discussed as one possible measure to combat climate change through a mechanism called reducing emissions from deforestation and degradation (REDD; Gullison et al. (2007)).

Rice is perhaps the characteristic crop of the tropics, particularly across Asia, where 90% of the population is dependent on it as a staple diet. It feeds more than 3×10^9 people worldwide, approximately 45% of the global population, and so is the world's most important crop (www.irri.org). It grows well in hot and moist climates with plentiful rainfall during the growing season (mean temperature >15°C for germination, >22°C for grain maturation,[2] but a maximum temperature of <33°C (Stansel & Fries 1980) and no frost). About 150 million hectares of land are devoted to rice production, more than any other crop. Two growth stages are normally used, the first exploiting the plant's water tolerance to restrict weed growth. However, this method of growth produces methane, a very powerful greenhouse gas. Up to three crops of rice may be grown annually, provided there is sufficient water. The need for water for the crop to grow – up to 5000 l kg^{-1} – can be critical in areas of periodic drought.

Even in areas where crops are grown extensively (e.g. fruit, oil palms or coconut for copra), the damage to the environment can be serious, removing the habitat of the great diversity of flora and fauna of these forests.

13.3   Agriculture in the savannas

The savannas are very important for humankind and many crops have replaced the climax vegetation of this climatic zone, particularly in India, the Americas and Africa.

Perhaps most important of the crops grown in many savanna lands are the grains: rice, wheat, barley, oats, maize and millet. These plants, developed from grasses, feed a large proportion of the population of the tropics, although the cultivated area remains relatively small compared with the natural grasslands used principally for pastoral farming. The crop type depends partly on climate, but also on the familiarity of the population. Among the cereals, oats and barley are resistant to periodic cold, including frosts, whereas maize requires warmer wetter conditions and millet is adapted to drier environments. Barley is salt-tolerant and oats can grow in very wet climates. Many of these grains are ground to form flour for baking or bread making. In general, tropical grains (apart from rice) contain little gluten, so bread does not rise well. Many grains are also used as animal fodder supplements. In countries where alcohol may be produced, malted barley is used to make beer.

To germinate, the mean temperature (in a sufficiently moist soil) must be above ~4°C for oats, ~9°C for wheat and ~11°C for maize (Petr 1991).[3] These temperatures are only a limit at altitude in most of the tropical zone or at the periphery of the tropics in winter.

Beans are also important crops, helping to preserve soil fertility by fixing nitrogen from the atmosphere, whilst providing valuable vegetable protein. The most significant root crop is cassava, grown mainly in Africa.

In the savannas animals and plants must be adapted to both dry and hot weather for part of the year, whilst able to withstand the very wet weather of the summer.

13.4   Dry-land agriculture

Most desert settlement is near rivers that, in some cases, are ephemeral (see Fig. 13.3), but where there is sufficient water, agricultural productivity can be high, since the lack of rainfall restricts leaching of the soils. The soils may be thin and poorly formed, so that their

Figure 13.3 The Hugh River, south of Alice Springs in the Simpson Desert, Australia. Although most rain falls during the summer (monsoon) season, this was taken following winter rain. The scrub vegetation is sustained by the rains of the previous summer.

mineral content has not been broken down to be readily used by plants, but still held within immature clays or bedrock. However, nutrients may be rapidly consumed or leached when water is added, as soils are usually loose and friable. Irrigation water is readily transpired in the low humidity and daytime warmth of this zone.

Irrigation was traditionally seasonal, provided by great rivers, such as the Nile. Alluvial soils, largely of fine porous but impermeable clays, store water and help make agriculture productive (FAO-UNESCO 1989). More recently, the building of dams and irrigation has allowed the greening of drier areas. Desert soils, not having been subject to much rainfall, are, in general, fertile, although their fertility is dependent on the surrounding geology. Where this is limestone, poorer soils are likely to form than from, say, claystones (Gass et al. 1986). Other areas may be largely bare rock, unsuitable for anything other than hardy animals that can live on the scrubby vegetation growing in

cracks in the rocks. Almost everywhere in the dry environment soils are poorly developed and thin.

The effects of desert irrigation over a long period can have a serious effect in dry lands. Soils naturally contain minerals, including salts. Addition of water, along with plant growth, gradually denudes the soil of essential phosphate and nitrate by leaching, whilst increasing many salts, in particular sodium chloride and calcium carbonate, due to evaporation. Sodium chloride is poisonous to many plants, so tolerant species may need to be grown.

In many parts of southern Asia scrublands have long been irrigated. An important crop grown in central India, between the Eastern and Western Ghats, as well as in Indo-China, is groundnut, usually for its oil.

The Sahel of Africa has come to prominence, mainly due to the effects of a growing population and the ephemeral nature of the rainfall. The human effect on water resources has been important in these marginal zones, in particular for agriculture, where inhabitants have become increasingly settled, rather than nomadic. However, there are also signs that these zones have become drier during human history. This change may well be due to changes in the environment of the Sahel and of the neighbouring savanna, as well as larger-scale atmospheric changes. As studies continue, no doubt our understanding will increase.

The great benefit of mountains in arid areas is demonstrated by the growth of the staple cereals (rye, oats and barley) on the mountains of Saudi Arabia and Yemen. Elsewhere, the orographic rains supply water to surrounding low ground, some of which may be in rain shadow. Notably, potatoes first grew on the western slopes of the tropical Andes. Seasonal rain, enhanced by mountains, is often sufficient to support crops such as sugar cane, which is a major crop in Nepal, the northern Philippines and coastal Queensland, Australia.

Although humidity is low on the highest ground, these areas are seen often to be cloudy in satellite imagery. The settlements of upland valleys are dependent on rain- and snowfall from these clouds. This is particularly important in north-east Afghanistan, Iran and Pakistan. Relatively large populations live in these countries of dry climate, where hardy crops are grown by irrigation from great rivers such as the Helmand, Indus, Chenab and Sutlej. Rain and snow fall on the Reshte-ye Alborz–Paropamisus–Kuh-e Bābā–Hindu Kush–Karakoram, Brahui and Sulaiman ranges to feed these areas (see Fig. 11.3).

13.5 Weather and locust swarms

The locust – a form of grasshopper of the arthropod phylum – is an ephemeral threat to crops and livelihoods in many parts of Africa, Asia, North America and Australia, capable of stripping vast areas of crop, especially in its early non-flying stage of life. The development and spread of bands or swarms and the

likelihood of decimation of croplands are highly dependent on the weather. Locusts are generally found in semi-arid areas with an annual rainfall less than about 200 mm (Food and Agriculture Organization 2009) and the stages of locust development are dependent on a cycle of moist and dry weather.

Locust infestation has a serious effect on agriculture in marginal desert areas and is related to the climate of these areas. Although different areas are affected in different years, locusts hopping and then flying in their millions destroy crops and thus livelihoods in the areas they land. Their pestilence is dependent on both weather and wind in an area where insects develop. Rain, usually associated with incursions of westerly winds, provides suitable conditions, promoting the growth of plants and the successful hatching of locusts (Dubey & Chandra 1991). Plagues may then develop following insect maturation once the wind returns into the east. This change of wind direction is typically seen in the transition from winter to summer in the monsoon zones (Chapter 8). However, many factors determine the occurrence of locusts, on all meteorological scales. Among ancient sources, the Bible readily suggests the serious effect of such plagues (e.g. Exodus 10:13–15).

Females lay their eggs only on bare sandy soil, close to a source of green crops. The crops are a primary food source to the growing insect and large numbers of full-grown locusts will only result if there are sufficient crops to support the young insect, which hops from plant to plant, gathering to form a 'band'. Thus there needs to have been adequate rainfall for crop growth along the route that the insects take. Once the locusts have matured to adulthood, a swarm may form as the grown insects take advantage of the winds to move into areas again likely (in good years) to support them. They eat the equivalent of their own weight in a day and, flying with the wind, may cover some 500 km. A biological trigger causes swarming and has its greatest effect early in the morning. The largest known swarm covered

1036 km², comprising approximately 40 billion insects. Although able to eat vast swaths of crop as adults, their damage is now more local, dependent on the distance they have travelled by night and the distribution of crops.

The adults' flight is relatively weak and the animals cannot fly any distance up wind so follow a route largely determined by winds near the surface. In north and west Africa crops are available at varying times of year, favouring migration when the winds are favourable and food is available. Wind-borne migration moves the insects to the margins of the tropics late in the summer and then returns them, over the course of several generations, to the Sahel. Similar migration patterns also occur in Australia and southern Asia.

The arid parts of the migration route often prevent further spread of the swarms, but irregular rains in these arid areas may cause an unusually prolonged outbreak, as occurred in 2004 in north Africa (Wikipedia 2014c).

Given the devastation that locusts may cause – in particular in years when sustaining crops would otherwise be expected, often in areas that are subject to periodic drought – many forms of pest control have been tried. Poisoned food can be very effective, but poses a threat of contamination and may be wasteful if the locusts do not land on treated farmland. Spraying with pesticide is most commonly used, but again this poses a risk to production: the crop cannot be sold, even if the swarm or band of locusts does not eat it. A biological pesticide to control locusts was tested in Africa in 1997. Dried fungal spores sprayed in breeding areas pierce the locust exoskeleton on germination and invade the body cavity, causing death. The fungus is passed from insect to insect and persists in the area, making repeated (chemical pesticide) treatment unnecessary. However, it has not entirely removed the threat from these creatures, although it greatly reduces the risk to human health. In Europe control has been very effective and swarms are very rare on the northern coast of the Mediterranean. However, political and economic factors can limit the effectiveness of locust control (Food and Agriculture Organization 2009).

13.6 The effects of agriculture in the tropics

The Neolithic Revolution saw the beginning of the cultivation of crops by humans and, as a result, of human settlement, saw the first changes in the environment linked to climate changes (Vedel 1978; Perkins 2010). In this respect, the tropical region is no different from any other part of the world. In general, the removal of natural vegetation (mainly forest) causes changes reductions in rainfall, an increase in temperature range and an increase in the likelihood that wind or rain will remove soil (Burroughs 2005). Atmospheric concentrations of the powerful GHG methane (CH_4) have changed rapidly in recent years, largely due to the agricultural need to grow more crops and animals to feed the world's growing population. Much of the increase in CH_4 is from rice cultivation and animal husbandry, both of which have expanded rapidly as the tropical population has grown. A secondary source of GHGs in the tropics has been the felling of rainforest, either for timber alone or for clearance to provide agricultural land. This includes re-planting with cash crops, which can rarely absorb the quantity of CO_2 that rainforest can.

Figure 13.4 shows the distribution of the natural vegetation types in the tropics. Humans have altered this environment, mainly by agriculture (Turner & McCandless 2004). The appearance of tropical lands, as modified by humankind, is shown in Fig. 13.5. At first sight the modification appears minimal, as the area under agricultural production is small (less than 10%). However, a closer look, comparing Fig. 13.5 with Fig. 13.4, reveals that the areas of natural vegetation vary significantly, particularly the area covered by tropical woodland.

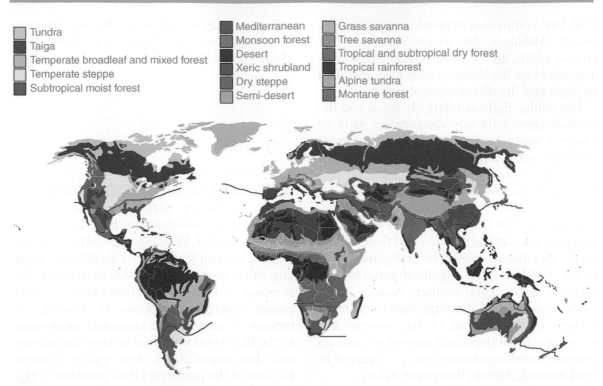

Figure 13.4   The natural vegetation of the tropics. The red lines indicate the northern and southern limits of fruiting palm-tree growth (approximately equivalent to the area within which frosts are rare or absent, cf. Chapter 1). Where these lines are closest to the equator, there is either high ground or the incursion of dry winter continental air. Courtesy of NASA-GFDL/Sten Porse.

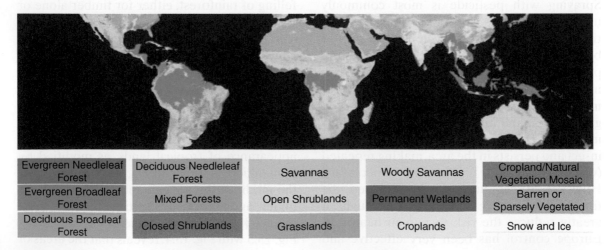

Figure 13.5   Present-day land cover of the tropics, as seen by the TERRA/MODIS satellite. Courtesy of NASA/University of Boston.

This is likely to have been a direct result of agriculture over the centuries.

It is now generally accepted that areas formerly forested, but at the margins of arid areas, have become drier since the last ice age, some 10,000 years ago. This drying is probably caused, in part, by the spread of agriculture, although importantly the drying in itself reduces the tree cover. This, in turn, reduces rainfall, since evapotranspiration is reduced, and so on (Burroughs 2005).

Clearly, humans must have the ability to grow sufficient food in the areas where it is needed: types that will grow well and that are familiar in an increasingly stressed world (see International Rice Research Institute 2014). There is also a need to grow crops that will bring an income (e.g. tea, coffee, cocoa, copra, fruit) to provide cash and supplement the staple diet, thus generating sufficient wealth to allow education and health care. Although this has an effect on the environment (as seen very clearly in Europe or North America), this is generally small in the tropics (notwithstanding the changes discussed above). The lowland humid tropical zone has the potential to produce vast crops from a small area, given its plentiful rainfall and sunshine, although supplements in the form of fertilizers (natural or manufactured) will be needed to sustain this growth and the growing population.

13.7 Agriculture and climate change

As we have seen, the interaction of humans with the environment has important effects on the climate in the tropics, every bit as much as in the industrialized nations. Here the possible adaptation strategies are discussed.

If the climate and the oceans, in particular, are warmer, more rainfall may result. However, some parts of the tropics are likely to become drier and others wetter. In general, the humid tropics are likely to become wetter, although drier parts of the tropics are likely to suffer more water shortages than occur at present (IPCC 2014). In Asia it seems likely that the south-west monsoon will become wetter and that monsoon rain is likely to be heavier (Lowe et al. 2005). However, on a regional scale, patterns of rainfall are likely to become more complicated. The Hadley Centre mesoscale model, running with a 25-km grid length, suggests that there will be more rain on the highest ground, but areas in rain shadow in southern India, as well as the relatively cool ocean areas, are likely to become much drier (Met Office 2004b). Changes, although subtle, are likely to favour some crops over others.

Increasing rainfall is likely to increase river flow, although in many drier areas any increase is likely to be minimal and the extra rainfall used by plants, particularly in areas of marginal agriculture (Lowe et al. 2005). Although generally of benefit to the population of these zones, there is a risk that the additional rainfall will be mainly in the form of heavy rain and hail, posing a threat of damage to crops, either directly or from floods and landslides. Some other, unexpected, changes may also occur, as already occurs in the western Pacific during the La Niña phase of the Southern Oscillation. In South-East Asia, when increased rainfall runs off high ground, rice terraces are often inundated, the rice seed washed away or young rice plants drowned in deep water. Plants such as maize may not be able to germinate or may suffer fungal infections when fields become waterlogged.

Rice is a significant source of methane gas (CH_4) emissions and rice emits increasing CH_4 as atmospheric carbon dioxide (CO_2) increases, the effect multiplied as the demand for rice grows. Although some efforts to move rice production to higher levels would reduce the yeild-related emission of CH_4, the cost of production would increase significantly, thus demanding an increase in price (Groeningen et al. 2013). Similar increases in CH_4 emissions are likely to be seen in response to the demand for meat as the taste for meat rises across the world. The high warming effect of CH_4 already makes this a significant input that is only likely to increase in coming years.

In areas of increased run-off and river flow, nutrients are likely to be more easily washed away, reducing natural fertility, as well as poisoning the sea, reducing oceanic productivity and fish stocks.

A warming world is also likely to bring changes to the El Niño–Southern Oscillation, the main control of climate variability in the tropics. However, whilst some research suggests that El Niño will become more common, ice-core studies for 10,000–5000 years ago, when the global climate was warmer than at present, suggest that El Niño will be less common, decreasing tropical climate variability (Burroughs 2005). Overall, the eastern equatorial Pacific is expected to warm, as shown in Fig. 12.6 (IPCC 2014). If tropical climate variability reduces, some of the hazards of tropical agriculture will either reduce or can be allowed for.

Warmer oceans might reasonably be expected to produce a greater likelihood of tropical storms and there is some evidence of an increase in storm activity over warming waters (Saunders & Lea 2008). However, many studies suggest overall a reduction in storm number; some areas seeing fewer storms while others experience more. Where they occur, they are likely to become more intense (Emanuel 2005). Coastal and island populations are likely to bear the brunt of a warmed world as sea level rises because of melting ice caps and ocean expansion (Grosvenor et al. 2005; Shutts 2005; Willett & Milton 2005). In turn, rising sea level (and an increase in coastal storms) will cause an increase in salination as sea water floods inland (Le Treut et al. 2007).

Given the key role played by the oceans in the development of weather systems bringing vital rainfall for agriculture, it is important that we use the best possible system to predict the likely SST changes in a warming world. In the Hadley Centre, ensemble and deterministic climate modelling are improving our understanding, using OSTIA (Met Office 2011a), the Met Office Operational SST and Sea Ice

Analysis, which incorporates climate-change feedback into ocean surface temperatures at a horizontal resolution of 5 km.

There are already signs of climate change in the tropics and effects can be significant in areas where rainfall is seasonal. The savannas of Australia have suffered recent drought, coastal zones in particular seeing much reduced rainfall totals, perhaps above and beyond what might be expected from El Niño–La Niña variability (see Bureau of Meteorology 2014). Uttar Pradesh in northern India is an important cereal-growing region, the cereal sown in November and harvested in March. It is dependent on rain from the westerly disturbances described in section 10.5. Rainfall (and snowfall), although always of very variable amount year-to-year, has declined to a lower level in this state in recent years as the depressions (when they occur) progress east or northeast into Jammu and Kashmir, rather than south-east. The early-winter season of early growth always has the greatest variability, but these months have seen the greatest decline in precipitation and rain may not be seen until January or February (Sheldrick 2005).

13.8   Question

1. Consider the range of pressures on agriculture in the tropics and their importance as regards development into the future.

Notes

1 Soils were described briefly in Box 6.2.
2 These temperatures are 'typical' values and vary according to grain cultivar. They must be sustained over a period of weeks, although the plants can usually withstand brief colder weather over a period of a few days. The maximum temperature most grain crops can endure is around 32°C and most need plentiful soil water for growth.
3 This Special Report on Emissions Scenario assumes a balanced emphasis on all energy sources in likely future energy consumption and technological changes that improve efficiency.

14

The Importance of the Tropical Ozone Layer

14.1 Background

It is now well established that a significant rapid reduction in stratospheric ozone during the late 20th century is related to the use and release of refrigerants that unexpectedly produced highly reactive chlorine and bromine that interfere with the production of ozone by sunlight in the stratosphere (Farman et al. 1985).

Although there has been much discussion of the effects of ozone depletion in the high latitudes, the effects of ozone are more significant in the tropics. Most of the protective layer of ozone is formed in the lower stratosphere (between altitudes of about 20–32 km in the tropics; Fig. 14.1) under the influence of near-constant daytime input of solar radiation in the tropics (WMO 1999). Whilst the effects of ozone depletion are most notable near the poles, where cooling of the upper stratosphere promotes the formation of polar stratospheric clouds (PSCs), the release of chlorofluorocarbons (CFCs) and halons (containing bromine) has also depleted the tropical stratosphere (where temperatures are consistently close to that required for PSC formation; see section 1.4).

The formation and destruction of ozone (O_3) is a perpetual process in near-constant balance. Three forms of oxygen are involved in the ozone–oxygen cycle: O (atomic oxygen), O_2 (oxygen gas) and O_3. Ozone forms when oxygen gas absorbs ultraviolet photons to produce two oxygen atoms. In the presence of a catalyst, the reactive atomic oxygen combines with O_2 to create O_3. Fortunately, the process works in both directions, allowing it to continue without any supply of fresh gaseous oxygen; the oxygen atoms join up with oxygen molecules to regenerate O_3. The process is complete when an oxygen atom 'recombines' with an ozone molecule to make two O_2 molecules: $O + O_3 \rightarrow 2O_2$, the splitting of the O_2 molecules by short-wavelength ultra-violet light continuing the cycle (A more detailed description of the chemistry of ozone depletion is given in Box 14.1.)

Even though anthropogenic ozone depletion is relatively small in the tropics (and the area is not generally affected by the ozone 'holes' of the high latitudes) because the catalytic effects of chlorine and bromine occur on PSCs, we must consider the effect of air motion in the lower stratosphere on the concentration of ultra-violet radiation that reaches the surface in the tropics. The motion of air acts in two ways: first, by transport of ozone away from the tropics and, second, by the effects of vertical motion on temperature.

Because ozone uses potentially harmful ultra-violet light with wavelengths less than 310 nm (1 nm = 10^{-9} m)[1] to produce a molecule

An Introduction to the Meteorology and Climate of the Tropics, First Edition. J F P Galvin.
© 2016 John Wiley & Sons, Ltd. Published 2016 by John Wiley & Sons, Ltd.

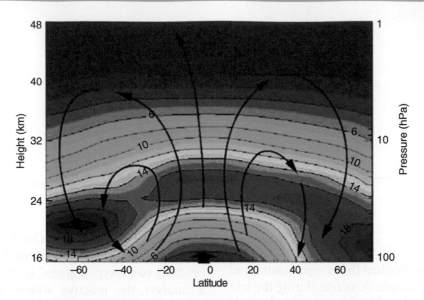

Figure 14.1   The Dobson–Brewer circulation and corresponding mean ozone concentrations (DU km^{-1}) measured by satellite Nimbus-7 between 1980 and 1989. The total amount of ozone over a single point on the earth's surface is measured as Dobson units (DU), 1 DU corresponding to a thickness of 0.01 mm of ozone at 1000 hPa. A typical column-ozone depth of 300 DU would therefore have a thickness of just 3 mm at the earth's surface. © NASA/GSFC.

Box 14.1   The chemistry of ozone depletion

PSCs are the medium on which chlorine compounds are converted into ozone-destroying chlorine radicals. Generally, there are two types of PSC: Type-1 PSCs are believed to be composed of nitric acid and water, which form at about –78°C. The clouds can be either solid or liquid, depending on the conditions. Type-2 PSCs are less common and composed of water-ice crystals at even lower temperatures: approximately –85°C near an altitude of 25 km.

Various chemicals promote the destruction and production of ozone in the stratosphere. Atmospheric oxygen molecules (O_2) are photo-dissociated in the stratosphere to form oxygen atoms (O). In association with the main atmospheric molecules (M), these oxygen molecules recombine to form ozone (O_3):

$$O + O_2 + M \rightarrow O_3 + M \qquad (14.1)$$

Much of the ozone produced in this manner is itself quickly photo-dissociated, the result reversion to diatomic and monatomic oxygen (1). The oxidation of so-called 'free radicals', particularly chlorine (Cl), bromine (Br) and nitrogen (N), is important in the longer-term reduction in the amount of ozone in the stratosphere. The oxidizing agent is usually the hydroxyl radical (OH), itself formed in the stratosphere from water vapour (H_2O). In this process, longer-living chemicals (e.g. ClO_x, NO_x) are produced which interfere with the balanced reversible cycle of reaction (14.1). These chemicals reduce the stratospheric concentration of ozone (Burkholder & Orlando 1998).

Although many of the reactions are reversible, particularly those involving nitrogen, it is the change of the balance between destruction and formation of O_3, as well as the transport of ozone in the stratosphere, which determine whether there will be an ozone 'hole' in the stratosphere.

The trend of rapid growth in the use of compounds containing Cl, Br or N (e.g. $CFCl_3$, known as CFC-11) produces an increase in stratospheric species such as ClO_X, especially around the poles. The production of Cl_2 from CFCs requires sunlight, so that ozone depletion requires air to be sunlit as well as cold (i.e. in spring, as observed, rather than in winter when the polar cap is continuously dark or in summer when the ozone layer is warm). However, some ozone loss can take place even in the polar winter due to atmospheric wave activity moving polar air into sunlit areas at lower latitudes for brief periods.

It was the anti-correlation of this trend with the concentration of ozone at high latitudes that provided the 'smoking gun' of proof that the reduction in stratospheric ozone has an anthropogenic cause (Fig. 14.4). This is despite the fact that chemicals such as CFCs, formerly used as refrigerants or aerosol propellants, are unreactive near the earth's surface. Indeed, their 'inert' character is what makes them such a potent source gas in the stratosphere, since they have a very long atmospheric lifetime (Table 14.1). The initial production of ClO, for instance, occurs in the following way in the presence of short-wavelength sunlight ($h\nu$):

$$CFCl_3 + h\nu \rightarrow CFCl_2 + Cl$$

$$Cl + O_3 \rightarrow ClO + O_2$$

Typical reactions that can occur in polar regions as a result of this production of ClO to bring about O_3 reductions are:

$$ClO + ClO + M \rightarrow Cl_2O_2 + M$$

$$Cl_2O_2 + h\nu \rightarrow ClO_2 + Cl$$

$$ClO_2 + M \rightarrow Cl + O_2 + M \qquad (14.2)$$

$$2(Cl + O_3) \rightarrow 2(ClO + O_2)$$

Overall:

$$2O_3 \rightarrow 3O_2 \qquad (14.3)$$

It can be seen from these reactions that free radicals may 'interfere' with the normal chemistry of the stratosphere during both the destruction and production of ozone, the overall effect a conversion of ozone to gaseous oxygen (14.3). Thus an 'excess' of gaseous oxygen is produced, while the overall ozone concentration is reduced (although the eventual outcome need not be a pro-rata change, as sunlight will continue the process of ozone production and depletion).

Table 14.1 Atmospheric lifetimes and depletion potentials of various artificial chlorine-containing source gases (from Table 3.1 of Warr 1991).

Compound	Cl atoms per molecule	Lifetime (yr)	Ozone depletion potential
CFC-11 ($CFCl_3$)	3	60	1.0
CFC-12 (CF_2Cl_2)	2	120	0.9–1.0
CFC-113 ($C_2F_3Cl_3$)	3	90	0.8–0.9
CFC-114 ($C_2F_4Cl_2$)	2	200	0.6–0.8
CFC-115 (C_2F_5Cl)	1	400	0.3–0.5
Carbon tetrachloride (CCl_4)	4	50	1.0–1.2
Methyl chloroform ($C_2H_3Cl_3$)	3	6.3	0.1–0.2
HCFC-22 (CHF_2Cl)	1	15.3	0.05

The lifetime of halons covers a similar range to those of CFCs.

The special situation of the polar regions

The polar regions are unique within the stratosphere. Not only are these parts of the atmosphere somewhat isolated from the flow within the rest of the stratosphere, but the large seasonal variations of insolation allow more complex chemical reactions to occur, especially during winter and spring.

Most importantly, these extreme northern and southern latitudes have prolonged periods of darkness through the winter, during which stratospheric temperatures fall (since, although there is O_3 present, there is no solar radiation for it to absorb, thus no 'greenhouse' warming). As temperatures fall below $-78°C$ nitric acid trihydrate – $HNO_3(H_2O)_3$ – condenses to form PSCs and temperatures below $-85°C$ yield water-ice PSCs. Sulphate particles are always present within the stratosphere, in part due to the addition of sulphur dioxide (SO_2) by volcanic eruptions. These may combine with water vapour and nitric acid at low temperatures, forming STS PSCs (Burkholder & Orlando 1998). Heterogeneous reactions can occur on the surfaces of PSC particles which activate the free radicals held within reservoir species (such as $ClONO_2$) into active forms (e.g. ClO_x), typically:

$$HCl + ClONO_2\,(PSC) \rightarrow Cl_2 + HNO_3$$

(Borman et al. 1997)

On the surface of these clouds nitrogen is also deactivated, reducing the chance of reactions that would deactivate ClO_x. The following reaction occurs:

$$N_2O_5 + H_2O\,(PSC) \rightarrow 2HNO_3$$

The nitric acid produced by this reaction stays on the PSCs, which fall under the force of gravity from the stratosphere, reducing the possibility of nitrogen reactions with reactive chlorine species, which would reduce their quantity, i.e. reaction (14.4) cannot occur:

$$ClO + NO_2 + M \rightarrow ClONO_2 + M \quad (14.4)$$

These reactions become particularly important when the sunlight returns each spring, rejuvenating the photolysing reactions. It is during this period that the greatest loss of O_3 occurs, before there is sufficient warming of the stratosphere to sublimate the PSCs (Sheldon et al. 1997). Until late spring, reactions such as (14.2) locally proceed unchecked within the polar vortex, forming the so-called ozone 'hole'. Most PSCs form in the lower stratosphere at altitudes between 12 and 25 km, where they can promote a large effect on ozone, which has a relatively high partial pressure in this height range.

In recent years the concentration of PSCs has increased in the Arctic stratosphere, indicating probable increases in the destruction of O_3 at high northern latitudes (e.g. Hofmann et al. 1989).

The presence or absence of PSCs will possibly have the greatest effect on stratospheric O_3 abundance and there remain many uncertainties about whether more or fewer will be seen in the near future because of uncertainties about the effects of global warming on the stratosphere.

of O_2 and an oxygen atom, it provides protection to plant and animal life. This is crucial, since ultra-violet light causes growth aberrations in cells and prolonged exposure – particularly of unprotected fair skin – can form melanomas, squamous cell carcinomas, basal cell carcinomas (Setlow et al. 1993; de Gruijl 1995; Abarca & Casiccia 2002; Fears et al. 2002) or cataracts (Dobson 2005). Plants and animals are also susceptible to the destructive effects of ultra-violet radiation, although most of the flora and fauna of the tropics have adapted to the high level of solar radiation (including relatively high levels of ultra-violet light) by developing pigmentation or another form of protection from cell damage.

14.2 The role of the tropics in replenishing extra-tropical stratospheric ozone

The production and depletion of ozone are nearly in balance so overall quantities, although small, change little. Indeed, the concentration of ozone in the tropics is generally below the global mean, as shown in Fig. 14.2. Throughout the year a proportion of the ozone produced in the tropics is carried slowly poleward by outflow above the deep convective tropical troposphere and the Brewer–Dobson circulation (Holton 1990). This pulls ozone from a level near 26 km in the tropical atmosphere to a lower level, around 21 km, in the extra-tropics,[2] where there is a low level of destruction (as well as production) of stratospheric ozone in winter (Fig. 14.1).

The upward and poleward motion is extremely slow in the dynamically stable stratosphere. Vertical motion is around 9 m day^{-1} above the tropical tropopause and poleward motion takes 4–5 months (Boucher 2010).

Because of the mean poleward motion of ozone in the stratosphere, tropical ozone production is essential for the flora and fauna of the higher latitudes, where depletion occurs in spring as the sun returns to the summer hemisphere.

14.2.1 Effects of the quasi-biennial oscillation

The QBO, described in section 12.4, has a significant effect on the amount of ozone in the tropical stratosphere. Figure 14.3 shows the total ozone mapping spectrometer (TOMS) mean column-ozone concentrations below 70°S recorded by satellite over a period of 14 years. In comparison, the westerly component of the equatorial zonal wind at Singapore

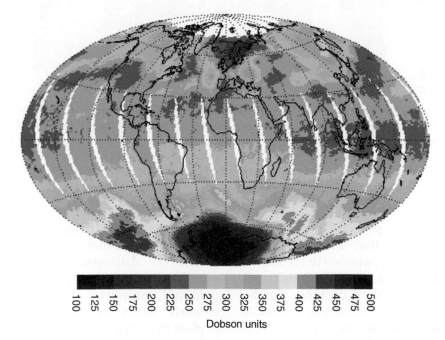

Dobson units

Figure 14.2 Total column ozone (DU) on 17 October 2005 sensed by the TOMS instrument on the AURA satellite. It is clear that whilst the tropics has a relatively constant level of ozone – contrasting with large variations around the South Pole in the southern spring – levels are below the global mean, so that there is an increased danger of exposure to short-wave radiation compared with the mid-latitudes, especially considering the elevation of the sun. Courtesy of NASA/GSFC.

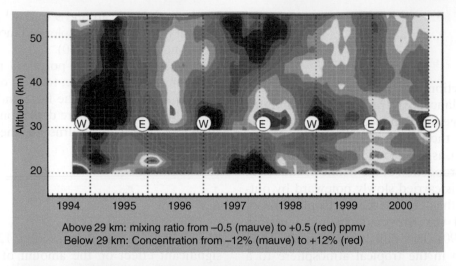

Above 29 km: mixing ratio from −0.5 (mauve) to +0.5 (red) ppmv
Below 29 km: Concentration from −12% (mauve) to +12% (red)

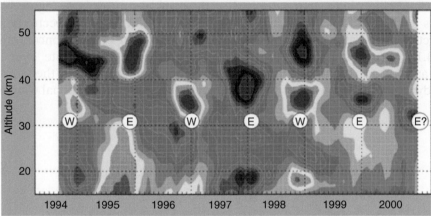

Figure 14.3   Seasonally synchronized ozone-column/QBO correlation observed by the Milo lidar, Hawaii, between 1994 and 2000. The maximum correlation is at 31 km, indicating a positive anomaly during the easterly phase (E) and a negative anomaly during the westerly phase (W). A weak signature around 45 km is out of phase with that at 31 km. Another, at 23 km, is in phase with anomalies at 31 km. The QBO signature is strongly disturbed by El Niño below 27 km (in ozone-poor air). Deviations in ozone mixing ratio from −3% (mauve) to +3% (red). Courtesy of NASA/GSFC.

is shown for the same period. The zonal-wind plot effectively shows that the oscillation of ozone in the Antarctic polar vortex is related to the QBO. The correlation is striking (coefficient 0.85) and led to the acceptance that the severity of the Antarctic ozone hole is partially caused by the QBO in its cooling phase. The effects are also seen in the northern hemisphere, but here the natural variation of ozone levels in the stratosphere is complicated by comparatively large vertical motions caused

by the mountain chains that surround the Arctic. Much smaller vertical motion is seen around the Antarctic. However, the effects are also important in the tropics, since the poleward transport is greatest when there are westerly winds in the tropics helping to deplete the tropical ozone layer.

Since the initial evidence was presented, two-dimensional models have shown the dependence of ozone-hole severity at the Antarctic on the phase of the QBO. Butchart & Austin (1996) found their simulated ozone hole was more dependent on the phase of the QBO and the transport of ozone from low latitudes than on stratospheric chemistry. The transport advects air of different temperatures into the vortex, thus affecting PSC formation. However, the modelling is imperfect and experiments with coupled three-dimensional models are currently being used to explain the inter-relationship fully (Lesley Gray & Lahoz 2002; Scaife et al. 2002). However, it is known that the tropical QBO exerts an influence on high-latitude planetary waves and can thus affect the timing of rapid stratospheric warming at the poles. (As we saw in section 12.4, the QBO has some influence on the monsoons and the formation of tropical storms. There is also some link to the variability in rainfall from mesoscale weather systems at the poleward edge of the ITCZ.)

14.3 The effect of 'global warming' on stratospheric ozone destruction in the tropics

Maintenance of the global radiation balance – the total amount of incoming solar to outgoing terrestrial radiation at all wavelengths – means that temperature increases in the lower troposphere, caused by an increase in GHG concentrations, must be matched by a cooling of the upper atmosphere. This cooling is greatest in the lower stratosphere, although there are many difficulties measuring temperature accurately between 20 and 50 km, complicated by the differing methods of assessment, including satellite radiometry. Thus the magnitude of the cooling of the stratosphere is not known accurately, but it is probably somewhat less than the magnitude of surface warming.

Cooling of the lower stratosphere in the tropics increases the potential depth of instability and the likelihood of deep convection above the tropopause, especially if surface temperatures increase (but see section 1.4).

The deepening of convection would entrain an increasing amount of water vapour into the tropical lower stratosphere.

14.3.1 The effects of an increase in stratospheric water vapour

The presence of water vapour upsets the balance between absorption of upwelling infrared radiation and increased emission of infrared radiation by the stratosphere, resulting in an overall cooling. Figure 14.4 indicates that this change is expected to be significant in the lower stratosphere with a maximum cooling of 0.3 K per decade at 18 km (Forster & Shine 1997). Throughout the upper stratosphere, most other non-ozone greenhouse gases, particularly CO_2, are still a cause of cooling.

A long-term trend of increasing stratospheric water vapour has been measured by radiosonde in Boulder, Colorado (Oltmans & Hofmann 1995). The trend has also been observed by the Halogen Occultation Experiment (HALOE) satellite, but over a much shorter time period. Hence, there is as yet no conclusive evidence that the stratospheric water vapour increase is a global long-term trend. The reason for the increase is also uncertain. It is thought that 40% of the increase could be accounted for by increases in methane oxidation (Forster & Shine 1997), its conversion to carbon dioxide resulting in water vapour formation as hydrogen is released. This would mean that the increase in water vapour would be a feedback of the increase in non-ozone greenhouse gases in the stratosphere, rather than a direct cause of temperature reduction (Forster & Shine 1997).

The deepening of convection also increases the likelihood of PSC formation. In the tropics, only a small reduction of temperature would support Type-1 PSCs within the lower stratosphere of the tropics and even where PSCs do not form, water vapour may be carried poleward by the Dobson–Brewer circulation. Thus a relatively modest cooling would increase the probability of imbalance in the

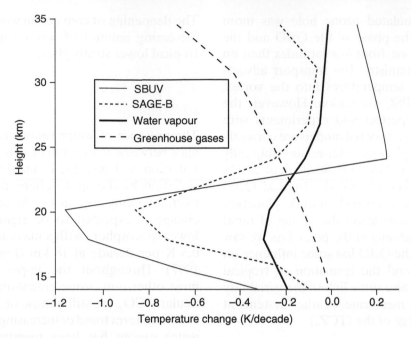

Figure 14.4 Annually averaged stratospheric temperature change (K per decade) as a function of height at 40°N. Results are shown for solar backscattered ultraviolet (SBUV) and stratospheric aerosol and gas experiment (SAGE) ozone changes, well-mixed GHG changes and stratospheric water vapour changes (from Forster & Shine 1997, Fig. 13). Changes are similar in tropical latitudes.

process of formation and destruction of stratospheric ozone. Given that the ozone layer of the middle latitudes is replenished in part by transport from the tropics, the implications are serious.

14.4 The effects of exposure to short-wave radiation

The depletion of the stratospheric ozone layer increases the risks associated with exposure to short-wave solar radiation in the tropics (Warr 1991; Abrahams et al. 2012). The risks are notably high in the tropics because of the high angle of the sun in the middle of the day and the relatively constant sunshine year-round. Solar radiation has a very short path through the atmosphere in the middle of the day.

Indeed, the ultra-violet index in the tropical zone remains high to extreme (due to the comparatively low levels of total ozone), mainly between 9 and 11, but locally 14 or more in summer, particularly in the tropical deserts and at high altitude (WHO 2002). For levels of index above 8, the World Health Organization (WHO 2014) recommends: 'Avoid being outside during midday hours; make sure you seek shade; shirt, sunscreen and hat are a must!'

Notably, the risks are present in oceanic areas as well as on land, despite the ready absorption of ultra-violet light by water (Häder et al. 1991; Sinha et al. 1999; Medicalecology. org 2006). Because water vapour absorbs, and clouds both reflect and absorb, solar radiation (in particular ice clouds, defined in Appendix 3) risks are somewhat lower in the humid zone than in the dry tropics. However, sunburn and sun stroke remain significant hazards

Table 14.2 Model-calculated steady-state lifetimes in years (from Butchart & Austin 1996, Table 1.4)

Species	AER	GSFC	CSIRO	Harvard 2-D	SUNY-NPB	LLNL	UNIVAQ 2-D	LaRC 3-D	GISS-UCI	MIT 3-D
N_2O	109	130	117	122	106	125	122	175	113	124
$CFCl_3$ (CFC-11)	47	61	53	68	49	49	44	57	35	42
CF_2Cl_2 (CFC-12)	92	111	100	106	92	107	105	149	90	107
$CFCl_2CF_2Cl$ (CFC-113)	77	101	83	55	81	87	81	70	79	
CCl_4	41	53	46	64	42	39	36	42	28	30
$CBrClF_2$ (H-1211)	16	12	36		21	29	21			
$CBrF_3$ (H-1301)	63	78	69		61	93	52			

AER, Atmospheric and Environmental Research; GSFC, NASA Goddard Space Flight Center; CSIRO, Commonwealth Scientific and Industrial Research Organisation; SUNY/NPB, State University of New York – NPB; LLNL, Lawrence-Livermore National Laboratory; UNIVAQ 2-D, Universita l'Aqulia 2-Dimensional; LaRC 3-D, NASA Langley Research Center 3-Dimensional; GISS-UCI, Geographical Information System for Science – University College Irvine; MIT 3-D, Massachusetts Institute of Technology 3-Dimensional.

throughout the tropical zone. Where the air is clearest and the humidity is least, the risk is greatest. In this respect Australia has the greatest risks as its mean humidity (and, for the most part, cloudiness) is somewhat lower than other similar parts of the world.

14.5 Current state of the stratospheric ozone layer

Although there has been some stabilization of the amount of ozone in the stratosphere since 1990 (WMO 2011; NASA-GSFC 2014), levels remain low and the emissions of both CFCs and halons before the adoption of the Montreal Protocol will continue to have an effect in the coming decades (Table 14.2). As can be seen from the discussion of the effects of ozone depletion, this is critically important in the tropics, where there is already an increased likelihood of effects on health and any depletion will reduce a level of stratospheric ozone that is already low.

14.6 Question

1. Much effort has been put into the observations and prediction systems of the ozone layer since the ozone 'hole' was discovered. Why do you think that most investment has been government funded, especially in extra-tropical countries?

Notes

1 All UV-C radiation (λ = 100–280 nm) is absorbed high in the atmosphere and most of the radiation absorbed by the stratospheric ozone is in the UV-B range (280–315 nm).

2 The difference in the level of the ozone layer between the tropics and the higher latitudes might also be used as a way to define the tropical zone (section 1.2.2), although the gradual change in altitude would make this difficult (and somewhat arbitrary) in the northern hemisphere.

15

Remote Sensing of Tropical Weather

15.1 Background

As was revealed in Chapter 1 and Appendix 1, observations are few and far between in much of the tropical zone, so remote sensing suggests an alternative method of observation, potentially on a small spatial scale. Notably, there are very few observations from oceanic areas, the sparsely populated deserts and, often, the poorest countries (despite the attempt to establish comprehensive surface networks towards the end of the colonial period), where weather observation may be seen as an expensive luxury. Techniques have been developed in recent years that may well fill the gap between observations in this important area. However, their use is not straightforward, so they cannot be seen as a comparatively cheap panacea and the required range of sensors is not fitted to every weather satellite.

15.2 Satellite remote sensing

Remote sensing has developed into a very important tool in the description, forecasting and nowcasting of weather throughout the tropics.

Satellite sensors, including imagery and microwave probes that can sense wind speeds, are possibly the most important tools. Using a variety of wavelengths, it is possible to image more than the reflected brightness (visible imagery), emission temperature (infrared imagery) or absorption (water-vapour imagery). Dust storms, volcanic ash, the intensity of convection, airmass characteristics and even areas where precipitation may be expected can be 'seen' using the differences between the wavelengths (Eumetsat 2014). The results are often displayed as 'false-colour' images, where the radiances are assigned to selected colours to enhance the aspect to be shown. In a similar way, 'true-colour' imagery is portrayed, often combining one or more infrared wavelength with visible imagery (which is sensed over a very narrow wave band: 0.6 or 0.8 μm, respectively, in the orange and near-infrared parts of the spectrum). This type of image is seen throughout this book.

Developments in satellite-image processing have allowed wind fields (within a comparatively deep layer) to be derived. These may be used to supplement observed data or validate numerical weather prediction background fields.

These observed winds may allow development to be predicted (as a nowcast). Because motion vectors may be derived from water-vapour imagery (as well as from infrared emissivity), upper-tropospheric divergence may be used to indicate areas of rapid upward motion. Derived divergence fields are now available

An Introduction to the Meteorology and Climate of the Tropics, First Edition. J F P Galvin.
© 2016 John Wiley & Sons, Ltd. Published 2016 by John Wiley & Sons, Ltd.

from some geostationary satellites for the tropics (Schmetz et al. 2004).

15.3 Precipitation

In much of the developed world – in particular the USA and Europe – radar has been a valuable addition to the forecaster's toolkit. Precipitation is shown in near-real time at high resolution (1 km within about 100 km of the radar), so that localized areas of heavy rain (or hail or snow) may be seen (Fig. 15.1). This gives the forecaster (really a nowcaster) a good

chance of warning customers of the likelihood of heavy rainfall, so that plans may be made to mitigate any effects from it.

Spatial observations also allow mesoscale models of the atmosphere to be verified and the model used to extrapolate the position of an area of rainfall up to around 12 hours ahead, giving the customer a longer lead time for preparation. In recent years, the value of these techniques has allowed a great improvement in nowcasts in many parts of the tropics. Temperatures are high in the lower troposphere and the troposphere is deep, so most tropical precipitation is convective, bringing a risk of

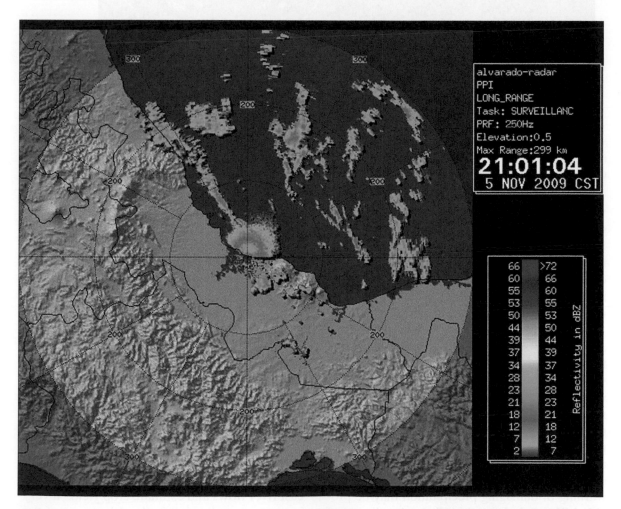

Figure 15.1 Rainfall-radar image from Alvarado, Mexico at 2101 CST on 5 November 2010, showing moderate showers over the Gulf of Mexico. © Servicio Meteorológico Nacional, Mexico.

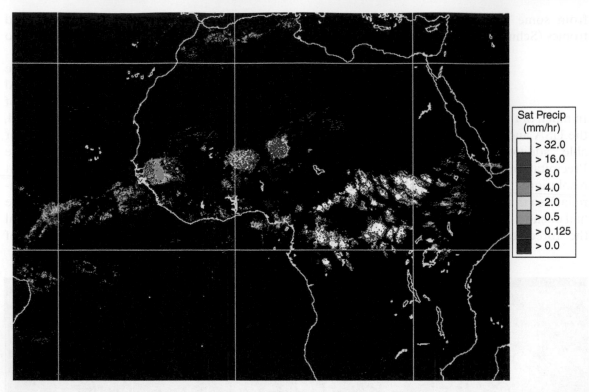

Sat Precip (mm/hr)	
	> 32.0
	> 16.0
	> 8.0
	> 4.0
	> 2.0
	> 0.5
	> 0.125
	> 0.0

Figure 15.2 Precipitation field during the West African monsoon at 1800 UTC on 2 July 2010, derived from Meteosat-9 imagery and deduced cloud thickness. Numerous MCSs can be seen over land, although the rain is sporadic along the coast, where there are occasional cumulonimbus clouds. © EUMETSAT/Met Office 2010.

very heavy rainfall. Great efforts are now going into the establishment of radar networks. However, where there are no suitable places for radar to be developed, particularly in remote or less-developed regions, techniques have been developed to allow satellite imagery to indicate areas of rainfall (Fig. 15.2).

Initially, polar-orbiting satellites were used and showed some success imaging the rainfall from tropical revolving storms, thus helping to verify their position and stage of development, which cannot be indicated with precision by low-resolution atmospheric models. More recently, geostationary satellites have been able to allow probable areas of precipitation to be tracked in areas where there are few observations and these show a great deal of success. In this way, some warning of damaging precipitation may be provided. The technique also allows the climate of relatively small areas to be studied.

However, there remain some problems with the use of radio waves that look down on cloud. Without returns from telemetering rain gauges, the true rainfall reaching the surface cannot be verified. Thus the amount of precipitation may be over-estimated (only occasionally under-estimated), a safer, if occasionally misleading, error. Since cloud depth is used in this technique, precipitation falling from high-based cloud may also be sensed. Over desert regions the base may be so high that the precipitation evaporates before reaching the ground (Chapter 7), but it will still be shown as measurable (perhaps heavy) precipitation to the surface.

More importantly, spatial rainfall estimates, using satellite techniques, can be used to forecast river flows and estimate crop yields, a very important application in the drier parts of the tropics (Grimes 2003; Diop & Grimes 2003;

Grimes & Diop 2003; Wheeler et al. 2005). However, there are also large errors in these estimates. Whilst hydrological forecasting is well developed in the UK (Grahame & Davies 2008; Werner et al. 2009) using a wide range of variables, including land use and soil type, neither of these variables is well mapped in most of the tropics. Land-use changes, such as deforestation and the rapid growth of plants following rainfall, can change these variables on as little as a weekly basis. As plants grow, the characteristics of a soil change: it loses less moisture from the surface as it is shaded from the sun (although there may be depletion of moisture below the surface), it is bound together by moisture and it becomes more likely to be able to absorb and retain moisture as its surface breaks up (although it may be at least a little less absorbent in depth).

15.4 Wind profilers

Wind profilers use microwave radiation (emitted from a 'nest' of aerials) to observe wind continuously in the vertical and some tropical countries have added these sensors in recent years, particularly to supplement radiosonde observations. The sensors (dependent on wavelength used) are able to obtain wind profiles to middle or upper tropospheric levels. Their use assists short-term forecasting (nowcasting) of weather systems. At times mesoscale systems may be observed by these systems, revealing detail for which only estimates were available previously. In 2013 Typhoon Usagi was observed in detail crossing the south-eastern coast of China (Li et al. 2015).

15.5 Thunderstorm observation

Since the Second World War various systems have been developed to use the emission of radio waves by lightning strikes. Each time lightning is produced, the signal it produces

may be sensed by a radio receiver. The use of very low frequency (VLF) radio receivers allows storms many thousands of kilometres away to be detected.[1] Once a network of three or more VLF receivers is established, arrival-time difference can locate the point of emission to within a few kilometres, even with stations at a large separation. Large networks, such as the sferics system established by the Met Office or similar arrival-time delay triangulation techniques used elsewhere, can detect thunderstorms across much of the world, although the accuracy of the location of any strike decreases with distance to around 30 km at a distance of more than about 5000 km (Met Office 2011b). Data are usually displayed as points or crosses on map displays (Fig. 15.3).

The large number of emissions from most thunderstorms can easily overwhelm any system covering a large area, so many thunderstorm-detection systems are selective, their aerials receiving data in short bursts. Smaller storms therefore may not be detected. There are also small errors in direction-finding systems, sometimes as a result of atmospheric effects. This can lead to errors in positioning, especially where the number of receiving stations is small.

15.6 Monitoring surface cover, fires and volcanic eruptions

In another 'colour-slicing' technique, the surface cover may be estimated (down to a resolution of about 1 km), for example the Normalised Difference Vegetation Index (NDVI) (Sellers 1985; Myneni et al. 1995). This method uses the differences in emission at various wavelengths to assess the surface cover: grass appears different from dense forest, dense forest appears different from farmed forest, vegetation can be easily differentiated from bare ground (Fig. 15.4). However, care is needed in areas where the surface cover varies

Figure 15.3   The location of lightning strikes (coloured crosses and squares) overlaid on false-colour imagery of north-eastern South Africa from Meteosat-9 at 1230 UTC on 7 October 2010. An MCS can be seen over the city of Johannesburg with an isolated storm further south-east, near Newcastle, Kwazulu-Natal. © EUMETSAT 2010.

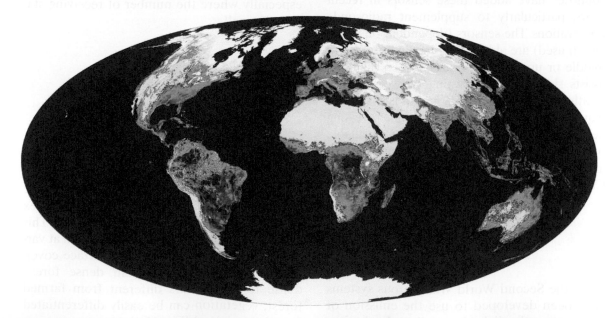

Figure 15.4   Mean global NDVI values in July. Vegetated areas show up as green shades, whilst predominantly-bare soil is yellow. Red areas are zones of transition. Courtesy of NASA/GSFC.

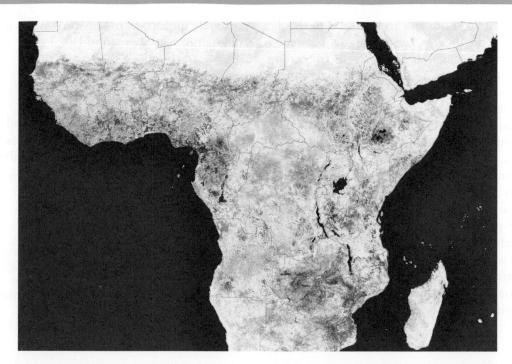

Figure 15.5 Anomaly of NDVI in central Africa for July 2005. The image is created from data collected by the SPOT Vegetation satellite. The anomaly is a measure of the density of vegetation compared with the long-term average. In regions where vegetation is less dense than normal, the image is brown, while areas of greater vegetation growth are shown as green. In July 2005 the vegetation was much denser than normal in many areas north of the equator, but there is a significant area affected by drought further south, as well as along much of the coast of the Gulf of Guinea. The good rainy season revitalized both agriculture and native plants, but also triggered deadly floods where the ground was sun-baked in the previous summer's drought. Areas that have suffered drought are more likely to be subject to fire, although the risk will vary according to vegetation type and normal density. Courtesy of NASA/GSFC. Image created by Jesse Allen, Earth Observatory, using data obtained courtesy of USDA FAS and processed by Jennifer Small and Assaf Anyamba, NASA GIMMS Group.

over a small scale. In recent years, maps of evident high resolution, showing surface type, have been published in atlases.

Although this is not a technique that may be applied directly to forecasting, it allows rapid analysis of the effects of weather and climate on a small scale in areas where there may be few observatories. This in turn allows responsive planning by governments, aid agencies and, where there is access to data in near-real time, farmers. Perhaps the largest challenge will be the speed of response to changes: governments, agencies and even farmers are notorious for their conservative response to changes.

Over time changes in vegetation cover may be assessed (Fig. 15.5), an important application in climatology. However, changes in surface

cover also have a close link to weather. Changes in vegetative cover are linked to day-to-day weather and recent changes, due to deforestation or agriculture, may be large. Where there is more vegetation, transpiration feeds more moisture back into the lower troposphere, helping to destabilize the atmosphere and produce more convective cloud. This in turn reduces (potential) evaporation and the cloud depth may be sufficient to produce more rainfall. Any additional rainfall will increase the amount of moisture available to evaporation, in turn possibly increasing precipitation still further.

Fire monitoring and fire risk are also useful features available from satellite imagery. The heat of large fires can be detected readily; a combination of surface cover and surface dryness allows a fire-risk index to be derived. Whilst no distinction can be made between fires started by mankind and those triggered by lightning strikes, the use of image time-series may allow this to be deduced. It may be possible to take action with support from these images either in fire control or as part of the monitoring of climate and demographic changes.

Volcanic eruptions are also readily monitored by satellite. The spread of ash and reactive gases from explosive eruptions can have local or global effects on the climate and it is extremely dangerous for aircraft to fly in ash clouds, which may have an effect on engines and airframes. Given the large number of volcanoes in the tropics and the paucity of monitoring from the surface, the satellite is an extremely important tool. More on the uses of satellite imagery is given in Box 11.2.

15.7   Question

1. What changes in the observing and forecasting systems of tropical (and other) nations might result from the introduction of remote sensing systems?

Note

1 The Met Office sferics system uses 12 receivers, most of them in Europe, to detect lightning strikes as distant as the east coast of North America, much of South America, Africa and western Asia.

16
Tropical Weather and Health

16.1　Introduction

The tropical environment supports an enormous variety of life and the plentiful sunshine allows areas with an adequate water supply to teem with life (Chapter 6). However, life includes a great variety of parasites and predators that prove a challenge. In addition, dry lands are a major source of atmospheric dust. The tropics have more than their fair share of these hazards.

16.2　The effects of tropical sunshine and warmth

Perhaps the most significant health effect of the tropics is on the metabolism. High temperature causes a high loss of fluid and salt from the body, but at high temperatures this process becomes inefficient and may result in overheating and heat exhaustion. Perspiration needs to be replaced if heat stroke and dehydration are to be avoided and it is necessary to drink much more water in the tropics than in the high latitudes. In the hottest climates, additional salt is also needed since this is present in sweat and is an essential component of the blood (at a concentration near 0.7%) (Voluntary Aid Societies 1997). When accompanied by high humidity and light winds, the risk of heat stroke and overheating are increased since it is more difficult to lose heat by the evaporation of sweat.

As described in Chapter 14, the high sun at midday also increases the exposure to damaging ultra-violet radiation in the tropics. Whilst the effect is most significant under clear skies (and cloudy skies significantly reduce the risk of exposure), the World Health Organization (Lucas et al. 2006; WHO & WMO 2012) advises those with lighter skin pigmentation to avoid the midday sun. Indeed, the greater exposure of the body to sunlight as the sun's elevation decreases probably means that the risk of skin damage and cancer remains for much of the day (Met Office 2014d).

16.3　Lifted dust and its effects on health

Wind-blown dust is relatively common in and around the tropical deserts. However, much of the dust has its origins in three localities: (i) the northern Sahara of Libya, southern Tunisia and central Algeria, (ii) the valley of the River Euphrates in Syria and Iraq, (iii) and the Sahel. These areas are a ready source of small particles of clay that can be lifted and carried long distances (section 7.2). Dust may be concentrated where mountain ranges block the low-level flow.

An Introduction to the Meteorology and Climate of the Tropics, First Edition. J F P Galvin.
© 2016 John Wiley & Sons, Ltd. Published 2016 by John Wiley & Sons, Ltd.

Suspended clay affects the health of many suffering from respiratory ailments – notably chronic obstructive pulmonary disorder (COPD) – and may also affect the eyes (Sarkies 1967), even at what may appear to be low concentrations. Trachoma is frequently found affecting the eyes of those living or working in dry dusty environments for many years (Sarkies 1967). By the time the horizontal visibility drops to 10 km, the suspended dust may have a significant effect on anyone suffering from respiratory disorders and when the visibility drops below 3500 m there are significant effects on aviation. At high concentrations the dust can be tasted and respiration becomes more laboured. The acceptable limit of exposure to suspended particles is 50 µg m^{-3} measured as PM10 load, i.e. all particles less than 10 µm in diameter – through a 24-hour period (Cohen et al. 1999). The dust load may exceed 1000 µg m^{-3} at times (Republic of Cyprus, Department of Labour Inspection 2015).

There is also a significant link between lifted dust and meningitis outbreaks, and the African Centre of Meteorological Application for Development (ACMAD) issues monthly bulletins to its member nations to ensure preparedness, although control of the disease remains difficult.

In recent years computer models have been produced that can predict significant dust events, which we know can have serious health effects. These models use standard weather-prediction models with a small grid length (around 10 km) coupled with the physics of dust particle ascent, advection and deposition. Warnings may be disseminated in the press, by radio or online (although this may leave large numbers of people without knowledge of an expected outbreak). The issue of warnings may help medical professionals to prepare for the likelihood of additional hospital admissions and GP appointments. Those with respiratory ailments can make suitable arrangements to ensure they remain healthy.

16.4 Industrial and smoke pollution

Industrialization is an increasing threat in the tropics. Rapidly developing industry in South Asia produces considerable pollution and the winter monsoon period in particular is characterized by moderate or poor visibility. The industry of northern China also produces considerable pollution that is frequently carried south and east in the winter period and may seriously affect much of South-East Asia at times.

The deliberate clearance or natural setting of fires in the tropical rainforest also produces serious pollution at times. However, the widespread distribution of smoke haze in these areas is a comparative rarity, requiring greater low-level tropospheric stability than is common in this zone. Nonetheless, there have been several significant events in recent years, not least in 1997 (WHO & WMO 2012) and 2013 (Roberts 2014; Sherwin 2014).

Such pollution can cause serious damage to health due to the inhalation of fine particles produced by combustion (WHO & WMO 2012).

16.5 Parasitic and infectious diseases

Areas of constant warmth and high rainfall are characteristically associated with serious diseases, many of which, as a result, are confined to and thrive in the humid tropics (e.g. WHO 2013). Almost all are parasitic (sometimes bacterial) infections. These diseases are usually associated with high rates of mortality or severe incapacity. The types of disease can be split into two groups: (i) those carried by flies or other insects (e.g. malaria, dengue fever, river blindness, philariasis, Leishmaniasis, African trypanosomiasis) and (ii) those of water bodies (e.g. bilharzia or schistosomiasis, helminth worms, hæmorrhagic fever). The strong link to climate is evident, but those of the first group may be carried widely (away from the wettest areas

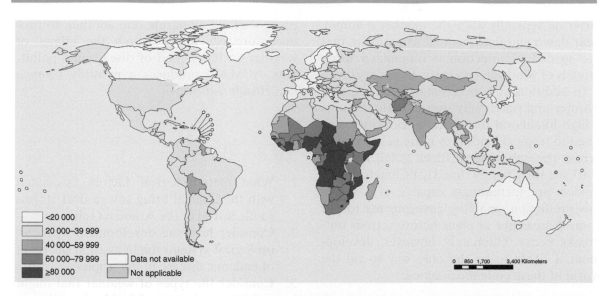

Figure 16.1 The effect of tropical diseases, displayed as disability-adjusted life years per 100,000 population, a measure that adds the number of years suffered with a disease to the number of years of life lost due to the disease in 2012. Values are largely above 40,000 years per 100,000 people in south and South-East Asia (pale–mid greens) and are mainly above 60,000 years per 100,000 people in much of tropical Africa (dark green), although there are lower values (mainly below 40,000 per 100,000) in Central and South America. Whilst it may be expected that low values would be seen in the higher latitudes, there is clearly a much higher rate in tropical Africa than in most other tropical lands. (The boundaries and names shown and the designations used on this map do not imply the expression of any opinion whatsoever on the part of the World Health Organization concerning the legal status of any country, territory, city or area or of its authorities, or concerning the delimitation of its frontiers or boundaries. Dotted and dashed lines on maps represent approximate border lines for which there may not yet be full agreement. The borders of the map provided reflect the current political geographic status as of the date of publication (2014). However, the technical health information is based on data accurate with respect to the year indicated (2012). The disconnect in this arrangement should be noted but no implications regarding political or terminological status should be drawn from this arrangement as it is purely a function of technical and graphical limitations.) © WHO.

and, sometimes, into the sub-tropics), whereas those of the second are strongly linked to local climate, in particular that of lower-lying regions (Prüss-Üstün & Corvalán 2006; WHO & WMO 2012). However, this assumption masks details; many of the diseases are endemic in only some parts of the tropics and infection rates are generally greatest in Africa, where control measures are poorly developed (Fig. 16.1). Elsewhere, control measures have been effective, particularly against malaria (WHO & WMO 2012).

Control measures are usually on a local scale, but can range from immunization (where practicable) and provision of suitable facilities (such as mosquito nets or new wells and sterilization tablets) to the somewhat extreme measure of broad-scale eradication of disease

vectors (notably flies). This last measure is carried out with little or no regard for the ecological balance and probable long-term effects, and does not discriminate disease vectors from, say, pollinators. The draining of wetlands also falls into this group of extreme measures, possibly also removing livelihoods dependent on the local ecology, a likely reason for initial settlement in the area. Nevertheless, such measures have proved effective.

However, there is clearly a role for the forecaster in the control of diseases carried on wind-borne vectors once all suitable improvement measures have been carried out on a local level. The need for insecticide spraying can be minimized by careful selection of probable times that infected insects may be airborne, typically when the weather is humid, near dawn and dusk, as well as ensuring that the spread of infection is minimized using models of wind velocity in the boundary layer.

In addition, it is becoming clear that, in a warmer and potentially wetter world, there is a high likelihood that diseases currently confined to tropical lands may more easily spread across the sub-tropics (Black 2006; Kington 2007; see also Chapters 6 and 12). Climatological models are increasingly capable of assisting this prediction, enabling governments to put control measures in place before serious outbreaks occur. Ultimately, however, development is likely to be the only way to rid the world of these endemic diseases.

16.6　Response of the meteorological community

The serious effects of weather on health and thus the economies of whole nations has led to increased efforts by national weather services,

often working alongside one another, to produce forecasts and research the effects of weather on the spread of disease and pollutants. ACMAD produces a monthly *Climate and Health Bulletin*.

16.7　Questions

1. What meteorological factors associated with the Shamal bring severe dust storms south east over the Arabian Gulf?
2. Consider how the development of meteorological systems might assist in the relief of endemic disease in the tropics.
3. Consider the types of weather that might assist in an increased incidence of locust swarms.
4. Which wind system is responsible for carrying pollution across southern and South East Asia and why is the associated pollution such a significant hazard?

17
Conclusions and a Look to the Future

17.1 A summary

Throughout this book there are descriptions of the variations in the climate and weather in the tropics on the synoptic scale and, to some extent, the mesoscale. Many of the local variations have been described from a forecaster's perspective. However, it was not possible to cover the subject in detail and this is only a taster for this vast subject, the tropics covering about half the world. Many specific texts are available for readers whose interest has been spurred by this series. These cover many details and often deal with the mathematics of the tropical atmosphere.

Throughout, there is an attempt to show the way that the weather affects mankind and, in section 12.5, how humans have modified the natural environment. What can be more important than to understand the weather across the globe and to predict its serious effects? As we have seen through the book, there is great potential for severe weather in most of the tropical zone. From parched deserts to torrential downpours, marked seasonal changes to tropical revolving storms, to squall lines with hail and thunder (as described in Chapters 5, 7, 8, 9 and 10) the tropics has perhaps more than its fair share of severe weather. Associated with the weather are plague and pestilence (Chapters 13 and 16).

By improving our understanding, we can forecast the weather and minimize its effects to improve our prospects in this productive, but often fragile, environment.

More than 70% of the world's population live in the tropics (i.e. about 4900 million people in 2010),[1] many of them along coasts or rivers, where transport and water resources have allowed populations to grow crops and flourish (Wikipedia 2014e; United Nations, Economic and Social Affairs Division 2012). The extreme character of most tropical climates, for example the discomfort of humidity, large desert temperature ranges, the cold of high mountains and copious rainfall, are well represented by the climate data in Figs 6.2, 6.4, 6.6, 6.8, 6.9, 6.10 and 6.13. (Further examples of tropical climates can be found in Pearce & Smith (1984).)

The land area of the tropics is about 40% of the world total, although the habitable area is much less than half of this. Urbanization is increasing rapidly and 12 of the 20 largest urban areas in the world lie between 30°N and 30°S, each with a population greater than 14 million (Demographia 2014). In more developed countries, development is extending into desert areas, resulting in a profound effect on the climate. Clearly, there is a large and growing need for increased knowledge of the effects of the weather in all tropical countries.

An Introduction to the Meteorology and Climate of the Tropics, First Edition. J F P Galvin.
© 2016 John Wiley & Sons, Ltd. Published 2016 by John Wiley & Sons, Ltd.

17.2   Forecasting the weather

In recent years there has been increasing research into the weather of the tropics, both to aid our understanding and to improve forecast models. As this research proceeds, our understanding grows and numerical models improve, particularly as model resolution is increased (Grosvenor et al. 2005; Shutts 2005; Willett & Milton 2005; Met Office 2014e,f), yielding benefit to all. However, even with new mesoscale models at a resolution of 4 km or less, the random character of tropical convection (even in areas where deep convection might reasonably be expected to develop) cannot be captured, so rainfall patterns cannot be forecast in detail even a day ahead. Nonetheless, there is increasing ability to forecast significant rainfall events over comparatively large areas. The effects of the weather on a local scale can be forecast several hours ahead using experimental models at a resolution around 1 km. Ensemble output is now available out to a period of 15 days or more, giving a range of likely outcomes (Richardson 2011; Mayes 2012) and shows considerable reliability in the prediction of both temperature and rainfall in the tropics. Increasingly, this reliability is reflected in products such as the Ten-day Climate Watch Bulletin of ACMAD.

The Met Office provides many new services to assist tropical countries in forecasting and dealing with the weather (Met Office 2013a, 2014g,h,i), adapting to the climate and assisting national development (Met Office 2014g,j). These include relocatable mesoscale models, warning systems for tropical revolving storms, direct assistance to national meteorological services, and forecasts for medium- and long-haul aviation at medium and high levels from its World Area Forecast Centre (Met Office 2012, 2014k). The clear need for forecasting services has also led to the rapid development of weather and climate prediction, not least in West Africa (Parker & Diop-Kane 2015).

Climate prediction and adaptation strategies are produced by the Met Office Hadley Centre for Climate Prediction and Research.

In many ways the future of forecasting in the tropics looks bright. With new forecasting systems, accompanied by developments in the internet that allow data and imagery to be exchanged rapidly, as well as new observing systems, our understanding of and ability to predict tropical weather is increasing rapidly.[2] On the other hand, as outlined in Chapter 12, the increasing population of the tropics, much of it in coastal areas, makes the threat of climate change serious right across this zone. Clearly, tropical nations have their part to play in reducing the threat posed by global warming and may be instrumental in the necessary reduction of greenhouse gases, despite the largest contribution to GHG loading having its origin in the extra-tropics.

17.3   Questions

1. Consider how a mixture of increasing ability to observe and forecast the weather, and international collaboration can be of benefit to both developing- and developed-world nations.
2. Thinking again about the threat posed by increasing GHGs in the earth's atmosphere, as well as increasing populations and the wealth of the nations involved, which countries' populations are most threatened by global warming and its effects?

Notes

1   This depends (to a small extent) on the way we define the tropics, as discussed in Chapter 1. However, as the tropical margins are predominantly dry areas, populations are mainly small.

2   This clearly poses the problem in this book that reference sources, especially those on the internet, will rapidly become obsolete, so readers should check for updates to keep abreast of developments.

Appendix 1
Observations from the Tropics

Observations are a necessary component of the forecasting system, as well as providing a wealth of weather details, continuously observed in a local area. In days past, forecasters used detailed observations to produce forecasts for up to several days ahead and without them analyses at the surface and at significant levels within the atmosphere are difficult. These days computer models assimilate observed data (at main observation hours 0000, 0600, 1200 and 1800 UTC), adjusting the forecast produced 6 hours ago (often with several hours of data around the nominated analysis time) to accommodate observed weather. This allows increased accuracy in the shorter term, as well as the production of forecasts into the medium term, giving guidance up to 10 days ahead or more. The data from radiosondes are most important for the adjustment of forecast parameters, since models maintain balance by correspondence of data throughout the atmospheric column (Lorenc 2006; WMO 2009a). Increasingly, models also assimilate data at other observation times to improve fine tuning (4d-VAR methodology; ECMWF 2002). Mesoscale models with a grid length around 10 km generally use a greater variety of observed surface data reported in the SYNOP code (WMO 2009b), but the amount of data within these models restricts the length of forecast that can be realistically produced.

The limit of prediction for these models is usually around 48 hours.

Ultimately, the small errors inherent in the observations themselves are magnified in numerical prediction models, which is why any observations too different from the T+6 forecast field are excluded. In general, the errors grow so large by day 10 that forecast accuracy is seriously impaired and no more than a general type can be forecast. This is usually based on a range of forecast solutions (an 'ensemble'; John 2006), with output produced as a high likelihood of wetter than normal weather or of higher temperatures than normal.

Before gaining independence from rule by the nations of western Europe and the USA, many tropical countries had established a good network of observatories, including radiosonde stations that could launch a balloon twice per day (at the standard hours 0000 and 1200 UTC). However, the increasing costs of weather observation, its transmission and the declining per-capita income in the developing world have greatly reduced – in some cases almost removed – the observing network, in particular that of radiosondes. This decline has occurred despite substantial efforts by the WMO to maintain acceptable networks. In many cases observatories need to have a secondary role, in particular the provision of data for local air operations, so areas that have no air service

An Introduction to the Meteorology and Climate of the Tropics, First Edition. J F P Galvin.
© 2016 John Wiley & Sons, Ltd. Published 2016 by John Wiley & Sons, Ltd.

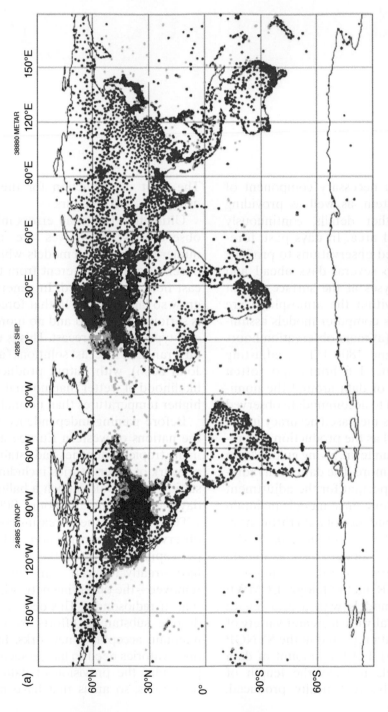

Figure A1.1 (a) Surface observations assimilated into the ECMWF numerical model at 1200 UTC on 29 May 2015.

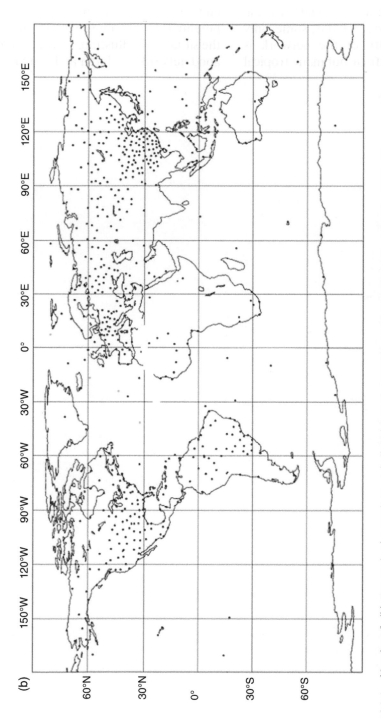

Figure A1.1 (*Continued*) (b) Upper-air observations assimilated into the ECMWF model at 1200 UTC on 29 May 2015. The networks are poor in much of the tropics, with only China and India having a reasonably dense coverage. Some stations declared to WMO are effectively dormant for financial or political reasons. The difficulty of making and recording ascent data over the open ocean makes these areas the least observed of all, largely dependent on ascents from island nations, but supplemented at times by commercial shipping. © ECMWF.

rarely have a meteorological observatory. The nominal radiosonde network of the WMO is shown in Fig. A1.1, although many of these stations report only once per day, some only occasionally. The paucity of the network is particularly notable in Africa, but many tropical countries have very widely spaced stations compared with those in Australasia, Europe and much of North America. Nevertheless, the potential for development to help to rectify the situation is illustrated by the comparatively good networks of India and China.

Appendix 2
Named Winds of the Tropics

This appendix details local and regional winds and associated weather in the tropics, edited from a document produced by Nick Weight of the Met Office.

Winds are listed alphabetically along with a brief description, most from publications by Schamp (1964), Martyn (1992), Kendrew (1961), Atkinson (1981) and Levi (1963). Additional information was supplied by Dr Danuta Martyn. The causes and peculiarities are supplied from original descriptions or as provided by Nick Weight. The more general terms valley, katabatic and ravine winds are not described in this paper, as a more precise definition and description is provided by the numerous books on meteorology available through libraries (Mayes & Perry 1989; Meteorological Office 1991).

Whilst the list is not exhaustive, as many winds as are generally recognized in the tropical zone have been included. There are few books with a comprehensive listing of winds, so it was considered important to include this list.

Some of the winds (some of which were named in extra-tropical regions) have become part of the terminology describing particular winds that occur in many regions, such as Bora and Föhn. Readers should become familiar with some of these terms on reading the listing.

Abroholos
See Abrolhos.

Abrolhos
Location: Brazil
A squall – often violent – associated with showers, occurring mainly between May and August over east Brazil (the southern winter). Named after the rock of the same name on the east coast of Brazil on the hill of Caravelas.

Afghanetz
See Bad-i-sad-obistroz.

Afghanistani
See Bad-i-sad-obistroz.

Albany Doctor
See Doctor.

Aloegoe
See Nirta.

Aref
See Arifi.

Arifi
See Scirocco.
Location: Morocco.

An Introduction to the Meteorology and Climate of the Tropics, First Edition. J F P Galvin.
© 2016 John Wiley & Sons, Ltd. Published 2016 by John Wiley & Sons, Ltd.

Aziab
See Khamsin.
Location: Egypt and Red Sea.

Bad-i-kasif
See Tebbad.

Bad-e-Simur
See Bad-i-sad-obistroz.

Bad-i-sad-o-bistroz, Bad-i-sad-obistroz
Location: Iran, Afghanistan.
Persistent ravine winds generally from north or north-west. The name means 'wind of 120 days', associated with the summer monsoon, often significantly enhanced by daytime heating to become strong or gale force, the intensity often reinforced by local topography, bringing raised dust and sand, reducing visibility significantly. Usually has low humidity and is gusty; can reach hurricane force.

Bagio
See Baguio.

Baguio
Location: The Philippines.
Native name for tropical cyclone or typhoon winds that affect the north of the islands bringing copious rain, mainly between June and December, literally 'wind from Baguio', a mountain city in central Luzon.

Bagujos, Baguo, Bagyo
See Baguio.

Bahorok
See Bohorok.

Bai
Location: Egypt.
A wind from the north in ancient Egypt, called 'the ram'.

Bali Wind
Location: Indonesia.
A strong easterly wind affecting the eastern end of Java, produced by local funnelling, a feature of the archipelago, that enhances the synoptic-scale gradient winds as they intensify.

Barat
Location: The Celebes.
A strong, often squally west or north-west wind affecting the north of the island of Sulawezi mainly from November or December until February. The wind is deflected by the many islands of the archipelago through the numerous channels between islands, and many rapid changes in direction and velocity may occur. These winds are frequently associated with the passage of tropical cyclones to the north of the Celebes.

Barico
See Baguio.

Barinés
Location: Venezuela, Brazil.
A strong westerly wind on the coasts of Venezuela and Brazil, this flow opposes the prevailing easterly trade-wind flow and is likely to be caused by drainage from the coastal highlands.

Barah Al
Location: Bahrain.
Dry north-westerly winds across the Arabian Gulf affecting the island between June and September, associated with low pressure over Pakistan and Iran in summer. It can be persistent and brings some relief from high temperatures.

Bat Furan
Location: Arabian Sea.
Persistent, but modest 'open sea' wind from the east during the winter months, permitting navigation under sail and associated with high pressure over Central Asia.

Bat Hiddan
Location: Arabian Sea.
Persistent strong 'closed sea' southerly to south-westerly winds during the south Asian

summer monsoon, bringing dangerous seas over shallow waters and difficult or impossible navigation.

Bayamo
Location: Cuba.

A violent northerly wind on the south coast of the island, especially near the Bight of Bayamo and over the northern slopes of the Sierra Maestra, linked to extra-tropical developments moving north-west across the Caribbean Sea, bringing severe thunderstorms and squalls.

Belar
See Belat.

Belat
Location: Southern coasts of the Arabian Sea.

A strong dry northerly or north-westerly wind, akin to a Bora between December and March, reaching 15 m s^{-1} or more in places, linked to the passage of weather systems over the Arabian Gulf and associated with the Shamal (see Chapter 7). It is often dust-laden.

Belats
See Belat.

Belot
See Belat.

Berg (Wind)
Location: South Africa.

Hot, dry, sometimes dusty winds with Föhn characteristics that carry continental air to the coasts, associated with high pressure over the South African plateau. Their mean frequency over western coasts is 50 days a year. Over the south-west they blow from the east, on the south coast, especially Cape Province, from the north and from the north-west over Natal. Most frequent during the winter months, they may last 2–3 days, bringing high temperatures, often damaging crops. On coasts, the Berg may be neutralized by the sea breeze.

Bhoot
Location: India.

A dusty wind, probably with dust devils.

Bhut
See Bhoot.

Biliku
Location: Andaman Islands, India.

A dry wind from the north affecting the Andaman Islands, associated with the north-east (winter) monsoon. However, showers sometimes occur as moisture is entrained during its passage across the Bay of Bengal.

Bist-rot
See Bad-i-sad-obistroz.

Black South Easter
Location: South Africa, notably Cape Town.

A dry generally easterly wind with Föhn characteristics, blowing off Cape Town at the beginning of summer. This name is also used in the winter months in New Zealand and New South Wales, Australia (not described here). Because of the steep orography of the region, its effects are very local.

Boe
See Broeboe.

Bofu
Location: Japan.

Name for a gust of wind (applied to typhoons).

Bohorok
Location: Indonesia.

This Föhn-type wind across the north-east backbone of Sumatra, blowing from the Karo Plateau between May and September, may cause damage to crops.

Bohorot
See Bohorok.

Bokorot
See Bohorok.

Bolon
See Nirta.

Bora
Location: The Adriatic Coast of the Balkan Peninsula (similar winds occur in many other mountainous coastal locations).

A strong east to north-east katabatic wind. Dry and cold and often very gusty, most noticeable around the Trieste region and associated primarily with winter and spring. Can occur along the length of the Adriatic coast, but most prominent between Trieste and Split. Under the influence of persistent high pressure (over central and eastern Europe), a cold pool of air builds over land. This eventually breaks through the numerous passes that lie along the mountain barrier and sweeps (westwards) towards the coast, enhanced by funnelling and topography. If there is a depression over the Adriatic, the Bora is accompanied by thick cloud and snow. Gravity waves in the lee of the Dinaric ridge provide an explanation for the gusts, along with some features of a hydraulic jump. Often very dry with very low humidity on the coast. The gusts of the Bora may reach 40 m s^{-1}, but can attain 60 m s^{-1} and last 1 or 2 days, at times as much as a week. Visibility can be reduced to fog conditions by dust over the sea called Fumarea or Spalmeggio. The vertical extent of the Bora is less than 2 km. The equivalent of the Bora occurs in many locations around the world (e.g. as described in Galvin (2004)).

Braw
Location: Papua New Guinea.

A south-westerly (dry) Föhn-type wind associated with the Australian winter monsoon over the Schouten Isles, north of New Guinea and lasting 7–8 days.

Brickfielder
Location: Australia.

A hot, dry and very dusty north or northwest wind, mainly in Victoria and western New South Wales. It forms due to the pressure gradient between the interior of Australia and the Tasman Sea. Temperatures can often exceed 37°C. The mountain barrier of the eastern coasts protects the coastal region somewhat from the effects of the Brickfielders. (Similar to the Zonda of Argentina.)

Bricklayer
See Brickfielder.

Bris
See Brisa.

Brisa
Location: South America (a similar wind also occurs in Puerto Rico – the Briza – and the Philippines).

A north-east wind blowing onto the coast of South America (possibly includes sea breezes).

Brisas
Location: Uruguay.

Strong south to south-east sea breezes near Montevideo, reaching greatest strength during the afternoon.

Brisole
See Brisote.

Brisote
Location: Cuba (Caribbean).

Applied to the north-east trade wind when blowing with more than its usual strength of 9 m s^{-1}, this wind may be associated with tropical depressions passing to the north-east of the island.

Briza
Location: Puerto Rico (Caribbean).

Associated with the trade winds over northern Puerto Rico (see also Brisa and Brisas).

Broe
See Broeboe.

Broeboe
Location: The Celebes.

A strong, dry Föhn-style squally easterly wind affecting, in particular, the south-western part of the islands. The strength may be associated with local funnelling.

Broboe, Broebroe, Brubru
See Broeboe.

Buhrga
See Bad-i-sad o-obistroz.

Burster
Location: Australia (also South Africa).

These strong cool southerly winds follow the passage of eastward-moving depressions, bringing relief from hot dusty north or north-westerly winds, especially the Brickfielder. Rough seas can be expected, especially in shallow waters.

Buster
See Burster.

Cambûeiros
Location: Brazil.

This cool squally southerly wind on the east coast of Brazil near Salvador occurs especially in August, bringing enhanced convection as cool air travels northwards into low latitudes under the influence of a slow-moving TUTT.

Camsin
See Khamsin.

Cape Doctor
Location: South Africa.

A dry, often very stormy, pseudo-Föhn type easterly wind, blowing from False Bay over Cape Town to Table Bay, especially during the summer months. This wind is relatively cool and has the effect of purifying the air over Cape Town, bringing refreshment to its inhabitants.

Cat's Paw
Location: Australia.

A general term used to describe a wind just strong enough to ripple a water surface or sand surface.

Caws
See Kaus.

Chabascos
Location: Pacific Coast of Central America.

This wind blows off the ocean onto the Pacific coast of Central America between Cape Santa Eugenia and the Isthsmus of Panama, normally commencing at the beginning of May and continuing through the wet season, into September. These winds have their origin in the south-east trades of the southern hemisphere and are moisture laden, but have a capping inversion, leading to areas of mist and low cloud that envelop the coastal hills.

Chalca
See Yalka.

Challiho
Location: India.

Strong southerly winds commencing in March–April prior to the onset of the summer monsoon as intense heating over the Indian subcontinent forms a thermal low-pressure system. The wind may be very dusty and can last 40 days.

Chamsin
See Khamsin.

Chandui
See Chanduy.

Chanduy
Location: Ecuador.

A cool mountain breeze setting in during the afternoon during the dry season (July–November) around Guayaquil.

Chemal
See Shamal.

Chergui
Location: Morocco.

A very dry and dusty 'sharp' east or south-east wind from the desert. Hot in summer and cold in winter, this wind can last for 40 days.

Chibasco
See Chabascos.

Chi'ing Fung
Location: China.
   A gentle breeze.

Chichili, Chihili, Chili
See Scirocco.

Choiu
See Arifi.

Chom
See Scirocco.

Chocolate, Chocolate Gale, Chocolatero
Location: USA and Mexico.
   A dusty wind blowing over the Gulf of Mexico from the north.

Chocolatta North
Location: West Indies.
   A north-westerly gale. The name chocolatta may refer to a dark, heavily laden sky, associated with depressions or storms to the west.

Chubasco, Chubaxo
See Chabascos.

Chudras, Churada, Churadas
Location: Mariana Islands.
   North-easterly trade winds with frequent squalls during the winter monsoon (November–April/May) caused by the convergence of winds from east Asian and the easterlies of trade-wind belt, generating instability over warm seas. The sea state may be rough.

Colla, Colla Tempestada
Location: The Philippines.
   Particularly squally southerly or south-westerly winds, mainly around Luzon, during the summer monsoon, when a depression passes to the north of the islands, enhancing the pressure gradient. Rough seas are likely.

Collada, Collado
Location: Mexico.
   A strong north or north-west wind blowing in the upper part of the Gulf of California, but blowing from the north-east in the lower part of the Gulf, with a possible katabatic origin. It may attain a speed of 15–20 m s^{-1}.

Cordonazo (de San Francisco)
Location: Mexico.
   Strong southerly wind peaking around the Feast of St Francis of Assisi in early October, the 'Lash of St Francis'. The strong southerly wind indicates the presence of a storm system to the west of Mexico or over the eastern Pacific. (The name can be applied to winds over San Francisco.)

Coromell
Location: Mexico.
   The usual evening (southerly) land breeze in the vicinity of La Paz from high ground at the southern tip of Baja California between November and May.

Cowshee
See Kaus.

D'Aibafu
See Giba.

Dahatoe
See Nirta.

D'Aibafu
See Giba.

Dchaoui
See Haboob.

Demani
Location: Africa (East).
   A light breeze during April, the Swahili term for the 'favourable winds' that are a precursor of the onset of the northern summer monsoon.

Depeq

Location: Sumatra.

A strong wind occurring over Loet Tawar, Sumatra during the south-west monsoon that may have Föhn characteristics after passage over the mountains of the island.

Dirty Northerly Wind

Location: The Philippines.

North-easterly winds of winter over eastern Luzon ascend the windward slopes of the Sierra Madre and the mountains of the Bicol peninsula during the winter monsoon, bringing copious cloud and rainfall.

Diver's Storm

Location: Egypt.

Cold, often stormy northerly winds towards the end of winter, around late January, in the vicinity of Alexandria. A marked temperature drop is common and the sea state is often rough or very rough, especially in shallow bays or exposed harbours.

Doctor

Location: Tropics.

A general term for cooling sea breezes in the tropics that bring relief from coastal humidity and heat. In Australia the following words precede Doctor: Fremantle, Albany, Esperance and Perth. The term is also applied to the Harmattan on the shores of Guinea and to a south-east sea breeze in South Africa (see also Cape Doctor).

Dschani, Dzhani

Location: Africa, mainly Libya and nearby western and southern Sahara.

A desert wind of the Sahara, it is at its most intense around midday and may be associated with lifted or blowing dust and sandstorms.

Egyptian Current

See Harmattan.

Elephant, Elephanta, Elephanter

Location: India.

This strong south or south-easterly wind on the Malabar Coast of south-west India occurs mainly in September and October, at the end of the summer monsoon. Thunderstorms, squalls and heavy rain may occur.

Embata, Embate

Location: Canary Islands.

This is a wind that is anti-trade in direction, often due to eddies south-west of the islands forming in the north-east trade wind. Von Karman vortices may be seen to leeward of the islands.

Esperance Doctor

See Doctor.

Evgey

Location: China.

A strong tunnel-like katabatic wind of the Bora type, especially frequent in winter, blowing from the Junggar Gate, implying a wind from the eastern quadrant, possibly connected to the circulation around the intense Asiatic high-pressure system of winter.

Fakatiu

Location: Melanesia.

North-west winds.

Feh

Location: China.

Light breeze over Shanghai (may be a sea breeze).

Föhn (or Foehn)

A relatively dry and warm wind downwind of mountain barriers. Moisture-laden winds deposit precipitation on the windward slopes, allowing the air to dry and warming adiabatically on its descent on the leeward side of the mountain barrier. Although formally a European Alpine wind, the name characterizes winds of this type. Rising temperature can initiate avalanches and desiccate growing crops.

Foh-Rse-Brune

Location: Japan.

A light and rather variable wind.

Foh-Rse-Jonn
Location: Japan.

A wind favourable for the reproduction of frogs. It is usually a springtime wind, preceding the summer monsoon, heralding milder damp conditions, associated with drizzle and extensive low cloud on windward slopes.

Foh-Rse-Torouj
Location: Japan.

A (summer) wind that is hot and humid.

Fremantle Doctor
See Doctor.

Friagem
Location: Brazil, Bolivia.

Cooling southerly to westerly winds over the southern Amazon Basin during the rainy season, mainly between April and June, may be associated with disturbances running along a TUTT.

Fung Chiao-hsueh
Location: China.

A cool, dry north-east wind associated with the winter monsoon and the circulation of the winter Asiatic high-pressure system.

Gending
Location: Java.

A dry Föhn-type wind that blows across the plains of northern Java, this south-westerly wind is highly modified as it crosses the belt of high peaks of the southern part of the island. Local gustiness is caused by funneling through mountain gaps.

General Santa Ana
See Santa Ana.

Gebli
See Ghibli.

Gerbui
Location: Algeria.
  See Scirocco.

Ghibli
Location: Mediterranean coast, mainly Libya and Malta.

A dry wind from the Sahara desert, usually originating hundreds of kilometres to the south, carrying much dust or sand, prominent during March–May and August–October, south-east of depressions over the Mediterranean. The low humidity can occasionally cause damage to vegetation. In the cold part of the year, this southerly flow from the Sahara is called a cold Ghiblis (Khamsin). There are many alternative spellings.

Giba
Location: Japan.

A wind that 'seizes the horses with enough strength to send the rider backwards in his mount'.

Gibla, Gibli
See Ghibli.

Glacier wind
See Nevado.

Guba
Location: Papua New Guinea.

Term applied to strong squally winds occurring at night over the south coast of New Guinea, accompanied by heavy rain and thunderstorms. There is an interaction of the prevailing winds and the night-time katabatic drainage winds from the surrounding high ground. Local steep topographic features and narrow valleys can enhance the intensity of these winds (see also Sumatras).

Guebli, Guibli
See Ghibli.

Gully-squall
Location: Central America.

A nautical term applied to a violent squall of wind in north-easterlies from the mountain ravines of the Pacific coasts of Central America, most often associated with the winter months.

Hababai
Location: Sudan.

A hot, dry easterly or north-easterly wind blowing over the western coast of the Red Sea near Port Sudan from Saudi Arabia, mainly during October and November, associated with the end of the summer season.

Habbub, Haboob, Haboub, Habub
Location: Sudan and Egypt.

Severe dust storms of northern Sudan and Egypt that can raise a wall of dust to hundreds of metres, especially in summer. The dust wall can be 25 km long and advance at 15 m s^{-1}. Strong instability is required where south-west winds near the surface interact with the dry Harmattan aloft. Many local names are given to dust storms, including many alternative spellings of the name, which means 'a strong wind'.

Haleakala
Location: Maui, Hawaii.
See Naalehu.

Harmatan, Harmattan, Harmetan, Haur
Location: North and West Africa.

The dry north-east trade wind of West and North Africa, most common in the winter season, its extent usually greatest in February, when it is the predominant wind along the coast of the Gulf of Guinea. It is dry and often dusty, with its origins in the central Sahara. The name comes from 'harmata', the name of West Africa in Fanti. It may also be known as the Doctor or the Egyptian Current. See Chapter 8 for details of its association with the West African monsoon.

Hava Janubi
Location: Arabian peninsula.
A southerly wind.

Hava Shimali
Location: Arabian peninsula.
A northerly wind.

Hayate
Location: Japan.
A gale.

Hermitan
See Harmattan.

Hubbob, Hubbub, Hubub
See Haboob.

Hupe
Location: Tahiti

A land breeze bringing some relief from high humidity, having Föhn-type characteristics.

Ihamsin
See Khamsin.

Irifi
Location: Western Morocco, Western Sahara and the Canary Islands.

A hot, dry, often dusty, north-east wind from the Sahara, especially during spring and autumn, when it is sometimes strengthened like a Föhn and often reaches the Atlantic coast, overcoming prevalent sea breezes.

Jalca
See Yalka.

Jaloque
See Scirocco.

Junk Wind
Location: South-East and east Asia (especially in Thailand, Japan, China and Malaysia).

A southerly or south-easterly monsoon wind favourable for the sailing of junks.

Junta, Junte
Location: Bolivia.

A strong stormy wind through mountain passes in the Andes that can reach hurricane force.

Kâchchan
Location: Sri Lanka.

A hot, dry Föhn-type wind from the west or south-west along the east coast of the island. Associated with the summer monsoon between May and August, this wind overcomes the prevalent sea breeze on the east coast, lifting the temperature to 38–40°C.

Kadja
Location: Indonesia.
   Any steady breeze from the sea.

Kai
Location: China.
   A light southerly wind.

Kal Baisahki, Kal-Baishakhi, Kala-Andhi
See Nor'Wester.

Kandahari
Location: Pakistan.
   A fierce hot, dry wind, usually from Kandahar, Afghanistan, that prevails in Baluchistan (see also Afghanetz).

Kapalilua
Location: Hawaii.
   A sea breeze.

Karaburan, Karadarimsky Storm
Location: Mongolia and east Turkestan, Kazakhstan and Kyrgyzstan.
   Easterly 'black storms' blowing in daytime between early spring and the end of summer. Speeds may reach gale force, carrying clouds of dust, darkening the sky and bringing unbearable conditions. The sand deposited can rapidly change the local course of rivers in the desert. As the anti-cyclone of the winter months over Asia breaks down due to increasing insolation and heating over the Tibetan plateau, this wind develops. The smooth undulating topography and differences in heating of this elevated region cause variations in wind speed.

Kaous
See Kaus.

Karif
See Kharif.

Kaus
Location: The Middle East and Arabian peninsula.
   The south-east wind occurring in the winter months mainly 1–3 days ahead of eastward

travelling depressions across the Arabian Gulf. The weather associated with this wind is typically gloomy, damp and squally, with thunderstorms that ultimately give way to rain and drizzle (see Chapter 7).

Khamaseen, Khamsin
Location: North Africa, mainly Egypt.
   Hot, dry southerly dusty winds, most prevalent between April and June ahead of eastward tracking depressions over the Mediterranean Sea. The name is also applied to south-easterly gales over the Red Sea (see Ghibli).

Kharif
Location: East Africa, mainly coastal Somalia and Sudan.
   Applied to often very hot, dry, dusty south to south-west winds during the summer months that can reach gale force, associated with the summer monsoon flow over the Indian Ocean and intense heating over the highlands of the nearby land mass. In Sudan the term denotes rain-bearing winds heralding the start of the rainy season during the summer monsoon, direction south to south-west. This wind has Bora characteristics over the Gulf of Aden (not to be confused with the Khareef, a local microclimate of the Oman).

Khemsin
See Khamsin.

Kibli
See Ghibli.

Kite Wind
See Junk Wind.

Klod
Location: Bali.
   A Föhn wind that may be associated with the monsoonal airflow of the summer months.

Klood, Kloof
Location: South Africa.
   A cool south-east to south-west wind of Simon's Bay, Capetown, associated with low

pressure, introducing air from higher latitudes across the south-west of South Africa. It is described as damp, with extensive low stratiform cloud cover, mist, drizzle and hill fog.

Koembang
See Kumbang.

Kohala, Kohilo
Location: Hawaii.
A gentle breeze.

Kolawaik
Location: Argentina.
A southerly wind affecting the Gran Chaco of northern Argentina.

Kona, Kona Storm
Location: Hawaii.
A south to south-west wind of January and February associated with rain and connected to the passage of depressions north of the islands, 'Kona cyclones', at a time when the Hawaiian anti-cyclone is at its southernmost position. This wind brings rain to normally sheltered locations in the lee of the prevailing north-east trades.

Krakatao, Krakotau, Krakatoa
Location: Asia.
A deep layer of easterly winds that top a 3.5 km deep westerly layer 1.2 km above the tropopause (associated with the QBO, included for reference).

Kü, Kü-Fun
Location: Taiwan.
Strong winds in small tropical cyclones over the Gulf of Taiwan, baby versions of the tai-fun (typhoon), bringing rough seas.

Kuban, Kubang
Location: Java.
A Föhn wind.

Kumbang
Location: Indonesia.
A dry southerly or south-easterly wind over

the central Pembarisan mountains with Föhn characteristics at Tjiriban and Tegel on the north coast of Java.

Laawan
Location: Middle East, Morocco.
A gentle west wind of Arabia, the 'helper' that aids farmers to winnow grain.

Laheimar
Location: Arabian Gulf.
Squalls of cold air from various wind directions, whose passage is followed by unusually good visibility over Iraq and the Arabian Gulf in October and November. Rough sea conditions can be expected over shallow seas.

Lakawa
Location: South America, Argentina.
A wind of the Gran Chaco, Argentina.

Lan San, Lansan
Location: New Hebrides.
A strong south-east trade wind, enhanced locally, but sometimes persisting for many days.

Laveche
See Scirocco.

Laxwaik
Location: Argentina.
A westerly wind of the Gran Chaco.

Leung
Location: China.
A cold northerly wind along the coast of China during winter. Associated with disturbances tracking around the extensive winter Asiatic high-pressure system.

Leste
Location: Madeira, Canary Islands, Morocco.
A hot, dry, dusty easterly wind of scirocco type in Madeira and North Africa, most common in winter and spring, associated with the passage of depressions over the eastern Atlantic at an unusually southerly latitude. The

temperature can rise to 50°C as the humidity falls to between 15% and 20%.

Levanto
Location: The Canary Islands.
An easterly pseudo-Föhn wind blowing from the Orotava Valley to Tenerife.

Leveche(s)
See Scirocco.

Loehis
See Nirta.

Loo, Loo Marma, Loo Sindhi, Loo Urdu, Lu
Location: India.
A hot, dry westerly Föhn wind between the months of March and May.

Maoi Feung
Location: China.
A north-east wind.

Marajós
Location: Brazil.
Strong squalls associated with a north-east wind during the rainy season in the first half of the year, indicated by 'popcorn' development on satellite imagery.

Mauka
Location: Hawaii.
A land breeze of Hawaii.

Mawsim
See Monsoon.

Merisi
See Simoom.

Mezzer-Ifoullousen
Location: Morocco.
A cold violent wind from the south over the east of Morocco that 'plucks the birds' feathers'. This winter wind cools over the snow-covered plateau of the Haute Atlas.

Mfi
See Arifi.

Midnight Wind
See Night Winds.

Minuano
Location: Brazil.
Cold, often strong west to south-west winds along the eastern coast of Brazil near Rio Grande do Sul, blowing from the interior between March and September (see Pampero).

Mistral
Location: France (included to describe this generic wind, associated with mountain barriers).
A north wind, most prominent in winter and spring, blowing with force down the Rhône Valley. A depression over the Gulf of Lyons or Gulf of Genoa and high pressure north of the Alps. A large pressure gradient develops and the Alps force a stable air stream through the narrow Rhône valley. Sometimes blowing for days, it is often at its most intense during the afternoon, but there can be periods of lighter winds. The effects of a strong mistral event can be felt far to the south, even as far south as the Algerian coast. Some mistrals are broad, forming a current as much as 700 km across.
Other names: Gending, Gully-squall, Junta, Junte, Yalca, Yalka.

Monçãn
See Monsoon.

Monsoon, Monsun
Location: The Indian sub-continent and Indian Ocean, South-East Asia, northern Australia, tropical Oceania and the tropical eastern Pacific Ocean, West Africa and the tropical eastern Atlantic Ocean.
A seasonally reversing continental-scale wind system accompanied by corresponding wet and dry seasons, associated with the asymmetric heating of land and sea. Strong to gale-force winds are common along the Somali, Yemeni and Omani coasts during the summer monsoon (see Chapter 8).

Murwa
Location: South Africa.
   A southerly wind of the Bavendas.

Muscat
Location: Oman.
   A very hot humid mid-summer wind affecting the littoral of Oman, described as enervating with high humidity, this is a pseudo-Föhn from the Hajr mountain range to the south of the Gulf of Oman. This wind often blows in strong gusts down the ravines that funnel it.

N'aashi, N'Aschi
See Nashi.

Naalehu
Location: Hawaii.
   A land breeze, especially over south-western Hawaii.

Nafhat
Location: Arabia.
   A light wind of the Arabian desert (also applied to a squall).

Nairutya Maarut
See Monsoon.

Namib
Location: Namibia.
   A very persistent wind, often accompanied by sandstorms blowing along the coast of the Namib desert as a result of the contrast in temperature between the heat of the land surface and the cool waters of the Benguela current. Local funnelling between the large sand dunes may increase the wind at times.

N'aashi, N'Aschi, Naschi, Nashi, Nashim
Location: Arabian Gulf.
   An east or north-east wind of bora type affecting the coast of Iran and the southern chain of mountains near the entrance to the Gulf. Quite gusty, with frequent lulls, but persisting for 3–5 days. The weather is dull, cloudy and wet, often with thick haze prior to arrival of this wind of winter months (see Chapter 7).

Nasim
Location: Saudi Arabia.
   Possibly a variant of the Naschi.

Nevada
See Nevado.

Nevadas de San Juan
Location: Bolivia.
   Snow-bearing winds of winter in the Bolivian Andes.

Nevado(s)
Location: Ecuador.
   A cold katabatic glacier wind descending from a mountain glacier or snowfield, especially from the higher valleys, often occurring after snowstorms.

Newhall Winds
See Santa Ana.

Night Winds
Location: Congo Basin.
   Dry, squally winds that occur at night in south-west Africa and the Congo. The term is loosely applied to other diurnal local winds.

Nirta
Location: Sumatra.
   One of many names for a local wind of northern Sumatra on Lake Toba.

Norder
See Norte.

Nortada(s)
Location: Philippines.
   Strong northerly winds occurring during the winter monsoon due to the interaction with a depression, bringing rough sea conditions.

Norte
Location: Argentina.
   A warm, damp northerly wind, often preceding a Pampero over Argentina.
   Location: Mexico
   A cold north–north-east wind over northern Mexico and the Gulf of Vera Cruz.

Nortes

Location: Mexico.

Occurs off the Pacific coast during January and February. A strong, very dry north-east wind with Föhn characteristics, blowing from the interior plateau. May sometimes be observed as far south as the Galapagos Islands.

North-Easter

Location: Australia (and New Zealand).

Stormy squally, sometimes rainy north-east winds occurring over eastern Australia (and along the east coast of the North Island of New Zealand) as depressions pass to the north.

Norther(s)

Location: Mexico California, Caribbean Sea.

Strong cold northerly winds of winter affecting mainly the Gulf of Mexico and Caribbean Sea with Föhn characteristics, especially along the Sacramento Valley, associated with the establishment of a pressure gradient between the continental interior and the warmth of the tropics. Local funnelling may also be a factor, especially along north–south orientated valleys.

Location: Australia.

A hot, dry northerly desert wind affecting the south coast of the continent, linked with the passage of depressions to the south.

Nor'Wester

Location: India and Bangladesh (also Pakistan and Afghanistan).

Strong thundery squally north-west winds, sometimes associated with dust storms, occurring over the Ganges Delta and Plains of the Ganges, prior to the onset of the summer monsoon and as the summer monsoon retreats (see section 10.5).

Ouari

Location: Somalia, Djibouti.

A Khamsin-type sandstorm in Somalia.

Pali

Location: Hawaii.

A local name for strong winds, enhanced through the Pali Pass above Honolulu.

Pampeiro, Pampero, Pampero Secco, Pampero Sucio

Location: Argentina

A west to south wind of polar origin associated with cold-frontal passage, the strong wind lasting several hours after the passage of the squall line of the (dirty) Pampero Sucio, occurring mainly in spring and summer.

Panas Oetara

Location: Indonesia.

A strong warm, dry north wind during February over Indonesia with Föhn characteristics, occurring on the south side of the island chain to the lee of its high mountains, through the centre of the archipelago.

Papagayo(s)

Location: Central America.

A strong, often stormy, unstable northerly wind of the Pacific Coast of Mexico, Nicaragua, Guatemala, San Salvador and Costa Rica, occurring in winter. Sea level is raised when water is piled up in bays facing the wind flow. Temperatures are often lowered by 5–8°C for about 2 days in low-latitude locations. On descent from the high interior, it produces fine and clear weather on the coast. Known as the Norte or Norther over the Gulf of Mexico (see also Tehuantepecers).

Paramito

Location: Colombia.

Cold easterly wind affecting Bogota, accompanied by extensive fog and drizzle between July and September.

Passat

Location: Oceanic.

A general term for steady winds blowing from the sub-tropical high-pressure areas towards the ITCZ. In the northern hemisphere they are north-easterly and in the southern hemisphere they are south-westerly (see also Bris, Brisa).

Perth Doctor
See Doctor.

Perth Eucla
See Freemantle Doctor.

Puff of Wind
See Cat's Paw.

Puna
Location: Peru and Bolivia.
A cold dry wind, possibly of katabatic origin, usually associated with the town of Puno on the shores of Lake Titicaca.

Qaws, Quas, Quaus
See Kaus.

Quexalcoatl
Location: Mexico.
A westerly wind (Aztec), likely to be dry, with Föhn-type characteristics descending onto the eastern lowlands.

Quibla
See Ghibli.

Raki
Location: Melanesia.
A persistent westerly wind typical of the island of Tikopia Melanesian Islands.

Reboyo(s)
Location: Brazil.
Cool south-west gales occurring over the east coast of Brazil, lasting 3–4 days, mainly during the rainy period of December–June. They reach their strongest after the initial rain has cleared.

Reshabar
Location: Iraq and Iran.
A dry, very hot dusty 'black' north-east wind over Iraq and Iran in summer. During the winter months this wind can bring snow to the upper mountainous regions.

Retôrno dos Aliseos
Location: Brazil.

Winds from the south-east trade-wind belt of the Atlantic that have backed north-easterly along the coast, occurring particularly between October and March.

Retour
See Vent de Retour.

Ribut
Location: Malaysia.
Short, sharp squalls preceding the onset of the summer monsoon over the South China Sea. This may be similar to the Sumatras of the Strait of Malacca, an interaction of the night-time land breeze, which opposes the prevailing wind.

Riseè
See Cat's Paw.

Rrashaba
See Reshabar.

Rrashabe, Rushabar
See Reshabar.

St Giles Wind
Location: Indian Ocean.
A south-east trade wind eddy that has veered westerly to the lee of the island of La Réunion, causing hazardous sailing conditions.

Safid Rud
Location: Iran and Afghanistan.
A ravine wind that occurs on most days during the summer, reaching gale force at times through gorges and low-lying basins.

Sahel, Samiel, Samum
See Scirocco.

Sam, Samun
See Simoom.

Sansar
See Shamsir.

Santa Ana
Location: USA.

A hot, dry, generally easterly wind with Föhn characteristics funnelling through the Santa Ana valley of southern California, occurring most frequently in winter. The very low humidity (<5%), can dry skin. In spring, blooms and young fruit drop off their trees after a short time. Windows sometimes have small round holes as a result of the coarse gravel carried by this wind.

Santana, Santanta
See Santa Ana.

Saoet
See Nirta.

Sarsar
See Shamsir.

Sarat
See Chergui.

Scheheli
See Scirocco.

Schergui
See Chergui.

Schobe
Location: The Levant.

An east or south-east hot, very dry pseudo-Föhn wind blowing quite strongly over the coast of the Levant in spring and autumn, associated with depressions over the eastern Mediterranean. Also known as 'fire wind' and 'the breath of death'.

Scirocco
Location: Mediterranean Sea and its coastal nations.

A hot, dry dusty, generally south-easterly wind ahead of depressions moving east over the Mediterranean Sea, most prominent in spring. The depressions draw air ahead of it from the deserts to the south. It is very hot and dry on the north African coast and can cause damage to crops. It picks up moisture over the Mediterranean Sea and becomes moist, depositing copious dew and occasional heavy rain, sometimes containing dust from the Sahara. When strong, the Scirocco can cause a rise of several feet in sea level on northern shores of the Mediterranean. Many wind names are grouped under the term Scirocco: Ghibli (Libya/Malta), Khamsin (Egypt), Chichili, Chihili, Chom (Algeria), Simoom, Xaroco, Jaloque, Sahel (Morocco, Algeria), Samiel, Scheheli (Algeria), Xaloc, Xlokk (Malta).

Seca
Location: Brazil.

A dry wind (or a drought).

Seistan
See Bad-i-sad-obistroz.

Seletan
See Broeboe.

Shaluk
Location: Africa and the Middle East.

A hot, dry desert wind (Simoom used in the northern Sahara).

Shamal, Schamali
Location: Middle East, Arabian Gulf.

A hot, dry and dusty north-west wind of some persistence, prominent during the summer months in Mesopotamia (see Chapter 7). Many local names exist, but can collectively be put under the generic term Shamal: Chemal, Schemal, Shemaal, Shimal, Shumal, Szemal.

Shamsir
Location: Iran and the Near East.

A cold north or north-west winter wind.

Sharav
Location: Israel.

A hot, dry and dusty east, south-east or south wind, literally 'heat of the land' (a form of Scirocco), having pseudo-Föhn characteristics. In Tel-Aviv, the Sharav occurs on 20 days per year and in Jerusalem 113 days between the months of September–December and May–July.

Sharki
See Kaus.

Sharkiye
Location: Libya.
A fresh wind of the Libyan desert.

Sharqieh, Sharki
Location: Israel.
An east or north-east wind of Israel: hot in summer, cold in winter.

Sharqui
See Kaus.

Shawondasee, Shawosee
Location: USA.
A rather hot and humid southerly wind of late summer. Algonquin term for 'lazy wind'.

Sheleli
See Chichili.

Shemaal, Schemal
See Shamal.

Sherki
See Kaus.

Shi Lung
Location: China.
A north-east breeze.

Shimal, Shumal
See Shamal.

Shurkiya, Shuquee, Shurqee
See Kaus.

Si Girring Girring
See Nirta.

Simoom
Location: North Africa and the Middle East.
A hot, dusty, dry suffocating wind or whirlwind over the deserts of North Africa and the Middle East. Most frequent in summer and relatively short-lived (20 minutes). Term also applied to the Scirocco of the eastern Mediterranean region. Derived from ssim and samma, Arabic for poison. Temperatures on the North African coast can reach 50°C or more and humidity can fall well below 10%.

Siroeagang
See Nirta.

Slatan
See Broeboe.

Sondo
See Zonda.

Sonora (Storm)
Location: USA, Mexico.
A warm southerly wind, originating in Mexico and slowing across the south-western USA, especially southern California and Arizona, as well as the north-western Mexican states of Sonora and Baja California.

South Easter
Location: South Africa.
South-east pseudo-Föhn wind across the south-west of the country, associated with high pressure to the south-east during summer. These winds can often blow at gale-force strength for many days. The sky often clears, bringing bright sunshine and low humidity on west coasts. The wind is deflected by topography to become southerly over Cape Town, helping to form the banner cloud of Table Mountain ('tablecloth').

Southerly Buster
See Burster.

Su-estado
See Suestado.

Suahili
See Suahili.

Sudestades, Sudestado(s), Suestada
See Suestado.

Suestado(s)
Location: Uruguay, Argentina, Brazil.

A strong rainy gale, mainly from the south-east during the winter months, associated with cold-frontal disturbances moving north, corresponding to the Pampero. A Suestado-prolongada is often very wet, prolonged by interaction with a Norte, which is also present.

Suhaili
Location: Iraq, Qatar.

A strong south-west winter wind of some hours' duration with thick cloud, rain and sometimes fog, following a period of south-east winds (the Kaus) ahead of a depression or trough moving east over the Arabian Gulf region. It can attain gale force and cause problems for coastal sites of Iran (see Chapter 7).

Sumatra(s)
Location: Indonesia.

Strong south-west winds occurring from May to September over the east of Sumatra during the height of the northern summer monsoon. It is often accompanied by rain and thunder.

Sundowner
Location: USA.

A hot down-slope wind that occurs periodically along a short segment of the southern California coast around Santa Barbara. Typical onset is in the late afternoon or early evening, possibly as the sea breeze wanes, but can occur at other times. In extreme cases it reaches gale force or stronger and temperatures along the coast may rise to 38°C. This implies Föhn-type characteristics (see Santa Ana).

Suraçon, Surazo(s)
Location: Peru.

Cold winds off the high ranges of the Andes Altiplana, affecting especially the high passes of Peru and lowlands east of the Andes. Clear skies and often violent winds bring temperatures well below 0°C. This wind is especially pronounced in daytime. It may well be katabatic in origin, the occurrence during daytime possibly due to differential heating in the clear thin atmosphere of the high mountains. The violence of the wind may cause the closure of mountain passes.

Location: Bolivia, Brazil, Argentina.

Over the lowlands of Bolivia, northern Argentina and south-eastern Brazil, this is associated with stormy intrusions of cold air from the south during the winter rains. Outbreaks of cold air reach low latitudes due partly to the semi-permanent trough east of the Andes and frontal disturbances edging slowly north across the continent. Marked temperature falls occur (sometimes in excess of 20°C) and may affect sensitive tropical crops, such as coffee.

Swahili
See Suahili.

Syzyzy
Location: Papua New Guinea.

Westerly winds which usually precede the north-west monsoon of the summer months over the Arafura Sea and Torres Strait. There is an influence from continental heating.

Sz
Location: China.

The first breath of autumnal wind.

Szaraw
See Sharav.

Szarki
See Kaus.

Szemal
See Shamal.

Tamboen
See Nirta.

Tanga Mbili

Location: Tanzania (Zanzibar).

Changeable winds between seasonal (monsoonal) rains, denoting the transitional times of the year. Its meaning is 'the two sails', indicating one sail is insufficient. These winds are common in September.

Tebbad

Location: Iran.

A hot, dry and dusty wind.

Tehuantepec, Tehuantepecer(s), Tehuantepecero

Location: Mexico, Gulf of Tehuantepec.

A strong and relatively constant north or north-east wind of November–March that can lower temperatures by 10°C and reach gale force at times. Heavy rain falls on the north of the Isthmus of Tehuantepec, but the wind is dry when it reaches the south of the Gulf of Tehuantepec due to adiabatic warming. It is caused by rising pressure over the cold interior plateau.

Temporal, Temporale(s)

Location: Central and South America.

A wind bringing rainfall from the south-west to west onto the Pacific coasts of Central and north-western South America. (The term is also applied to a north to north-west wind in central Chile during the winter rainy period.) The wind is a deflection of the south-east trades of the eastern Pacific during the northern hemisphere summer when the ITCZ is at its most northern latitude over the South America. (Over Chile this wind is associated with eastward-travelling depressions of temperate latitudes.) These moisture-laden winds produce copious rainfall over windward slopes of the mountains of the Pacific coast. (In Chile, this wind associated with heavy rain on the coast and heavy snowfall over higher ground.)

Tenggara

Location: Celebes.

Föhn-type easterly winds along the Spiemonde archipelago to the lee of the southern peninsula of the islands. The drying and warming of these initially moisture-laden winds by passage over the high ground of the Celebes may bring some relief from the oppressive humidity of the equatorial zone.

Terral

Location: Chile, Peru (also Spain).

A land breeze on the coasts South America, opposing the Virazon sea breeze. In Chile, the term is more generally applied to night-time land and mountain breezes.

Terre Altos

Location: Brazil.

Down-slope gusty winds from the north-west off the mountains surrounding Rio de Janeiro.

Tezcatlipoca

See Norte.

Thar Wind

Location: India.

A hot, dry wind of Rajasthan, most likely associated with the summer monsoon (see Bad-i-Sad-o-bistroz).

The Witch

See Santa Ana.

Tiempo del Monte

Location: Canary Islands.

North-east trade winds heated and dried by Föhn or pseudo-Föhn effects on the lee side of Tenerife.

Tokalau

Location: Fiji.

A wind from the north-east.

Tokerau

Location: Melanesia, Tikopia.

A northerly wind.

Tonga

Location: Melanesia.

A south-east wind (originating in the direction of Tonga).

Tongara, Tongara Putih
Location: Indonesia.

A south-easterly wind meaning 'white south-easterly wind', occurring mainly at the end of the dry period (July and August) in the Gaspar and Karimata Strait between Sumatra and Borneo. The atmosphere at this time is greatly enriched by the presence of hygroscopic dust particles from combustion of tinder-dry rainforest. These particles are so dense that they form a grey mist, which also occurs over the Tanimber Islands in the Arafura Sea, near Timor.

Traversia
Location: Chile.

The name comes from a South American nautical term (derived from the Travesier of the Mediterranean) for a west wind from the sea, blowing directly into port, so causing rough seas and a hazard to shipping.

Tuaura
Location: Melanesia.

A southerly wind of Tikopia and the Melanesian island system.

Tung-Shang
Location: China.
    A north-east trade wind.

Turbonadas
See Pampero.

Vagio, Vaguio, Vario
See Baguio.

Vash-ki-ry
Location: Surinam.
    A humid northerly breeze.

Vent de Retour
Location: Gran Canaria.

A light wind in an opposite direction to the prevailing flow on the leeward side of the island. In a strong north-easterly trade wind, an eddy forms in the lee of the island, leading to a light south-westerly breeze at low level, the reverse of the predominant flow.

Vento di Cima
Location: Amazon basin.

A rather cool, moist westerly wind in the upper Amazon basin, often associated with rain. The westerly direction suggests a katabatic wind from the Andes interacting with the warm moist environment within the Amazon basin, producing pseudo-frontal characteristics. The name is derived from the Portuguese word for top or upper, indicating the mountain origin of the wind.

Ventos Gerais
Location: Brazil.

The generally weak variable easterly winds over the southern interior during the summer.

Viento Zonda
See Zonda.

Viraçao
Location: Congo and Angola.

Occurs over the mouth of the Congo and refers to the sea breeze. Corresponds to the Spanish Virazon 'turning wind'.

Virazon
Location: Pacific and Atlantic coasts of South America.
    Name for a sea breeze.

Vriajems
See Friagem.

Waazay
Location: Nigeria.
    A light and variable wind.

Waimea
Location: Hawaii.
    A sea breeze that clears morning mist.

Wam-Andai
Location: Papua New Guinea.

Strong gusty westerly winds of summer, associated with the moist summer monsoon. The gustiness may be due to pressure surges or local funnelling.

Wam Braw, Wambra, Wambru, Warm braw
Location: Papua New Guinea.

A warm. dry wind that persists for 4–8 days during the winter monsoon season over Biak, Schouten Is, off the north coast. This wind is part of the circulation that blows across the equator to become part of the summer monsoon system of South-East Asia. Pseudo-Föhn characteristics originate from the drying out of these winds as they cross the high ground of the centre of the island. These winds may offer some relief from the predominantly humid conditions and are part of the long dry season in that region.

Waryaraik
Location: Argentina.

A northerly wind of the Gran Chaco in Argentina.

Xaloc, Xaroco, Xlokk
See Scirocco.

Yalca, Yalka
Location: Peru.

A strong and gusty mountain wind occurring in the passes of the Andes in northern Peru, mainly during the winter months. May be a katabatic wind.

Yellow Winds
Location: China (and elsewhere in East Asia).

A loess-laden wind blowing from the west or north-west from the Gobi Desert and loess-steppes of eastern and northern China during the winter months. The Eurasian semi-permanent high-pressure system of winter is associated with local pressure surges as weak weather systems move east across the continent.

Zabaa, Zoboa
Location: Egypt.

A southerly wind; also the term for a dust devil. May be an alternative name for the Khamsin/Scirocco.

Zonda
Location: Argentina.

A hot, dry dusty westerly wind of western Argentina, most frequent in spring. It has pseudo-Föhn characteristics on descent from the Andes, associated with the passage of depressions across southern South America, and it often precedes the Pampero Sucio. These infrequent winds are derived from the upper westerly airflow and descent from altitude brings a significant temperature rise of up to 30°C, with relative humidity lowering to 5–10%.

Appendix 3
An Introduction to Cloud Types, Cloud Species and Precipitation

A3.1 Introduction

As in the classification of animals and plants, genus (type), species and variety, developed from a scheme devised by Howard (2011), define cloud forms. Not all cloud genera have species or varieties, but all meteorologists know the form of each cloud type, as prescribed by the international body for meteorology and hydrology, the WMO (1956).

Clouds are formed from water, but this may be found in three states: droplets above freezing point (0°C), water drops below freezing point (supercooled) or ice crystals. Between 0 and –20°C supercooled water is predominant in cloud. This is mainly because cloud droplets are very small, having a typical radius of about 0.1 μm, and occupy only a small proportion of a volume of air (<3%). In order to freeze, droplets must form a lattice around an initial ice nucleus. However, there are too few water molecules in a cloud drop and they are too active for this homogeneous nucleation to occur at temperatures above about –40°C. It is only when a nucleating substance that replicates an ice lattice is present (such as dust, salt or products of combustion) that ice forms in cloud when the temperature is higher, the process known as heterogeneous nucleation. Only very small amounts of ice are present at temperatures above –10°C, but there is a significant amount as the temperature falls below –20°C.

According to their typical composition, clouds are placed in three levels (étages): high (mainly ice clouds), medium (water – supercooled if above about 4500 m in the tropics) and low (water above 0°C in the tropics).

A3.1.1 Precipitation processes

Supercooled water is very important in the production of rainfall as ice crystals growing in the cloud are in a medium greatly supersaturated with respect to ice, even though the saturation with respect to water is low. Thus the ice crystals can grow rapidly, soon becoming heavy enough to fall and collect cloud drops to become a snow crystal, the process accelerating until as the snow melts (still collecting cloud drops as it falls) it forms a raindrop with a typical radius of 2 mm. This method of production of rainfall (or snowfall, if the temperature is not sufficiently high to melt the falling snow) is known as the Wegener–Bergeron–Findeisen process (McIlveen 1992). Precipitation from the tallest of cumulus (*Congestus* sp., described below) and cumulonimbus clouds is dominated by raindrops (or snow above about 4000 m in the tropics) formed in this way.

An Introduction to the Meteorology and Climate of the Tropics, First Edition. J F P Galvin.

However, in the tropics rain also falls from 'warm' clouds with temperatures between about 0 and $-10°C$ by a process of drop coalescence, slightly larger drops ascending less rapidly than small ones, ultimately growing large enough to fall and collect further drops to become raindrop size (McIlveen 1992). It follows that cloud must attain a depth of at least a few kilometres for rain or snow to form. Coalescence may also produce slight rain from layer clouds that have spread out from deep convective clouds.

Hail is an occasional product of cumulonimbus clouds. Its production uses both the Wegener–Bergeron–Findeisen and agglomeration processes. It is dependent on relatively large ice crystals ascending into cold cloud where collision with water drops (or snow or raindrops) builds up a large mass of ice, freezing of supercooled drops occurring due to the large size of the developing hail. It is only produced in cumulonimbus clouds. Large hail (defined as having a diameter > 5 mm) may have several layers of ice, reflecting several cycles of ascent and descent through cloud. Thus it can only form if these cycles of motion are possible. This requires first that there is a small amount of wind shear (change of direction) with height, so that one part of the cloud is ascending alongside a part of the cloud that is descending. Secondly, there needs to be a large amount of convectively available potential energy in the cloud, so that there are up-draughts strong enough to carry hail upwards through the cloud and to cause it to ascend following each cycle of convection (Browning 1963; Grant 1995). For this reason, it is only seen near the margins of the humid-tropical zone and is rare over the sea, where CAPE is comparatively modest.

A3.2   The high clouds

The high clouds (cirrus, cirrocumulus and cirrostratus) are found at levels above 8000 m in the tropics and are predominantly composed of ice crystals.

- Cirrus (Ci): Wispy streaks or clumps of ice, often seen in association with convection (Figs A3.1 & A3.2) or the STJ (Fig. A3.3). Halo phenomena may be seen in these clouds (Fig. A3.4).
- Cirrocumulus (Cc): Patches of cloud through which the sun can shine weakly (Fig. A3.5), often with the appearance of clumps of cotton (a 'mackerel' sky), composed partly of supercooled water, so not generally seen above 11,000 m in the tropics.
- Cirrostratus (Cs): A relatively thick layer of ice cloud through which the sun can usually be seen, the cloud often exhibiting optical phenomena (haloes: Dunlop 2004; Schipper & Mühr 2010). This cloud is most commonly observed in the ITCZ (Fig. A3.6).

In this book these cloud forms are generally clustered together as cirriform clouds. Precipitation in the form of snow may fall from these clouds if they are sufficiently thick, but will not reach any but the highest of mountains.

A3.3   The medium-level clouds

The medium-level clouds (altocumulus, altostratus and nimbostratus) are found at levels between 2000 and 8000 m in the tropics. Although composed mainly of water, the water is supercooled above the freezing level.

- Altocumulus (Ac, a common layer-cloud type of the tropics): Layers (Fig. A3.7) or clumps, sometimes of substantial vertical development (Fig. A3.8) with a definable shape, often occupying much of the sky over an observer at the surface.
- Altostratus (As, a comparatively rare cloud in the tropics): A poorly defined greyish layer cloud, usually occupying most of the sky observed from the surface, through which the sun may appears as if through

Figure A3.1   Dense cirrus (sp. *spissatus*) over the central Philippines. A characteristic cloud of the ITCZ, cirrus is possibly the most commonly observed cloud in the humid tropics, persisting after other clouds that have formed from convection have dispersed.

ground glass, but often thick enough to obscure the sun (Fig. A3.9).

- Nimbostratus (Ns): A dense, thick poorly defined layer cloud almost always covering the whole sky, as observed from the surface and from which rain (or snow) falls. The base and top of nimbostratus may extend into the high- and low-cloud étages. This is a rare cloud type in the tropics, but does occasionally form in tropical storms or near the poleward edges of the tropics.

Rain (snow over high ground) often falls from thicker layers of tropical altocumulus or altostratus, but most of this rain evaporates before it reaches sea level.

A3.4   The low clouds

These clouds can be seen in most parts of the tropics where there is sufficient moisture.

- Cumulus (Cu) clouds develop over a warmed surface and are usually dense enough to have a dark base, which is flat over level ground. Initially (or in a dry atmosphere or where there is a stable layer preventing deeper convection) cumulus clouds are shallow (Fig. A3.10). However, the tops of cumulus clouds commonly reach into the medium-cloud étage in the tropics, typically developing to their maximum depth during the afternoon over land

Figure A3.2   In this case, an extensive area of cirrus spissatus can be seen, evidently the top of a cumulonimbus cloud that has dispersed at low and medium levels.

Figure A3.3   Cirrus spissatus over Lanzarote. Cirrus is often the only cloud associated with the STJ and there is a semi-permanent upper trough in the STJ over the Canary Islands.

Figure A3.4   Vivid circumhorizontal arc in dense cirrus cloud over Orlando, Florida, USA, an example of the halo phenomena visible in cirrus or cirrostratus clouds. This form of halo is less often produced by ice clouds than other haloes, such as parhelia, requiring an unusual orientation of ice crystals (Dunlop 2004; Schipper & Mühr 2010). © G.R. Galvin.

(Fig. A3.11). Rain showers (their intensity dependent on the height of the cloud base and cloud depth) may fall from these clouds, but precipitation is short-lived, as are individual cumulus clouds, which have a lifetime rarely as much as an hour.

- As cumulus develops, it may reach the high-cloud étage, when it is re-classified as cumulonimbus (Cb). At this stage, ice is evident at the top of the cloud and further development may form a large icy top, spreading horizontally from the cloud tower (Fig. A3.12). Convection at medium levels may also form cumulonimbus clouds. Heavy rain (or snow), hail and thunder are all commonly seen in association with cumulonimbus clouds, although these are not a necessary accompaniment. Because of their size (tens to hundreds of kilometres across), cumulonimbus clouds have lifetimes greater than an hour and are most commonly seen over land in the tropics during the evening.

- Stratocumulus (Sc) is in the form of layers or elements that develop either due to the horizontal spreading of a convective tower or more commonly as a result of turbulent overturning of moist air at the surface (Fig. A3.13). The cloud is usually thick enough to obscure the sun and, if extensive, can make the weather appear dull. In the tropics, precipitation (in the form of drizzle) is only rarely formed by these clouds, most often where warm moist air drifts over cold water masses and the freezing level is low (Overton & Galvin 2005). Indeed, it is much more likely to be observed on mountain-tops (given that drizzle cannot fall through

Figure A3.5 Cirrocumulus over the Visayan Sea, Philippines. Perhaps the least common of clouds, these high-cloud layers are composed of a high proportion of supercooled water. It is unlikely that this cloud can exist above about 11,000 m in the tropics, the level at which the temperature is usually below the critical level of −40°C at which water spontaneously freezes.

more than about 300 m of unsaturated air (McIlveen 1992, Appendix 6.4).

- Finally, there is stratus (St), a thin grey cloud found only very near the surface. This is most often seen in the tropics in the early morning over tropical rainforest, where the temperature falls to its minimum at the end of the night (Fig. A3.14). Stratus also forms over cold seas under the influence of warm moist air. However, it is too thin ever to form precipitation (Overton & Galvin 2005), although it may prevent the evaporation of slight precipitation falling through it.

A3.5   Cloud species and varieties

There are many cloud species and many apply only to particular genera. However, the only defining species of cloud that are discussed in this book are castellanus, floccus and lenticularis (in these cases species of altocumulus, stratocumulus or cirrocumulus). Castellanus or floccus clouds (Fig. A3.15) form as a result of moist convection as cool air moves over warmer moist air by the process of advection. Castellanus clouds form towers of significant vertical extent, often with a moist layer at their base. Floccus cloud has the appearance of cotton-wool balls. Only in the case of altocumulus clouds forming over mountain ranges (or deserts) is this convection a result of sensible heating of the surface, producing a medium-level form of cumulus clouds, which are of the species castellanus. Lenticular clouds (Fig. A3.16) form in moist, but stable, air over mountains. As suggested by the name, they look like lenses, forming at the crest of one or more waves over and down-wind of the mountains forming them.

Varieties and supplementary features may also be used to aid the description of clouds, some applying only to particular cloud genera or species (Met Office 2006).

Figure A3.6 Cirrostratus at sunset over the Bohol Sea, Philippines. The ITCZ is often marked by extensive areas of this deep, rather fibrous high cloud, which may extend from about 12,000 m to 15,000 m or more. Although associated with ascent and deep convection, this cloud is much more persistent. Broken altocumulus can also be seen.

Figure A3.7 Layers of altocumulus are commonly seen in the middle-troposphere of the tropics. These clouds frequently exhibit some element of instability and the 'lamb's-wool' appearance of parts of this layer define it as sp. *floccus*. Less developed parts are sp. *stratiformis*. Occasionally, precipitation may fall from thick layers of altocumulus.

Figure A3.8   When medium-level cloud becomes unstable and rises rapidly to form the turrets evident in this picture, it is known as altocumulus castellanus. In this case, the altocumulus developed into cumulonimbus clouds over northern Mindanao, Philippines with tops near 16,000 m within about 15 minutes on 6 August 2008 as a short-wave upper trough crossed the island chain. Precipitation commonly falls from this cloud type, but if the base is above 3000 m it will not reach the ground.

Figure A3.9   Altostratus, the somewhat amorphous layer above a layer of altocumulus, is comparatively rare in the tropics as it is characteristic of a moist and stable atmosphere. It is most often observed over parts of South-East Asia in winter (Chapter 8) or within upper-tropospheric troughs (Chapters 3 and 4), although here it is seen over the Bohol Sea, Philippines within the ITCZ. Slight to moderate rain may fall from this cloud if its base is below 3000 m.

Figure A3.10 Cumulus clouds (sp. *humilis*) developing in the middle of the day over land in the southern Philippines. This cloud is entirely at temperatures above freezing point and rain does not fall from this species of cloud.

Figure A3.11 As cumulus clouds grow, they extend into the middle troposphere and have tops composed of supercooled water. Slight to moderate showers can fall from these large cumulus clouds (sp. *congestus*), which may grow into the upper troposphere, if the atmosphere is sufficiently unstable, to become cumulonimbus clouds. Cumulus humilis can be seen near the right-hand edge of this picture and patches of cirrus spissatus at the top and centre-right betray distant convection into the upper troposphere over the Philippines.

Figure A3.12 Cumulonimbus, in this case seen behind a large cumulus cloud. As the temperature in cloud drops below −20°C it is composed increasingly of ice, so that convective clouds that reach 8500 m or more in the humid tropics are of this genus. In most cases, tropical cumulonimbus clouds reach at least 12,500 m and in this case the cloud top was probably near 15,000 m, its species capillatus ('hairy'). Unless the base of a cumulonimbus cloud is higher than about 2500 m above ground level heavy precipitation can be expected from this cloud. This is the only type of cloud that produces hail, thunder and lightning. The rapidly rising currents within this cloud often produces severe turbulence and its density brings a high risk of severe icing.

Figure A3.13 Stratocumulus stratiformis over Lanzarote, Canary Islands. Although also seen as a bi-product of convection, this cloud is mainly found in windy weather over and near the cool-ocean areas on the western margins of continents under anti-cyclonic temperature inversions. It is a dense cloud and, if extensive, brings gloomy conditions. Wind shear near the cloud top frequently causes turbulence. However, icing is not associated with this cloud type in the tropics and as it is at temperatures entirely above freezing no precipitation falls from it.

Figure A3.14   Extensive stratus on the tree-tops of Sabah, Borneo. Over lower-lying tropical rainforest night-time cooling frequently causes this lowest of clouds to form, rather than fog, as radiation cools the air to a minimum at the top of the forest canopy. The presence of dense patches of high cloud does not appear to affect the radiation balance sufficiently to prevent stratus formation. Stratus may also form in precipitation or over higher ground as a result of ascent. © Faye Davies.

Figure A3.15   Altocumulus floccus over Melbourne, Australia. In this image, the cloud can be clearly seen to resemble cotton wool.

Figure A3.16 High altocumulus in the form of lenses (sp. *lenticularis*) in two layers (near 4500 m and 6500 m) above cumulus humilis and altocumulus castellanus, eastern Mindanao, Philippines. Cirrostratus can also be seen in the distant sky.

Appendix 4
An Introduction to Meteorological Diagrams, Stability, Instability and Aviation Weather Charts

A4.1 Temperature–pressure graphs

The tephigram, skew-T/log-p diagram (rotated emagram) and Stüve diagram are commonly used in meteorology to graph temperature T against pressure p (and, hence, altitude, which is inversely proportional to the logarithm of pressure). Examples of the first two appear in the text. In all cases, the temperature axis (the x-axis) is skewed 45° to allow for the rapid decrease in temperature with altitude (up to the tropopause). Further series of lines show equal potential for dry and saturated air.

In the tephigram, lines of equal pressure are slightly bowed upwards so that both lines of equal temperature and lines of equal potential temperature are straight. For this reason, equal areas contained by the curves have equal energies, leading to better comparisons of entropy (φ) within systems. This is the derivation of the name of this diagram, which compares T and φ.

In the emagram, lines of equal pressure and equal temperature are straight. In the tephigram, there is an angle of 90° between lines of equal temperature and of equal potential temperature for dry air. In the emagram, lines of equal potential temperature are slightly curved.

However, for most purposes, these diagrams are very similar and in each case it is possible to manipulate the plotted data to show what will happen when air ascends or descends, becomes saturated or unsaturated. Saturated air will move along the curved lines of equal wet-bulb potential temperature (θ_w) as it ascends, cooling at a variable rate (Γ_s). However, as it descends and warms to a temperature above that of the plotted curve of temperature, it becomes unsaturated and warms parallel to the lines of equal potential temperature (θ). Similarly, unsaturated air cools at the rate ($\Gamma = -0.098$ K m^{-1}). In free air, the lapse of temperature cannot exceed Γ (dry air, the dry adiabatic lapse rate (DALR)) or Γ_s (saturated air, the saturated adiabatic lapse rate (SALR)) as a greater fall of temperature with height in the environment results in 'instability' (see below). Descent generally causes stability, as a descending mass of air warms as it is compressed.

If detail allows, the graph may include lines showing the lapse of humidity mixing ratio (r), so that if we know the dew point of the air (usually plotted as dashed lines alongside those of temperature), we know the pressure level at which ascending air will become saturated. This is the point at which air cooling at a rate Γ reaches its saturation temperature,

An Introduction to the Meteorology and Climate of the Tropics, First Edition. J F P Galvin.
© 2016 John Wiley & Sons, Ltd. Published 2016 by John Wiley & Sons, Ltd.

intersecting the plot of r from the level at which ascent started. Above this level, air rises at rate Γ_s. An unplotted annotated tephigram is shown in Fig. A4.1.

The relationship of observed to dry or saturated potential can be used to assess stability.

In order to obtain height from these graphs, we can use the following equation, based on the universal gas law:

$$h = 67T\left(\log p_0 - \log p_1\right)$$

where h is height (m), T is the mean temperature (K) of the layer, p_0 is the pressure (Pa) at

the bottom of the layer and p_1 is the pressure (Pa) at the top of the layer. The effect of temperature on relative pressure can be seen easily in Fig. A4.1: pressure levels are considerably higher on the right-hand side of the diagram, where the temperature is higher, than on the cooler left-hand side.

Usually, wind fleches are added for significant levels, when these are available, allowing the assessment of cloud (or dry-air) motion within the area around the observed (or forecast) profile.

Temperature–pressure graphs can be produced from data observed instrumentally from

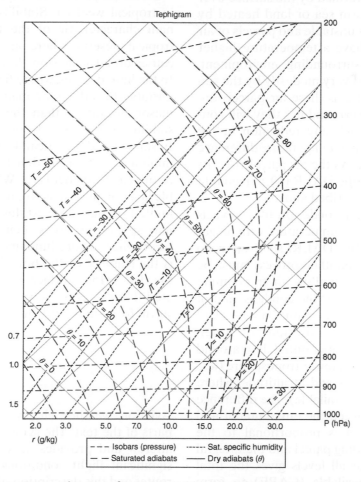

Figure A4.1   A tephigram. Intersecting axes of temperature (T), potential temperature (θ), wet-bulb potential temperature (θ_w), pressure (p) and humidity mixing ratio (r) are shown. The physical relationship between these curves allows the effects of mixing, convection and radiation to be assessed, as well as estimates of cloud development and formation. © RMetS.

ascending balloons (radiosondes), from falling instruments attached to parachutes (dropsondes), from data produced by numerical models of the atmosphere (a very useful adjunct of numerical forecasting) or from remotely sensed data obtained from satellite radiometry.

A4.2   Stability and instability

Using temperature–pressure diagrams, it is possible to assess the stability of an air mass. In the tropics, much of the rainfall is a result of instability. If air is warmed by the surface over which it passes (warm sea or land heated by the sun), it becomes unstable and parcels of air rise because they have a temperature higher than that of the surrounding environment, which is only heated very modestly by the sun (section 1.4). As long as these air parcels are warmer than their surroundings they will continue to rise, although any reduction in the air-parcel–environment temperature difference tends to halt ascent. As the air parcel rises, its temperature falls at the DALR (Γ), a fixed rate of -9.8 K km^{-1}, as described above.

Inspection of radiosonde profiles reveals that the mean lapse of temperature with height is much less than the DALR, so if air is to continue ascending another process must be involved. The energy to allow air to continue ascending by instability is the condensation of water vapour. This provides about 2.5×10^6 J kg^{-1} (the latent heat of vaporization) of water, reducing the rate of temperature fall to around 3 K km^{-1} in the tropical lower troposphere (less in the higher latitudes and at altitude, where there is less moisture available). The difference in temperature between the environment and the rising air parcel is proportional to the energy of the ascending parcel and the integral of this difference at all levels gives the total potential energy available (CAPE) to form convective cloud. The slow fall of temperature of saturated air as it ascends can give convective clouds in the tropics considerable energy

and vigorous convection may continue to rise beyond the point that the parcel and environment temperatures are equal for several thousand metres. In practice, moist instability can be assumed where θ_w falls with altitude.

As described in Chapter 3, ascent is most vigorous at the margins of the humid tropical zone where there is cool air in the middle troposphere, as well as in upper-tropospheric troughs (Chapters 4 and 10). The near-continuous ascent of air parcels in the ITCZ tends to warm the environment (so-called sensible heating by entrainment), so that CAPE values are relatively low in this zone.

Stable air also has an important part to play in tropical weather. Stability is the predominant characteristic of the troposphere over tropical deserts, where anti-cyclonic motion causes broad-scale slow descent and warming. In the humid zones, where the air is potentially unstable, the effects are mainly in the (<50 km) mesoscale, rather than the synoptic scale of stability in the extra-tropical frontal systems. In stable air, the vertical displacement of air is inhibited, so that air that has been displaced returns to its former level. Where there is a stable layer of air, convective ascent is slowed and limited, or may cease. At this level, layer clouds may spread out, as is seen at the top of the troposphere. More fundamentally, however, it is the stability of the boundary layer early in the day, before surface heating can be liberated that prevents shower-cloud and thunderstorm formation until late in the day.

A4.3   Aviation-significant weather charts

Aviation weather charts are reproduced in parts of the text. The symbology is familiar to aircrew and provides an easy way to assess significant flight conditions on long-distance routes and this description of the symbols used may assist the reader in the understanding of weather features commonly seen in the tropics and elsewhere. Please note that altitudes appear in multiples of 100 feet, assuming standard density at all levels of the atmosphere. Wind speeds in the jet-stream cores are shown in knots to a resolution of 10 kn.

The main features of aviation weather charts are (i) the jet streams (cores of winds near the top of the troposphere), defined by the International Civil Aviation Organization (ICAO) as having a speed greater or equal to than 80 kn, (ii) zones of significant cloud and (iii) areas in which there are likely to be pockets of significant (moderate or severe) turbulence. The only cloud zones indicated on the charts shown in this paper are those of cumulonimbus reaching an altitude greater than 25,000 feet (~7600 m) embedded in extensive layer cloud. Cloud boxes contain the following coded descriptions:

ISOL EMBD CB	Cumulonimbus clouds affecting less than 25% of the area, embedded in extensive layer clouds (high clouds) at an altitude exceeding 25,000 feet (~7600 m).
OCNL EMBD CB	Cumulonimbus clouds affecting between 25% and 50% of the area, embedded in extensive layer clouds (high clouds) at an altitude exceeding 25,000 feet.
FRQ CB	Cumulonimbus clouds affecting more than 50% of the area (usually assumed to be embedded in high clouds), seen only in tropical revolving storms.
450 XXX	Height of the top and base (XXX = below 25,000 feet) of cumulonimbus clouds in hundreds of feet (to the nearest 1000 feet) above mean sea level; in this case 45000 feet (~13,700 m).

Areas of significant clear-air turbulence (i.e. turbulence not encountered in cloud), enclosed by pecked lines, are labelled thus:

	Moderate clear-air turbulence.
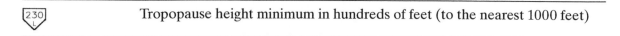	Moderate, occasionally severe clear-air turbulence.
460 290	Height of the top (in this case 46,000 feet (~14,000 m)) and base (in this case 29,000 feet (~8800 m)) in hundreds of feet above mean sea level of the layer in which turbulence of the specified type may be encountered.

Tropopause heights are also shown in rectangular boxes with tropopause minima indicated as shown below:

230 L	Tropopause height minimum in hundreds of feet (to the nearest 1000 feet)

The jet-stream core is indicated by a broad solid line and wind speeds indicated using conventional symbols: pennant = 50 kn (25 m s^{-1}), long narrow bar = 10 kn (5 m s^{-1}), each of which may be used in multiples.

Appendix 5
Snow in the Desert: A Case Study

A5.1 Introduction

Over the north and centre of the Sahara desert, as in many of the world's dry lands, in particular their poleward and central parts, almost all precipitation occurs in the winter months, associated with deep cold air masses extending towards the equator. Although there is a southward progression of cold (extra-tropical) air, the scant precipitation from these systems is not typically what would be seen from a mid-latitude front. Precipitation is mainly from convective clouds that develop along the boundary between relatively warm (tropical) air near the surface into anomalously cold air aloft.

However, because the air over the desert surface is very dry to a great depth, precipitation may not reach the ground and, where it does, small quantities are the norm. Over high ground, which provides uplift additional to that from free convection, sufficient rain may fall to support desert communities, particularly where precipitation can be stored in the rocks and is released from springs for those living around and on the uplands.

A5.2 Development of a depression over the desert

One particularly notable event occurred in early January 2008. As Fig. A5.1 shows, a strong north-westerly polar-front jet stream (PFJ) developed on 3 January 2008 at 10,000 m, with speeds reaching a maximum of 90 m s^{-1}. This extended an upper trough south (Miles 1959) across south-western Europe into north-west Africa over the following 24 hours. The divergent flow at the left exit to this jet stream developed a depression at the surface. During 5 January, the trough extension cut off, producing a deep upper low over Algeria (Fig. A5.2a). The temperature contrast between the air within this cut-off and that around its periphery may be judged by the tropopause height of only 8000 m – later 7000 m – (see section 1.4) near the centre of the low (Fig. A5.1) and the speed of the 300 hPa wind around it. The extension of the upper trough displaced the STJ south to 5°N over the Gulf of Guinea. The core of this jet stream can be seen at 12,000 m (FL390) in Fig. A5.1.

An Introduction to the Meteorology and Climate of the Tropics, First Edition. J F P Galvin.
© 2016 John Wiley & Sons, Ltd. Published 2016 by John Wiley & Sons, Ltd.

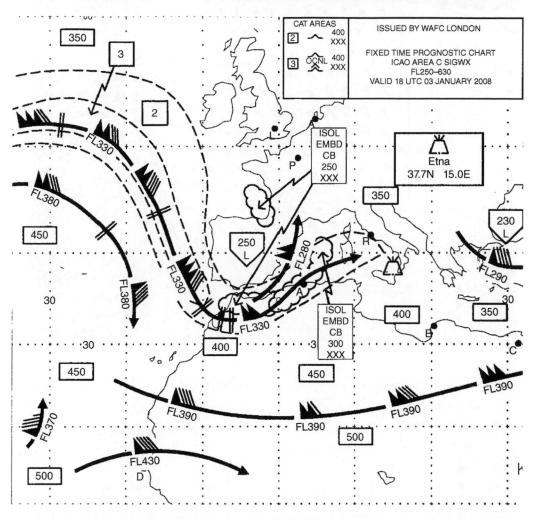

Figure A5.1 Cut-out of the AFI ICAO SigWx chart valid at 1800 UTC on 3 January 2008, produced by WAFC, London, showing the PFJ at FL330 (10,000 m) extending an upper trough over Iberia south-east into north Africa. Slight trough-ing is evident in the STJ at FL390 (12,000 m) between 20°N and 25°N ahead of the major upper ridge of tropical air in the North Atlantic. Well-separated cumulonimbus clouds, embedded in layer cloud (ISOL EMBD CB), were forecast ahead of and around the base of the trough (as shown within scalloped lines). These clouds brought snow to the Haut Atlas and Hoggar massif over the following 5 days. Areas where pockets of significant CAT could be expected are enclosed by pecked lines: moderate ⌒ and moderate, occasional severe ⌁. Turbulence would not have been expected at this time along the STJ, but was expected to be locally severe around the PFJ. Tropopause heights (at 5000 feet intervals) are shown in boxes and tropopause lows to the nearest 1000 feet, e.g. ▽. (Heights/flight levels are in hundreds of feet above mean sea level. Speed fleches follow the usual convention.)

As may be expected, deep convection was the result, both along the eastern edge of the system, ahead of the trough, and in the cooler air within the depression. As the low-level flow was dry, its origin over desert to the east (Fig. A5.2b), much of the cloud formed as alto-cumulus and cumulonimbus from instability in the middle troposphere. On 6 January, Fig. A5.3 suggests that, in general, cloud bases were near their characteristic height of 3600 m,[1] with tops reaching 6400 m or more in the cold air near the centre of the low, so that lower lying areas of the desert would have received little or no precipitation despite its thundery nature

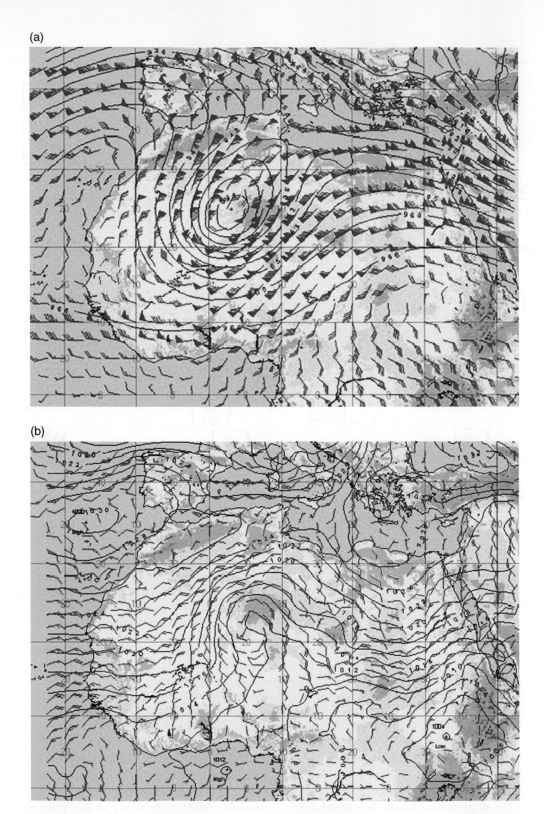

(a)

(b)

Figure A5.2   (a) Geopotential height (dam), wind speed and direction at 300 hPa at 1200 UTC on 6 January 2008 over north Africa. A cut-off low has developed just to the west of the Hoggar massif. (b) Mean sea-level pressure (hPa), surface wind speed and direction at 1200 UTC on 6 January 2008 over north Africa, showing a surface low at the southern edge of the Hoggar massif, Algeria with strong north-easterly winds around its north-western quadrant. Its position just ahead of the upper low would have aided the development of deep convective cloud, as described in the text. © Crown copyright (Met Office).

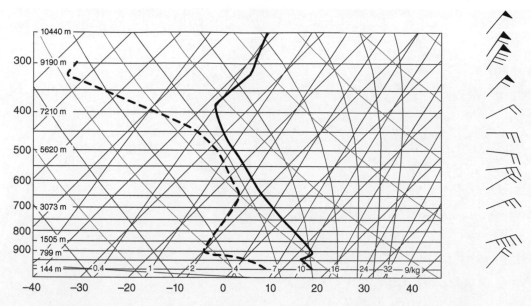

Figure A5.3   Radiosonde profile for In-Salah (Algeria) at 1200 UTC on 6 January 2008. Courtesy of University of Wyoming, Department of Atmospheric Science.

(Fig. A5.4). However, over the Hoggar massif of southern Algeria, which rises to more than 1500 m, with the peak of Mount Tahat at 2918 m, cloud bases were somewhat lower, and on 7 January cumulonimbus clouds were likely to develop *in situ* with a base near 2500 m from a surface temperature around 10°C (Fig. A5.5). These clouds had brought precipitation across higher ground in Morocco and northern Algeria on 4 January, snow falling on the Haut Atlas above about 1500 m. Between 6 and 8 January, as the system became slow-moving, convective clouds reached 10,000 m or more ahead of the low (area A in Fig. A5.6) and reached the tropopause in the cold air near the centre of the depression (area B in Fig. A5.6). With a freezing level below 2400 m near the centre of the low, snow could fall to about 1500 m.[2] Little or no snow could be seen on the Hoggar massif through 6 and 7 January, with temperatures on much of this area of high ground above freezing, but by 8 January, as the low migrated slowly north-east, snow accumulated on the highest ground. This is indicated in Fig. A5.7.

Separated from the source of deep cold air to the north, the cut-off upper low filled, warming out relatively quickly in this extreme

southerly position. Forcing mechanisms decreased and precipitation had died out by 9 January over the central Sahara as the temperature contrast forming the cloud was removed, the STJ moving north to form its more usual broad upper trough over north-western Africa (Atkinson 1971).

A5.3   The weather features associated with the upper low

Along the south-eastern edge of the weather system, the altitude of cirriform cloud associated with the STJ is notable. Figure A5.6 shows tops between 12,000 and 13,000 m (area C), close to the level of the jet core. The acceleration into this jet and (modest) anti-cyclonic curvature formed both this broad cirriform jet 'streak' and banded cirrus, indicating clear-air turbulence on the warm side of the jet (Ellrod 1989).

To the east of the upper low, the development of convection was aided not only by dynamical uplift over the mountains, but by strong divergence at the left exit of the remnant PFJ around the upper low. Thus the strongest development occurred north of the surface low, where ascent

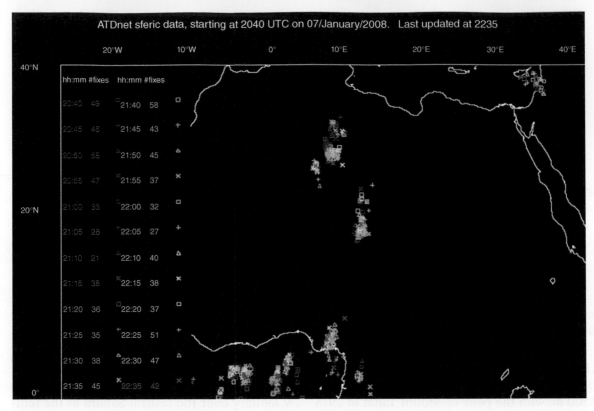

Thunderstorm discharges (sferics) over north-west Africa between 2040 UTC and 2235 UTC on 7 January 2008. The inter-tropical convergence zone can be seen over the Gulf of Guinea with significant zones of thunder evident over Niger from the cloud mass ahead of the upper low (A in Fig. A5.6) and the centre-east of Algeria, near the centre of the upper low (area B in Fig. A5.6). © Crown copyright (Met Office).

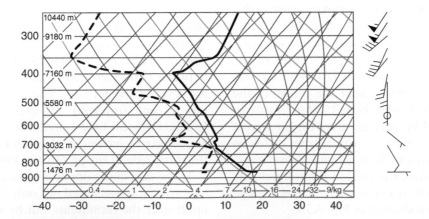

Figure A5.5   Radiosonde profile for Tamanrasset (Algeria) at 1200 UTC on 7 January 2008. Courtesy of University of Wyoming, Department of Atmospheric Science.

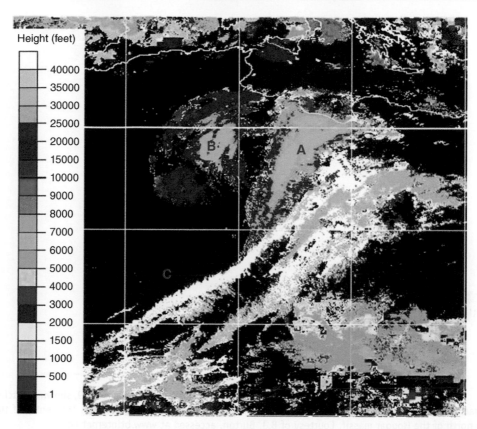

Figure A5.6   False-colour imagery from Meteosat-9 coloured using temperature and computer-model data to show cloud-top height over north Africa at 1700 UTC on 7 January 2008: 3000 m ≈ 10000 feet. Area A is the zone of altocumulus, altostratus and cirrostratus with embedded cumulonimbus developing ahead of the upper low in the left exit of the mid-latitude jet stream and may be assumed to have frontal characteristics in the upper troposphere. Area B is the area of medium-level cloud with occasional cumulonimbus tops reaching 7000–9000 m, near the centre of the upper low. Area C indicates the zone of turbulent cirrus and cirrostratus along the northern edge of the STJ. © Crown copyright (Met Office).

at lower levels rose into comparatively moist south-westerlies ahead of the upper low, which lay somewhat further west.

In the area of upper-level convergence and related descent of dry air over the world's 'hot' deserts, scant rainfall is dependent largely on the incursion of upper troughs in winter, which bring positive vorticity and associated ascent at medium and high levels (Knippertz 2007). This ascent is assisted by the displacement of the STJ (which is normally found near 30°N and 30°S), so that there is upper-level divergence where the jet accelerates away from the base of the trough.

High ground is also important in the development of weather systems, providing additional uplift, the ground level closer to cloud bases. Over the Sahara desert in early January 2008 unusually low freezing levels, locally as low as 1500 m, were modelled as far south as 25°N. Air with these characteristics (at upper levels) would be classified in western Europe as polar. Freezing levels are more typically near 3500–4000 m in this part of the tropics in winter. The result was significant snowfall on the Hoggar massif of southern Algeria.

Although there was clearly a change of air mass ahead of the upper low of early January 2008, frontal characteristics cannot be readily analysed at low levels, but could be seen near the cold side of the sub-tropical jet stream. At 850 hPa the wet-bulb potential temperature

Figure A5.7   High-resolution false-colour image from NOAA-18 at 1255 UTC on 8 January 2009, showing patches of snow on the Hoggar massif (red arrow) and extensive snow cover on the Haut Atlas (green arrow). The centre of the low can be seen just north of the Hoggar massif. Courtesy of B.J. Burton, accessed at www.btinternet.com/.

fell from 14 to 2°C across this upper-level baroclinic zone (to a minimum of –2°C near the centre of the depression).

A5.4   Summary

Whilst snow is commonly seen in autumn and winter on the Atlas Mountains that fringe the south-western Mediterranean and the Atlantic seaboard of north-west Africa, and is occasionally seen at low levels (such as that seen in Algiers on 27 January 2005), it is rare as far south as was the case on 7–8 January 2008 (but see the letter by Aymard in Kendrew 1937: 31).

Upper troughs and cut-off upper lows are an important source of precipitation over the world's deserts (Knippertz 2007), the additional precipitation available around high ground often supporting cities such as Tamanrasset in otherwise barren desert. Although upper troughs are semi-permanent features in the STJ, they are rarely seen over dry continental areas, so that precipitation is meagre and needs to be stored in aquifers of permeable rock if it is to be of value to local desert communities.

Notes

1   The lowest cloud layer in Fig. A5.3 has a base near 3000 m, which was characteristic of slightly warmer air sampled above Tamanrasset (Algeria) at 1200 UTC on 6 January. Cloud only rarely has a base below about this level over the deserts (as described in the main text). However, in the case of the Sahara desert (as much of the rest of Africa) the ground level is near 400 m.

2   The fall of snow at temperatures above freezing is described in Meteorological Office (1997).

Appendix 6
A Climatic Summary for Tropical Countries and States

This appendix summarizes the climates of countries and states, the majority of which lie between 35°N and 35°S. The list is in alphabetical order and includes a basic description of the main climatic type, the Köppen descriptor and a short description of local variations. Mountain climates are defined in this table as areas where the mountains have a significant effect on the weather, either producing rainfall not seen at lower altitudes and a similar latitude, or reducing temperatures below 0°C at some time during the year.

Some countries or states within latitudes 35°N to 35°S have climates that do not have tropical characteristics, even for part of the year, or may be affected by tropical weather for only a short period. These are indicated in italics.

Country	Climate type	Köppen descriptor	National and local climate variation
Afghanistan	Desert	BWh	Mountain areas experience frequent rains, especially in summer. Winters are cool.
Algeria	Desert	BWh	Atlas Mountains have autumn rainfall peak. Hoggar massif has some winter rain.
American Samoa	Wet	Af	Oceanic regime. Maximum rainfall in summer.
Angola	Mountain	H	Coastal plain is desert (BWh). Monsoon (Aw) conditions inland at modest altitude.
Australia			
New South Wales	Semi-desert	BShw	Coastal plain has a sub-tropical (Caf) climate. Mountain climate between regimes.
Northern Territory	Semi-desert	BShw	Monsoon (Aw) climate in north. Desert (BWh) in south. Occasional hurricanes.
Queensland	Semi-desert	BShw	Monsoon (Aw, Am and Caf) on coastal plain. Mountain divide. Occasional hurricanes.
South Australia	Desert	BWh	Coast has a Mediterranean (Csb) or semi-desert (BSh) climate. Winter rain peak.
Western Australia	Desert	BWh	Mediterranean (Csb) or semi-desert (BSh) climates in south-west. Hurricane risk in north.

(Continued)

An Introduction to the Meteorology and Climate of the Tropics, First Edition. J F P Galvin.
© 2016 John Wiley & Sons, Ltd. Published 2016 by John Wiley & Sons, Ltd.

Country	Climate type	Köppen descriptor	National and local climate variation
Bangladesh	Monsoon	Aw	Occasional tropical cyclones.
Belize	Wet	Af	Mountain climate (H) inland. Occasional hurricanes.
Benin	Monsoon	Aw	Semi-desert (BSh) in far north.
Bermuda	Oceanic	Caf	
Bhutan	Mountain	H	Monsoon (Caw) climate in the low-lying south.
Bolivia	Monsoon	Aw	Equatorial (Af) climate far north. Large area of mountain climate (H) in south-west.
Botswana	Mountain	H	Largely semi-desert (BSh). South-western areas within the Kalahari Desert.
Brazil			
Acre	Equatorial	Af	
Alagoas	Monsoon	Aw	
Amapá	Monsoon	Am	Marginal between wetter monsoon (Am) and equatorial (Af) climates.
Amazonas	Equatorial	Af	
Bahia	Monsoon	Aw	Mountain climate (H) in central areas.
Ceará	Monsoon	Aw	East and south has a semi-desert (BSh) climate.
Espírito Santo	Wet	Af	
Federal District	Mountain	H	Lies within the monsoon (Aw) zone.
Goiás	Monsoon	Aw	South-east has a mountain climate (H).
Maranhão	Monsoon	Aw	Coastal zone is wetter, having an Am climate.
Mato Grosso	Monsoon	Aw	
Mato Grosso do Sur	Monsoon	Ca	South-east has mountain climate (H).
Minas Gerais	Mountain	H	Coastal area is wettest, lying within the Af zone. The north-west is mountainous.
Pará	Monsoon	Aw	Coastal zone is in the Am climate zone, verging on Af locally.
Paraibá	Semi-desert	BSh	Higher ground inland brings a wetter (As) climate.
Paraná	Mountain	H	Coastal areas have a sub-tropical (Mediterranean) climate.
Pernambuco	Semi-desert	BSh	Mountain areas inland are relatively moist and have an As climate.
Piauí	Monsoon	Aw	Coastal strip has a wetter (Am) climate.
Rio de Janeiro	Wet	Af	Lies under the influence of a TUTT with a summer rainfall peak.
Rio Grande do Norte	Semi-desert	BSh	
Rio Grande do Sur	Monsoon	Caf	
Rondônia	Monsoon	Aw	North-west has an equatorial (Af) climate.
Roraima	Monsoon	Aw	South and west have an equatorial (Af) climate.
Santa Catarina	Mountain	H	Coastal and inland areas have a monsoon (Caf) climate.
São Paulo	Mountain	H	Coastal and inland areas have a monsoon (Caf) climate.
Sergipe	Monsoon	Aw	
Tocantins	Monsoon	Aw	
Burkina Faso	Semi-desert	BShw	South-west has a monsoon (Aw) climate and there is desert (BWh) in the far north.
Burma	Monsoon	Aw/Caw	Wetter coastal zone (Am) with occasional cyclones. Mountain (H) zone between Aw and Caw zones.

Country	Climate type	Köppen descriptor	National and local climate variation
Burundi	Mountain	H	Lies within the East African monsoon (Aw) zone.
Cambodia	Monsoon	Aw	Relatively dry climate.
Cameroon	Equatorial	Af	Central areas are monsoonal (Aw) and the north semi-desert (BShw).
Canary Islands	Desert	BWh	Lies under the influence of cool seas.
Cape Verde Islands	Monsoon	Aw	Occasional hurricanes.
Central African Rep.	Monsoon	Aw	South-west is equatorial (Af) and parts of the north semi-desert (BShw).
Chad	Desert	BWh	South-eastern and south-western parts semi-desert (BShw). Far south monsoonal.
China			
Anhui	*Sub-tropical*	*Cah*	*Some monsoon rains in late summer.*
Chongqing	Monsoon	Caw	Winter rains also seen, but a relatively dry cool inland climate.
Fujian	Monsoon	Caw	Relatively moist coastal climate, but cool in winter. Occasional typhoons.
Guangdong	Monsoon	Caw	Relatively moist coastal climate. Occasional typhoons.
Guangxi	Monsoon	Caw	Relatively moist coastal climate, but cool in winter.
Guizhou	Mountain	H	Lies within the monsoon (Aw) zone.
Hainan	Equatorial	Af	Occasional typhoons.
Henan	*Sub-tropical*	*Cah*	*Some monsoon rains in late summer.*
Hong Kong	Monsoon	Caw	Relatively moist coastal climate, but cool in winter. Occasional typhoons.
Hubei	*Sub-tropical*	*Cah*	*Some monsoon rains in late summer.*
Hunan	Monsoon	Caw	Winter rains also seen, but a relatively dry cool inland climate.
Jiangsu	*Sub-tropical*	*Cah*	*Relatively cool moist coastal climate. Occasional typhoons.*
Jiangxi	Monsoon	Caw	Winter rains also seen, but a relatively dry cool inland climate.
Macão	Monsoon	Caw	Relatively moist coastal climate, but cool in winter. Occasional typhoons.
Shaanxi	*Sub-tropical*	*Cah*	*Some monsoon rains in late summer.*
Shanghai	*Sub-tropical*	*Cah*	*Relatively cool moist coastal climate. Occasional typhoons.*
Sichuan	Mountain	H	Lower-lying areas are near the northern extreme of the monsoon climate.
Tibet	Mountain	H	Most rain or snow on highest ground in summer.
Yunnan	Mountain	H	Highest rainfall/snowfall on high ground throughout the year.
Zhejiang	Monsoon	Caw	Relatively moist coastal climate, but cool in winter. Occasional typhoons.
Columbia	Equatorial	Af	Mountain climate (H) inland. Climate is relatively dry in coastal areas.
Congo	Monsoon	Aw	Equatorial (Af) climate in the north.
Cook Islands	Monsoon	Am	
Costa Rica	Mountain	H	Monsoon (Aw) on west coast. Wet (Af) on east coast. Small risk of hurricanes.
Côte d'Ivoire	Monsoon	Aw	Coastal areas have a wetter Am climate.

(Continued)

Country	Climate type	Köppen descriptor	National and local climate variation
Cuba	Monsoon	Aw	Occasional hurricanes.
Cyprus	*Mediterranean*	*Cb*	*Mountain climate inland W*
Djibouti	Semi-desert	BS	Marginal with BWh desert climate.
Dominican Republic	Monsoon	Am	Mountain climate (H) inland.
Easter Island	Semi-desert	BS	
Ecuador	Monsoon	Aw	Wetter north of the equator (Am) and a mountain climate (H) inland.
Galápagos Islands	Desert	BWh	El Niño brings wet conditions.
Egypt	Desert	BWh	Semi-desert (BShs) in coastal north.
El Salvador	Monsoon	Aw	Mountain climate (H) inland.
Equatorial Guinea	Equatorial	Af	
Eritrea	Semi-desert	BS	Marginal with BWh desert climate.
Ethiopia	Mountain	H	Low-lying areas drier (BS) with desert (BWh) along the border with Somalia.
Fiji	Wet	Af	Lies under the influence of a TUTT in winter. Occasional typhoons.
French Guiana	Monsoon	Am	
French Polynesia	Wet	Af	Occasional hurricanes.
Gabon	Equatorial	Af	Some eastern and southern areas drier and monsoonal (Aw).
Ghana	Monsoon	Aw	South-east and far north drier semi-desert (BSh).
Gibraltar (UK)	Mediterranean	Csb	Convective precipitation is partly determined by the effects of surrounding high ground.
Guatemala	Wet	Af	Monsoon (Aw) climate on SW coast. Mountain (H) inland. Occasional hurricanes.
Guinea	Monsoon	Am	Drier (Aw) climate inland and along northern border.
Guinea-Bissau	Monsoon	Aw	
Guyana	Monsoon	Am	Drier (Aw) climate in south.
Haiti	Monsoon	Aw	Occasional typhoons.
Honduras	Monsoon	Aw	Mountain climate (H) inland.
India			
Akas	Mountain	H	Monsoon (Caw) regime.
Andaman Islands	Monsoon	Am	Occasional cyclones.
Andrha Pradesh	Monsoon	Aw	Semi-desert (BS) far west, wet in Eastern Ghats (far north-east). Occasional cyclones.
Arunchal Pradesh	Mountain	H	Monsoon (Aw) regime.
Assam	Monsoon	Caw	
Bihar	Monsoon	Caw	Occasional winter rains may be significant. Small risk of cyclones.
Chattisgarh	Monsoon	Aw	Significant rainfall enhancement on Eastern Ghats, but relatively dry in west.
Goa	Monsoon	Am	
Gujarat	Desert	BWh	Some semi-desert (BS) in south-east. Occasional cyclones.
Haryana	Monsoon	Ca	Semi-desert (BS) in west.
Himal Pradesh	Mountain	H	Semi-desert in lowland south-west.
Jammu/Kashmir	Mountain	H	Marginal monsoon (Caw) climate in lowland south-west. Spring rains may be significant.
Jharkhad	Monsoon	Caw	
Karnataka	Monsoon	Aw	Semi-desert (BSh) in far east. Wetter near coast. Mountain climate Western Ghats.

Country	Climate type	Köppen descriptor	National and local climate variation
Kerala	Monsoon	Am	Mountain climate (Western Ghats) inland.
Laccadive Is.	Monsoon	Am	
Madhya Pradesh	Monsoon	Caw	Semi-desert (BSh) in far north-west.
Maharashtra	Monsoon	Aw	Semi-desert (BSh) in far east. Wetter near coast. Mountain climate Western Ghats.
Manipur	Mountain	H	Monsoon (Caw) in lowland west. Occasional cyclones.
Meghalaya	Monsoon	Caw	Mountain areas are exceptionally wet. Occasional cyclones.
Mizoram	Monsoon	Am	Mountain (H) in far east. Occasional cyclones.
Nagaland	Mountain	H	Monsoon (Caw) regime.
Orissa	Monsoon	Aw	Significant rainfall enhancement on Eastern Ghats. Occasional cyclones.
Punjab (east)	Semi-desert	BSh	Monsoon (Caw) regime in far north-east.
Rajasthan	Desert	BWh	Semi-desert (BSh) in east.
Sikkim	Mountain	H	Monsoon (Caw) regime.
Tamil Nadu	Wet winter	As	Semi-desert (BS) in west.
Tripura	Monsoon	Am	Drier in north-east. Occasional cyclones.
Uttar Pradesh	Monsoon	Caw	Semi-desert (BS) in west. Occasional winter rains may be significant.
Uttaranchal	Mountain	H	Monsoon (Caw) in lowland south-west. Occasional winter rains may be significant.
West Bengal	Monsoon	Aw	Occasional cyclones.
Indonesia	Equatorial	Af	All larger islands have mountain climates (H) inland.
Iran	Mountain	H	Desert climate (BWh) in coastal areas and low-lying inland parts.
Israel	Semi-desert	BSh	Mediterranean (Cb) climate in north-west and desert (BWh) in far south-east.
Jamaica	Mountain	H	Monsoon (Aw) climate in coastal areas. Occasional hurricanes.
Jordan	Desert	BWh	
Kenya	Semi-desert	BSh	Desert (BWh) in north-east. Monsoon (Aw+H) in south and west.
Kiribati	Equatorial	Af	
Kuwait	Desert	BWh	
Laos	Monsoon	Aw	Mountain (H) climate locally close to Vietnam border.
Lebanon	*Mediterranean*	*Cb*	*Mountain climate (H) inland.*
Lesser Antilles	Wetter summers	Am	Occasional hurricanes.
Lesotho	Mountain	H	
Liberia	Monsoon	Am	
Libya	Desert	BWh	Narrow semi-desert (BSh) strip close to coast.
Madagascar	Monsoon	Aw	Wet (Af) climate on west coast. Mountain climate (H) inland. Occasional cyclones.
Madeira	*Sub-tropical*	*Ca/Cb*	*Pre-monsoon humid weather affects the island in high summer; there is a notable difference in climate between north-facing and south-facing sides.*
Malawi	Mountain	H	Monsoon (Aw) zone.
Malaysia	Equatorial	Af	Mountain climate inland.
Maldives	Equatorial	Af	
Mali	Desert	BWh	Monsoon (Aw) and semi-desert (BShw) climates in the south.
Mariana Islands	Monsoon	Aw	Occasional typhoons.

(Continued)

Country	Climate type	Köppen descriptor	National and local climate variation
Marshall Islands	Equatorial	Af	Occasional typhoons.
Mauritania	Desert	BWh	Semi-desert (BShw) climate in far south.
Mauritius	Wet	Af	Occasional cyclones.
Mexico	Mountain	H	Semi-desert (BS) or desert (BWh) lowlands, but monsoon (Af and Aw) in lowland south and east. Occasional hurricanes on both coasts.
Micronesia	Equatorial	Af	Occasional typhoons.
Morocco	Mountain	H	Semi-desert (BSh) around Atlas Mountains. Desert (BWh) in far south.
Mozambique	Monsoon	Aw	Semi-desert (BSh) north-east coast and inland south-west. Mountain (H) locally inland; cyclones.
Namibia	Mountain	H	Desert (BW) and semi-desert (BSh) in lowland south and west.
Nauru	Equatorial	Af	
Nepal	Mountain	H	Monsoon (Caw) in lowland south.
New Caledonia	Wetter summers	Am	Occasional hurricanes.
Nicaragua	Mountain	H	Wet (Af) climate along the coast.
Niger	Desert	BWh	Semi-desert (BShw) in far south.
Nigeria	Monsoon	Aw	Equatorial (Af) climate in south and semi-desert (BShw) in the far north.
Oman	Desert	BWh	Semi-desert (BS) uplands.
Palau	Equatorial	Af	Occasional typhoons.
Pakistan	Desert	BWh	Mountain (H) along north-western and northern borders. Semi-desert locally. Cyclones.
Panama	Equatorial	Af	Mountain climate (H) inland.
Papua New Guinea	Equatorial	Af	Mountain climate (H) inland.
Paraguay	Monsoon	Caf	Semi-desert (BS) in far west.
Peru	Equatorial	Af	Desert (BWn) along the coast and mountain (H) along Andes divide.
Philippines	Equatorial	Af	Mountain modification (H) in centres of larger islands. Occasional typhoons.
Pitcairn Islands	*Sub-tropical*	*Cah*	*Dry oceanic climate.*
Puerto Rico	Wet	Af	Occasional typhoons.
Rwanda	Mountain	H	Lies within the East African monsoon (Aw) zone.
Saudi Arabia	Desert	BWh	Semi-desert (BS) climate in mountainous areas of western coast.
Senegal	Semi-desert	BShw	Monsoon (Aw) climate in far south. Desert (BWh) in far north.
Seychelles	Monsoon	Aw	Occasional cyclones.
Sierra Leone	Monsoon	Aw	
Singapore	Equatorial	Af	
Solomon Islands	Wet	Af	Occasional hurricanes.
Somalia	Desert	BWh	Semi-desert (BSh) in southern coastal areas and parts of the mountainous north.
South Africa	Mountain	H	Semi-desert (BSh) interior. Desert (BWh) in west. Sub-tropical (Ca and Cb) in south and east.
Sri Lanka	Equatorial	Af	Monsoons (Aw) in northern and eastern coastal areas. Mountainous (H) interior.
St Helena	Desert	BWh	Lies under the influence of cool seas.

Country	Climate type	Köppen descriptor	National and local climate variation
Sudan	Desert	BWh	Semi-desert (BShw) in south, with monsoons (Aw) in extreme southern fringe.
Suriname	Monsoon	Am	Drier along western border.
Swaziland	Semi-desert	BSh	Mountainous (H) in west. Wetter along eastern border.
Taiwan	Monsoon	Caw	Typhoons common.
Tanzania	Mountain	H	Monsoon (Aw) climate with semi-desert (BSh) on south-east coast. Occasional cyclones.
Thailand	Monsoon	Aw	Equatorial (Af) in some southern areas.
The Bahamas	Monsoon	Ca	Occasional hurricanes.
The Gambia	Monsoon	Aw	Semi-desert (BShw) in north-east.
Togo	Semi-desert	BSh	Monsoon (Aw) in central areas. Dry weather in shelter of mountains.
Tonga	Wetter summers	Am	
Trinidad and Tobago	Wetter summers	Am	Occasional hurricanes.
Tunisia	Semi-desert	BShs	Desert (BWh) in south.
Turks and Caicos Islands	Wet	Af	Occasional hurricanes.
Tuvalu	Equatorial	Af	
Uganda	Mountain	H	Monsoon (Aw) to semi-desert (BSh) climate. Wetter around Lake Victoria.
United Arab Emirates	Semi-desert	BS	Desert (BWh) climate along inland fringes and locally elsewhere.
USA			
Alabama	*Sub-tropical*	*Caf*	*Occasional hurricanes.*
Arizona	Semi-desert	BS	Large areas of mountain (H) in east and locally desert (BWh).
Arkansas	*Sub-tropical*	*Caf*	
S. Carolina	*Sub-tropical*	*Caf*	*Occasional hurricanes.*
California	*Sub-tropical*	*Ca*	*Semi-desert (BS) or desert (BWh) inland. Only the extreme south of the state is equatorward of 35°N.*
Florida	Monsoon	Caf	Sub-tropical in north; occasional hurricanes. Under the influence of a TUTT in winter.
Georgia	*Sub-tropical*	*Caf*	*Occasional hurricanes.*
Hawaii	Wet	Af	Occasional hurricanes. Some extra-tropical characteristics at times in winter.
Louisiana	*Sub-tropical*	*Caf*	*Occasional hurricanes.*
Mississippi	*Sub-tropical*	*Caf*	*Occasional hurricanes.*
New Mexico	Mountain	H	Largely semi-desert (BSh).
Texas	Semi-desert	BSh	Sub-tropical (Caf) in east and north. Mountain (H) along south-western fringe. Occasional hurricanes.
Uruguay	*Sub-tropical*	*Caf*	
Vanuatu	Wet	Af	Occasional hurricanes.
Venezuela	Monsoon	Aw	Wetter (Am) on eastern coast. Equatorial (Af) in far south. Desert and semi-desert in far north.
Vietnam	Monsoon	Aw	Equatorial (Af) in central areas. Occasional typhoons.
Virgin Islands	Wet	Af	Occasional hurricanes.
Wallis and Futuna	Wetter summers	Am	Occasional hurricanes.
Western Sahara	Desert	BWh	
Western Samoa	Wetter summers	Am	Occasional hurricanes.

(Continued)

Country	Climate type	Köppen descriptor	National and local climate variation
Yemen	Desert	BWh	Semi-desert (BS and H) in mountainous inland parts (mainly north-west).
Zaïre	Equatorial	Af	Monsoon (Aw) in south with mountain (H) along southern and eastern borders.
Zambia	Mountain	H	Mainly monsoon (Aw), but semi-desert (BSh) in far south and south-east.
Zimbabwe	Mountain	H	Monsoon (Aw) regime, but semi-desert in north, west and south.

Appendix 7
Two Easterly Waves in West Africa in Summer 2009: A Case Study

A7.1 Introduction

The summer monsoon of West Africa brings much-needed rain to the Sahel between Chad and Senegal, where drought has been common since the late 1960s (Kennedy et al. 2009). However, the stability of the moist south-westerly flow[1] compared with the potentially unstable, but dry air inland – which has its origin in the trade-wind Harmattan from the Sahara desert[2] – usually limits rainfall. North of about 10°N, rainfall is sporadic, often occurring in heavy downpours from MCSs. In some years, monsoon rains do not occur north of about 13°N. In much of the monsoon zone, MCSs often develop into MCCs, the latter defined as having a cold-cloud (cirriform) area greater than $15 \times 10^4 \, km^2$ (Maddox 1980, 1986; Maddox et al. 1981; Augustine & Howard 1988). These large cloud masses develop within waves in the mid-tropospheric easterly flow, as described by Laing and Fritsch (1997).

The development of mid-tropospheric waves is associated with the upper-tropospheric easterly equatorial flow and their development is limited when these winds and associated large-scale forcing are weak. Initial formation appears to be the result of heating over the Marra Plateau of Darfur, Sudan, which strengthens potential vorticity, developing the mid-tropospheric waves (Thorncroft et al. 2008).

Within the waves, the development of MCCs requires the coincidence of favourable factors throughout the troposphere:

- the convergence of low-level winds, often around a low-pressure system
- a trough, usually seen near the 650 hPa level, that maintains convergence to mid-tropospheric levels, thus aiding development from the moist surface layer
- favourable wind shear that allows large-scale cloud development
- diffluence of the upper-tropospheric winds, often associated with the incursion of an upper tropospheric trough.

The role of the mid-troposphere is crucial: air may be lifted readily by convection in the boundary layer, but to continue ascending, convergence in the mid troposphere and modest wind shear are needed to force uplift into the upper troposphere.[3]

Maximum wave amplitude usually occurs late in the summer season, from mid-July onwards. At this time, the summer monsoon has reached its maximum northward progress, but upper troughs may still make progress southwards across the Sahara desert. Their development and potency are dependent on the alignment of factors favouring ascent, as well as time of day: vigorous convection into the upper troposphere tends to occur from around 1200 local time, lasting until around 0300 local time over tropical land masses.

Over the warm tropical oceans development is more gradual. Emanuel (2005) describes the intimate link between mid-tropospheric waves and the formation of tropical revolving storms in the Atlantic, although hurricanes form in fewer than 10% of these waves. In years when there are few well-developed mid-tropospheric waves and rainfall is lower than normal in the Sahel, fewer vigorous hurricanes develop in the Atlantic (Landsea & Gray 1992). Nonetheless, there is a strong link between easterly waves and major hurricanes (Landsea 1993). These factors are important in the formation and decline of large-scale tropical disturbances within easterly waves.

The two waves described in this paper vary in character. Development within them was linked to baroclinic forcing where the moist monsoon flow meets dry Saharan air (Chang 1993). The first brought flooding to a wide area of the Sahel, while the second, which was in the subsequent mid-tropospheric wave, formed a tropical revolving storm after moving over the Atlantic Ocean.

A7.2   The wave of 28 August to 5 September 2009

The development of deep convection within this wave occurred in two phases. Initial development commenced in a shallow mid-tropospheric trough over the Marra Plateau near the border of Sudan and Chad between 1200 UTC on 28 August and 1200 UTC on 29 August (Figs A7.1 & A7.2). Forcing was assisted by a 'dry slot' in the upper troposphere (Grahame et al. 2015), which had made slow progress south across the Sahara. The moderately dry air near 18°N assisted development within the trough as it was carried west, bringing ascent additional to that from the potential vorticity associated with convection. In the presence of low θ_w aloft, ascent is promoted

(Bradbury 1977) and evaporation around the ascending cloud aids cooling and moistening, thus reducing the static stability. The upper-tropospheric cooling around convective towers also may be a source of descending air that assists in the formation and maintenance of the MCS. Indeed, it may be important in the formation of squall lines (Roca et al. 2005). The cyclonic curvature of the trough also assisted development, positive vorticity assisting ascent.

The mid-tropospheric trough shown in Fig. A7.2(b) developed under the influence of upper-tropospheric diffluence and convection to form a cut-off low in the mid troposphere, moving as shown in Fig. A7.3 over the following week. During this stage of development, 700 hPa contour heights fell to a minimum around 313 dam early on 30 August, rising for a time thereafter.

It can be seen that the trough accelerated on 31 August from a mean speed of 6 m s–1 to 9 m s–1, its later speed typical for these waves (Diop & Grimes 2003). However, large-scale deep convection was only present when there was close alignment of the surface and mid-tropospheric depressions.

Deep convection is usually associated with the surface low, which often moves ahead of the mid-tropospheric trough. Although convection is usually enhanced just ahead of the trough axis, where modest directional shear above the surface and mid-level vorticity combine, ascent into the upper troposphere reduces when there is rapid directional shear with anti-cyclonic curvature in the mid troposphere. The system thus often degenerates after 18–36 hours as the depression fills ahead of the trough axis (Desbois et al. 1988). Development and decay of the system followed this pattern during the first phase, so despite the continued presence and westward motion of the trough at increased speed, the surface low filled as it accelerated, suppressing convection on 29 and 30 August until the development of a new surface low, close to the base of the mid-tropospheric trough, early on 31 August. The surface depression was closely

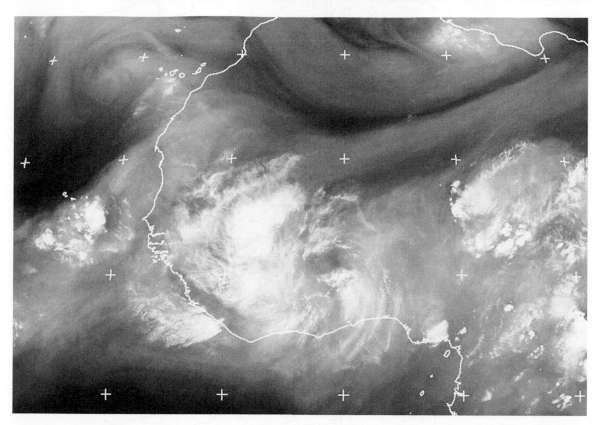

Figure A7.1   Water-vapour (6.85–7.85 μm) imagery at 1200 UTC on 29 August 2009 from Meteosat-9. 'Dry slots' can be seen north of the Sahel, their presence and the upper tropospheric trough they portray aiding development, including the initial convective phase of the first wave, which can be seen near 16°N, 12°E in this image. The previous wave has developed to a maximum near the base of the upper trough over Mali; a sequence of such developments were an important precursor to the flooding of early September, wave upon wave running into the region ahead of the axis of the slow-moving upper trough. © Crown copyright (Met Office).

aligned with the mid-tropospheric depression thereafter.

The new low formed as comparatively dry, but warm air at low levels moved south ahead of the mid-tropospheric disturbance, converging with southerly winds that fed moist (monsoon) air into its eastern side (Thorncroft & Blackburn 1999; Fink & Reiner 2003; Roca et al. 2005).[4]

The speed of development is indicated by the rapid expansion of the MCS to form an MCC of cloud-shield area ~24 × 10⁴ km² by 1500 UTC on 31 August (Fig. A7.4), which brought heavy rain across Niger, Burkina Faso, southern Mali, northern Côte d'Ivoire, Guinea,

Senegal, The Gambia and Guinea-Bissau between 31 August and the early hours of 3 September. Throughout this period of heavy rain, mid-tropospheric heights were falling, 700 hPa contour height declining to a minimum of ~311 dam over land around midday on 3 September near Bissau (12°N, 16°W).

The intense rain brought floods to Burkina Faso. Exceptionally heavy rainfall during August caused several rivers, including the Pendjari, Niger, Volta and Senegal, to break their banks, causing destruction of houses, bridges, roads and crops (ReliefWeb 2009a). Floods are reported to have affected 600,000 people (ReliefWeb 2009b) across six countries.

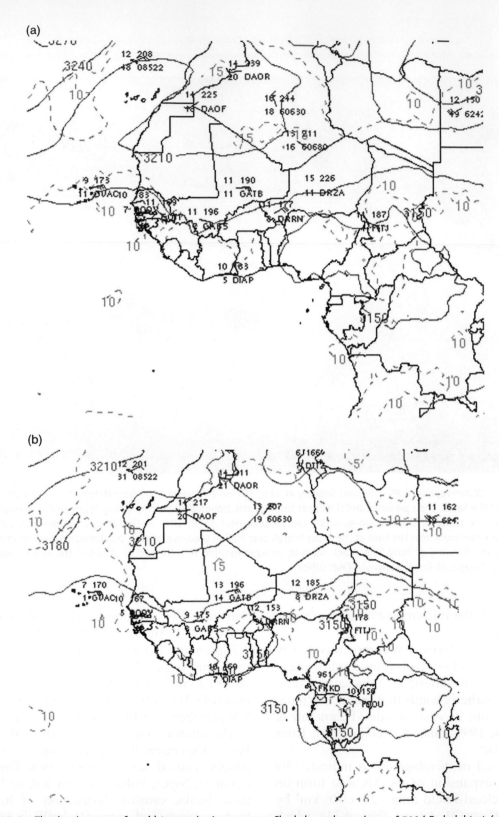

Figure A7.2 The development of a mid-tropospheric wave over Chad shown by analyses of 700 hPa height at (a) 1200 UTC on 28 August 2009 (a) and (b) 1200 UTC on 29 August 2009. The trough can be seen to have a steeper gradient on the northern flank of the mid-tropospheric low, which moved west-north-west. A complex area of lower contour heights on 28 August developed into a small mid-tropospheric low on 29 August with sharpening of the contour pattern on the north-western flank of the trough. Courtesy of University of Wyoming, Department of Atmospheric Sciences.

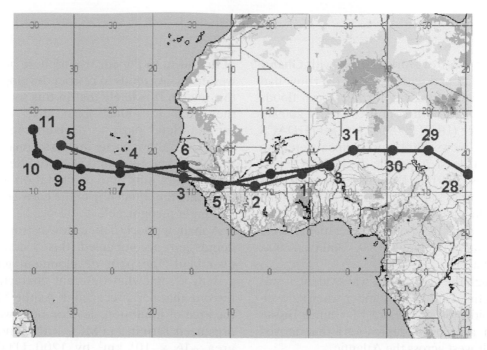

Figure A7.3   The progress of the two mid-tropospheric vortices in late August and early September 2009 across west Africa and the eastern Atlantic Ocean. Positions are at 1200 UTC on dates given.

Figure A7.4   Development of a cloud shield to form the MCC over the borders of Niger and Nigeria. False-colour image from Meteosat-9 at 1500 UTC on 31 August 2009. © Crown copyright (Met Office).

Near the border of Burkina Faso with Ghana, a dam threatened with possible overflow had to be opened, causing further local floods downstream in both countries (IRIN 2009a). Benin, Niger (IRIN 2009b), Senegal (GDACS 2009; BBC News 2009) and Guinea were also affected. At least 70 people died (IRIN 2009c; BBC News 2009). In Burkina Faso more than 150,000 people fled their homes, mostly in the capital Ouagadougou, where rainfall in one day was equal to about 25% of the normal annual total (BBC News 2009; Fominyen 2009; Schlein 2009).

Although the surface low was evident, it was poorly defined until 3 September, when it developed rapidly off the shore of Guinea-Bissau (to form a tropical depression) that moved westnorth-west across the Atlantic.

A7.3 The wave of 2–11 September 2009 and the development of Hurricane Fred

The next development occurred as a sharpening mid-tropospheric trough encountered the slow-moving upper trough on 2 September, a line of thunderstorms forming late in the day over Nigeria and Niger. A mid-tropospheric low had developed by 1200 UTC on 3 September near Niamey (Niger), although this was less well developed than that within the earlier trough (Fig. A7.5). Nevertheless, by this time an MCS had formed over the developing squall line.[5] Over the next 8 days, it moved westwards within the mid-tropospheric trough, as shown in Fig. A7.3.

By this time, the dry slot had become a complex feature, reflecting the development of a cut-off upper low. If the water-vapour imagery at 1200 UTC in Fig. A7.6 is compared with that in Fig. A7.1, it can be seen that the dry area in the middle and upper troposphere has progressed southwards. The base of the western portion of the dry slot can be seen near 15°N, 8°W. Roca et al. (2005) note that the progress

of dry air in the upper troposphere across the western Sahel tends to restrict development in this region. It is likely that in this second case there was too much convective inhibition for the cloud mass to develop rapidly into an MCC,[6] although its area had increased to ~13 × 10^4 km² by 1200 UTC on 4 September. However, by this time precipitation was scattered and the squall line of the early stages was no longer recognizable.

A shallow poorly defined lower-tropospheric depression accompanied this second MCS over land. The poor development gave this system a somewhat different character from the one that had preceded it. As the MCS left the area of inhibition, development was more organized, forming an MCC with cloud-shield area ~16 × 10^4 km² by 1200 UTC on 5 September. There does not appear to have been the rapid injection of warm air ahead of the system within the monsoon flow in this case and precipitation was comparatively modest. It is possible that this combination of circumstances had an important effect on developments within the second wave. In this case, the lower-tropospheric low deepened slowly over land, moving slightly ahead of the mid-tropospheric trough within which it had formed. Its mean speed was similar to that of the preceding system over land and it might not have lasted more than about 2 days had it formed further east.

However, in humid air it developed quickly into a tropical revolving storm during 7 September over an SST of 29°C before it reached the Cape Verde Islands. This storm – the sixth of the Atlantic hurricane season, named Fred – then moved rapidly west-north-west and north-west before stagnating and filling near 18°N, 34°W as it reached an area of weak forcing in the upper troposphere.

The speed of the surface low decreased as it moved across the ocean and it was able to move with the mid-tropospheric trough across the Atlantic at a speed of about 5 m s⁻¹.

There was a small, but significant, rise in tropopause height associated with the MCC of

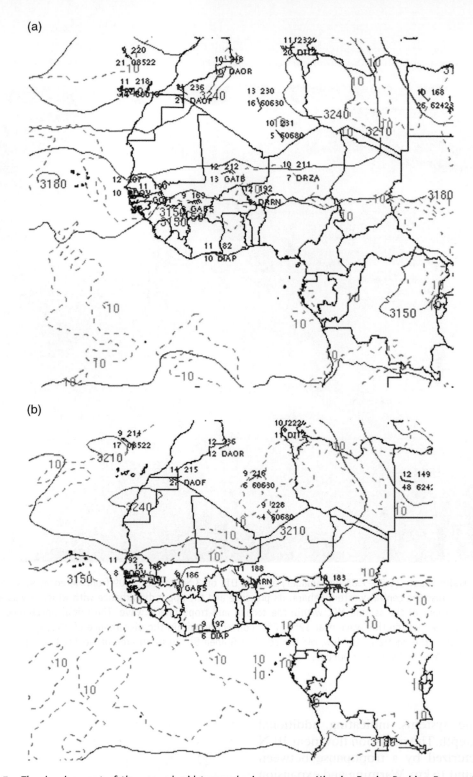

Figure A7.5 The development of the second mid-tropospheric wave over Nigeria, Benin, Burkina Faso and Mali shown by analyses of 700 hPa height at (a) 1200 UTC on 2 September 2009 and (b) 1200 UTC on 4 September 2009. In this case, development of the trough appears to have included similar strengthening of the gradient around the northern flank of the trough as was seen in the first trough on 29 August. However, the trough was in the form of two short-wave features north-west and north-east of its centre line, a possible reason for the lack of development of a squall line within this system. As these lobes became less developed on 6 September, close to the coast of Senegal, Guinea-Bissau and The Gambia, tropical depression formation commenced. The first trough can be seen over Mali and Guinea on 2 September. Courtesy of University of Wyoming, Department of Atmospheric Science website: weather.uwyo.edu/upperair/uamap.html.

Figure A7.6   Water-vapour (6.85–7.85 μm) imagery at 1200 UTC on 3 September 2009 from Meteosat-9. A complex dry slot can be seen across the central-western Sahara, its presence and the upper disturbance with which it was associated aiding the development of the second MCC within the second mid-tropospheric wave. The initial development of deep convection can be seen within this wave near 12°N, 2°E. Convection within the disturbance of 28 August–5 September has moved away from the centre of the low, off the coast of Guinea Bissau in an area of weaker forcing. © EUMETSAT, obtained from NEODASS, University of Dundee.

each of the systems, indicating additional warmth in depth. The monsoon flow near 10°N was characterized by a tropopause between about 15.5 and 16 km. Warming and expansion in the upper troposphere lifted the tropopause to about 16.5 km, a height maintained across each MCC. Immediately behind the systems, tropopause heights fell to their previous levels.

A7.4   A model for the development of squall lines in the Sahel

The systems described in this paper form typical MCCs. The first was associated with an intense squall line and satellite-derived rain rates[7] in excess of 32 mm h^{-1} (Fig. A7.7). Rapid development on 31 August formed a

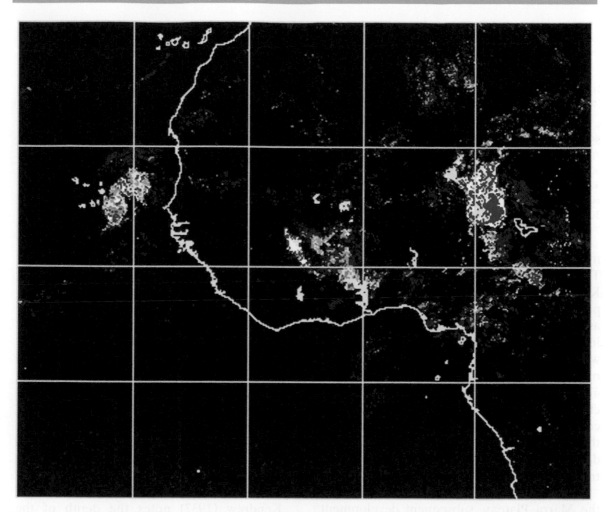

Figure A7.7   The development of a squall line of intense precipitation (estimated >32 mm h⁻¹) at 1500 UTC on 1 September 2009 over southern Mali, western Burkina Faso and northern Côte d'Ivoire. Satellite imagery from Meteosat-9 was used to produce this 'radar' image from a combination of the 0.8 μm (visible), 1.6 μm, 3.9 μm (both near-infrared) and 10.8 μm (infrared) channels. Assumptions about the relationship of rainfall to cloud-top temperature and cloud-drop size produce unrealistic 'showers' over the Sahara desert, where clouds are dense, but have a high base. Precipitation estimates may be considered reasonable in the moist monsoon zone. The area of less intense rain behind the squall line is likely to have fallen from comparatively modest convective towers and, to some extent, layer clouds forming the eastern part of the MCC. (The area of precipitation north-west of Lake Chad is likely to be an over-estimate, the convective cloud in this area developing mainly from mid-tropospheric levels above 3000 m, although the potential for development in this area is indicated by Fig. 10.14.) © Crown copyright (Met Office).

line of intense precipitation that moved steadily west, bringing 4–5 hours of rain or hail. (The squall line of the second system was not associated with MCC development, as discussed above. However, this fragmenting line of cumulonimbus brought heavy rain to Mali, Guinea, Senegal, Guinea-Bissau and

The Gambia on 5 and 6 September as the MCC formed.)

Many squall lines are observed in association with MCCs in West Africa (Leroux 2001), although short periods of very heavy rain, sometimes associated with squall lines, need not be from cumulonimbus embedded in

Table A7.1   Temperature (T), wet-bulb potential temperature (θ_w), geopotential height (H) and relative humidity (RH) data from Bamako, Mali at 1200 UTC

Level (mbar)	1 September				2 September				3 September			
	T (°C)	θ_w (°C)	H (dam)	RH (%)	T (°C)	θ_w (°C)	H (dam)	RH (%)	T (°C)	θ_w (°C)	H (dam)	RH (%)
925	23.6	23.3	81	77	21.4	22.5	80	88	21.2	22.4	81	89
850	18.8	21.7	155	77	17.6	22.9	153	95	18.4	22.0	154	81
700	11.2	20.0	319	54	9.0	20.6	317	82	13.2	20.4	320	47
600	1.4	18.0	446	50	2.8	20.6	445	80	2.8	20.3	445	71
500	−5.9	19.0	591	38	−5.5	21.0	589	88	−5.7	19.1	592	49
400	−16.1	19.8	762	11	−15.1	21.6	761	78	−15.7	20.3	764	38
300	−31.5	22.1	972	8	−29.3	22.0	972	56	−31.3	22.5	974	19

mesoscale cloud systems (Fink & Reiner 2003). They were well known by the 1930s, although Kendrew (1937) refers to them as 'tornadoes', perhaps reflecting the very strong winds associated with them, locally including rotating vortices. The squall line in the first easterly wave lasted less than 36 hours and the MCCs of other wave phases did not contain identifiable squall lines. A line of cumulonimbus clouds forms where air from the north converges with moist air that moves north and then west, often in association with a surface depression, under the influence of the mid-tropospheric trough.

Although the first MCS formed in the lee of the Marra Plateau, subsequent development within this mid-tropospheric trough does not seem to have been assisted by orographic forcing, at least in its initial phase. However, the squall line may have intensified as the trough passed over the rising ground in Burkina Faso and Mali. The system was notably weak over the lower ground of the plains of the River Niger (as the surface depression ran ahead of that at 700 hPa). Pronounced development is often observed when there is a coincidence of maximum insolation as a system encounters high ground. The Marra Plateau is noted as the main area of development of African easterly waves (Throncroft et al. 2008).

It is notable that the radiosonde ascent near the squall line (Fig. A7.8b), in which air was close to saturation between 700 and 11,000 m,

850 hPa θ_w was close to that of the moist southerly flow (~23°C). Instability was greater at low levels due to moistening and cooling just above the surface. Observed data across the trough at standard levels are shown in Table A7.1. It can be seen in Fig. A7.8(b) that the warm air ahead of the system (Fig. A7.8a) has begun to undercut the easterlies of the moist monsoon, assisting in the release of convective energy from the monsoon air in the region of mid-tropospheric convergence.

The poorly developed squall line in the second system may be explained by weaker convergence in the low-level flow and restricted development under the influence of dry air aloft.

Kendrew (1937) notes the depth of the humid monsoon layer over West Africa as reaching a depth of about 1250 m, as seen in this case (Fig. A7.8c). This high humidity at low levels feeds the development of thunderstorms, drier (low θ_w) air above the moist (high θ_w) flow associated with deep instability within the mid-tropospheric wave.

Various instability indices have been devised to predict deep convection and the likelihood of thunderstorms. The Potential Instability Index (PII; Bradbury 1977) and K index (George 1960) are commonly used in the UK, where PII values of −2°C or lower and K index values of 20 or more indicate probable thunderstorm development in summer.[8] Both indices use a combination of lapse rate and humidity as predictors.

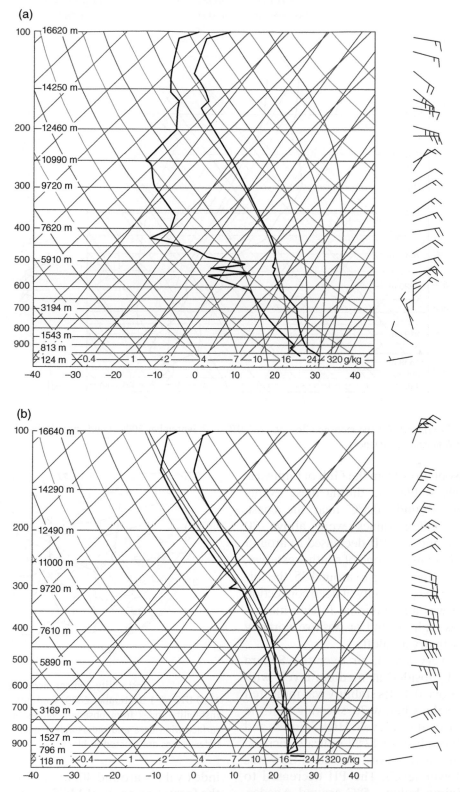

Figure A7.8   Radiosonde profiles for Bamako, Mali: (a) the Harmattan flow ahead of the first mid-tropospheric wave at 1200 UTC on 1 September 2009, (b) the squall line at 1200 UTC on 2 September 2009 and

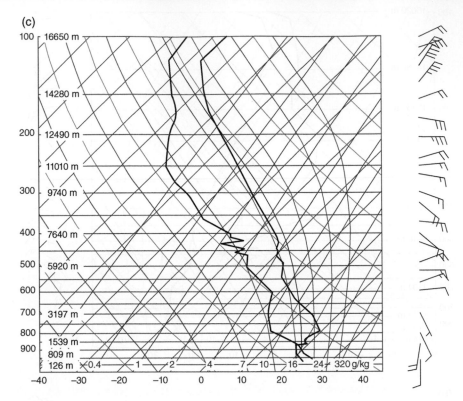

Figure A7.8 *(Continued)* (c) the monsoon flow at 1200 UTC on 3 September 2009. Courtesy of University of Wyoming, Department of Atmospheric Sciences.

The monsoon flow from the Gulf of Guinea is comparatively stable and during these events the PII was around –0.5°C near the coast, although *K* index values were above 30 (Fig. A7.9). Inland, the PII decreased to around –1.5°C over land in the zone of south-westerlies, with *K* index values rising to around 35, the former indicating the possibility of deep convection where there was an additional forcing mechanism, such as high ground or positive vorticity associated with dry air aloft.

Although very warm, air over the Sahara was dry and subsided at medium levels with PII values above 0°C. Its stability was indicated by *K* index values below 30, showing that the air was also dry at low levels. High *K* index and low PII values close to 12°N indicate increasing instability as the moist monsoon flow was heated at lower levels. The PII decreased to notable minima below –5°C around Agadez, Niger and Bamako, Mali, coincident with *K*

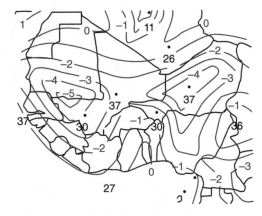

Figure A7.9 Instability indices across West Africa at 1200 UTC on 1 September 2009. Contours show Potential Instability Index from observed values, model products, surface wind data and radiosonde sequences. Observed *K* index values are plotted in black.

index values around 40, in particular ahead of the formation areas of MCSs in unstable air. In this zone mixing and convergence of the warm

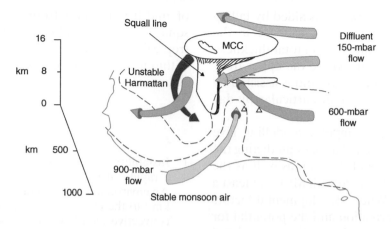

Figure A7.10   An idealized scheme for the development of a West African squall line: the cores of low-level (monsoon and Harmattan), medium-level and high-level flows are indicated by arrows.

and moist flows assist in the development of deep instability, spurring MCS formation.

The PII of the southward-flowing air ahead of the first surface low was below –3°C, strongly suggesting the potential for very heavy rain and damaging thunderstorms (Fig. A7.9), although a PII above –1°C is analysed to have been present behind this low. This suggests that gradients of PII are a good indicator[9] of the likelihood of squall-line development. These gradients agree well with observed areas of heavy precipitation over land in Fig. A7.7. Thus, whilst large-scale deep convection is not seen to coincide with areas of lowest PII, it coincides with gradients of PII where K index values are high. This has some similarities to the convective developments seen along gradients of low-level θ_w that coincide with convergence lines in the middle latitudes.

A schematic for the flow and cloud pattern seen in the first MCC as it crossed Burkina Faso is shown in Fig. A7.10.

Apart from heavy rain, squall lines may be associated with large hail. This is supported by winds veering and increasing with height just ahead of squall lines, in particular in the northern part of a squall line. The wind shear also supports the possibility that tornadoes could have formed. Hail and tornadoes are particularly damaging to crops and may affect food

production as a result, possibly negating the positive impact of the rainfall.

A7.5   Conclusion

Through studies in recent years we now have a greater understanding of the development and decay of squall lines in West Africa, some of them associated with MCCs. There is a strong link between mid-tropospheric waves, squall lines and MCCs. In this case, the persistence of waves (Diop & Grimes 2003) allowed several phases of convective development: MCS–latent–MCC–MCC with squall line–MCC–tropical depression in the first and squall line–MCS with squall line–MCC–tropical depression–tropical revolving storm in the second. However, as this suggests, prediction is not straightforward, since both MCCs and squall lines usually have a short lifetime within a mid-tropospheric wave and MCC formation may follow squall-line formation; in mid-tropospheric waves a squall line does not form.

A coincidence of factors is necessary for the development and persistence of MCCs: a mid-tropospheric wave must engage moist air in conditions where instability may be released, reducing pressure at the surface, with this area of lowered pressure moving at the same speed as the trough. Development is aided by falling upper-tropospheric heights, diffluence in the upper troposphere and a fall in total thickness. MCC formation occurs where the upper troposphere is sufficiently moist and is assisted by ascent over high ground, in particular by day (Ramage 1971).

Improvements in computer modelling give the forecaster a greater chance of predicting the heavy rainfall, hail and thunder, as well as the risk of flooding associated with MCCs, at least a day or so ahead. Where developmental factors coincide, MCC formation and the potential for damage may be expected, sometimes associated with squall lines. However, computer models need accurate data both from the surface and in the upper air to produce forecast fields sufficiently accurate to indicate the necessary combination of wind flow and stability. Where few observations exist, the accuracy of prediction suffers. A high-density observation network could not be expected to be established in the semi-arid Sahel, although there is a reasonably good network of radiosonde stations within the zone, indeed better than that further south, in the near-coastal zone.

Nevertheless, prediction is important, since much of the rainfall in the Sahel is from these occasionally damaging systems, which often bring heavy rain or hail during the summer monsoon, especially between July and early September. Even 18–24 hours' warning of such damaging weather would be sufficient to allow steps to be taken to mitigate the effects. Near its centre, Fig. A7.3 suggests that the squall line in the first wave was about 150 km across. Thus, although the mean rainfall from each West African wave-related system is around 10 mm (Diop & Grimes 2003), in this case rainfall totals are likely to have been generally in excess of 150 mm. This is reflected in the fall of more than 200 mm in Ouagadougou on 1 September, reported as the wettest day in the capital of Burkina Faso for more than 90 years (BBC News 2009).

As has been shown, not all mid-tropospheric waves produce squall lines. The development of mid-tropospheric troughs associated with squall lines over land does not indicate that a tropical depression or hurricane will form within it once it has reached the warm seas of Atlantic Ocean.

Notes

1   Instability is more dependent on moisture than a rapid fall in temperature above the surface. An environmental lapse rate of only about 10 K km^{-1} to the cloud base is capable of producing convective clouds when the air is very moist. Little or no convective cloud will form if humidity is low, even with a lapse rate much greater than 10 K km^{-1} in the lowest few hundred metres.

2   The air ahead of the areas of convection discussed in this paper is not a 'true' Harmattan; the driest air is usually confined to northerly winds in the lower troposphere, the flow more moist in the easterlies of the middle troposphere.

3   As air enters a trough, it must decelerate to maintain the balance between the pressure-gradient force and the Coriolis force. This deceleration causes convergence behind and across the trough axis. However, air accelerates and diverges ahead of the trough where curvature becomes anti-cyclonic. These factors are important in the formation and decline of large-scale tropical disturbances within easterly waves.

4   It is also probable that the dusty Saharan air provides condensation nuclei that assist in the development of deep convection (Zipser et al. 2009).

5   Squall lines may form and decay without the development of an MCC. This form often has a shorter lifespan than those that develop within an MCC.

6   This is reflected in dry subsided air above 870 hPa at Bamako (Mali) at 1200 UTC on 3 September.

7   Rainfall rates can be derived from satellite imagery; the MSG series of satellites uses cloud depth, liquid water content and cloud-top temperature data (Francis et al. 2006).

8   From data in this paper it appears that whilst a similar threshold of PII to that used for western Europe indicates deep convection in West Africa, higher values of K index are required, above about 30. The greater dependence of the

K index on humidity, compared with the dependence of PII on change of humidity and temperature with height, appears to make the former a weaker indicator than the latter.

9   Lifted index (not commonly used by UK forecasters) also gives good guidance and appears

similar to PII (although numerical values have greater magnitude). However, it is notable that the *K* index, whilst indicating regions of potential instability, is a weaker indicator of the areas in which MCCs are likely to develop within an easterly wave.

References

Abarca JF & Casiccia CC. 2002. Skin cancer and ultraviolet-B radiation under the Antarctic ozone hole: southern Chile, 1987–2000. *Photodermatol. Photoimmunol. Photomed.* **18** (6): 294–302.

Abdulazziz A-O & Essa R. 1994. A study of local thunderstorms (Sarrayat) in Kuwait. *J. Meteorol. (UK)* **19**: 150–155.

Abrahams J, Campbell-Lendrum D, Kootval H, Love G & Otmani del Barrio M (eds). 2012. *Atlas of health and climate*. WHO/WMO, Geneva.

Al-Maskari J & Gadian A. 2005. A study of orographic convection over the Hajar mountains in northern Oman. *RMetS Conf. 2005 Abs., Plenary Rev. Pap. Weather Air Quality Forecast.*: 34.

Alojado D & Padua DMV. 2010. *The twelve worst typhoons of the Philippines (1947–2009)*. typhoon2000.ph/stormstats/12WorstPhilippineTyphoons.htm, accessed 8 December 2014.

Andreae MO, Talaue-McManus L & Matson PA. 2004. Anthropogenic modification of land, coastal and atmospheric systems as threats to the functioning Earth system. *Earth system analysis for sustainability*: 245–264.

Andreotti B, Claudin P & Douady S. 2002. Selection of dune shapes and velocities. Part 1: Dynamics of sand, wind and barchans. *Eur. Phys. J. B*, **28**: 321–339.

Andrews DG, Holton JR & Leovy CB. 1987. *Middle Atmosphere Dynamics*. Academic Press.

Anthes RA. 1982. Tropical cyclones – their evolution, structure and effects. *American Meteorological Society Meteorological Monograph*, **19** (41). Boston, MA.

Aquino M & Ford D. 2008. *Peru bets on desalination to ensure water supplies*. http://uk.reuters.com/article/2008/03/11/environment-peru-water-dc-idUKN1161583720080311, accessed 23 June 2014.

Asif Q. 2008. Water shortage to hit drinking supplies. *Daily Times*, 29 February 2008: 7–34.

Asnani GC. 1993. *Tropical meteorology*. Pune.

Atkinson GD. 1971. *Forecasters' Guide to Tropical Meteorology*. University Press of the Pacific, Honolulu HI.

Atkinson BW. 1981. *Meso-scale atmospheric circulations*. Academic Press, London.

Attri SD & Tyagi A. 2010. Climate profile of India. *India Meteorol. Dept.* http://imd.gov.in/doc/climate_profile.pdf.

Augustine JA & Howard KW. 1988. Mesoscale convective complexes over the United States during 1985. *Mon. Weath. Rev.* **116**: 685–701.

Bader MJ, Forbes GS, Grant JR, Lilley RBE & Waters AJ. 1995. *Images in weather forecasting*. Cambridge University Press.

Baldwin MP, Gray LJ, Dunkerton TJ et al. 2001. The Quasi-Biennial Oscillation. *Rev. Geophys.* **39**: 179–229.

Barkan J & Alpert P. 2010. Synoptic analysis of a rare event of Saharan dust reaching the Arctic region. *Weather* **65**: 208–211.

Barrett EC. 1971. The tropical Far East: ESSA satellite evaluations of high season climatic patterns. *Geogr. J.* **137**: 535–555.

BBC News. 2009. *UN warns on West Africa floods*. http://news.bbc.co.uk/2/hi/africa/8239552.stm, accessed 10 December 2014.

BBC News. 2010a. *Fresh downpours hamper Pakistan flood relief.* http://www.bbc.co.uk/news/world-south-asia-10904903, accessed 6 December 2014.

BBC News. 2010b. *Asian monsoon's range of devastation.* http://www.bbc.co.uk/news/world-south-asia-10951517, accessed 6 December 2014.

BBC News. 2011a. *Queensland survives Cyclone Yasi with no known deaths.* http://www.bbc.co.uk/news/world-asia-pacific-12351647, accessed 7 December 2014.

BBC News. 2011b. *Brazil floods: more than 500 dead.* http://www.bbc.co.uk/news/world-latin-america-12187985, accessed 10 December 2014.

Bell GD, Halpert MS, Schnell RC et al. 2000: Climate assessment for 1999. *Bull. Amer. Meteorol. Soc.* **81**, S1–S50.

Berry G, Thorncroft C & Hewson T. 2007. African easterly waves during 2004 – analysis using objective techniques. *Mon. Weath. Rev.* **135**: 1251–1267.

Betts RA, Cox PM, Collins M, Harris PP, Huntingford C & Jones CD. 2004. The role of ecosystem-atmosphere interactions in simulated Amazonian precipitation decrease and forest dieback under global climate warming. *Theor. Appl. Climatol.* **78**: 157–175.

Betts RA, Malhi Y & Timmons Roberts J. 2007. The future of the Amazon: new perspectives from climate, ecosystem and social sciences. *Phil. Trans. R. Soc.* **363**: 1729–1735.

Bjerknes J. 1969. Atmospheric teleconnections from the equatorial Pacific. *Mon. Weath. Rev.* **97**: 163–172.

Black R. 2006. *Climate link to African malaria.* http://news.bbc.co.uk/1/hi/sci/tech/4827362.stm, accessed 16 January 2015.

Bond NA & Vecchi GA. 2003. The influence of the Madden–Julian Oscillation in precipitation in Oregon and Washington. *Weath. Forecast.* **18**: 600–613.

Bormann S, Soloman S, Dye JE, Baumgartner D, Kelly KK and Chan RG. 1997. Heterogeneous reactions on stratospheric background aerosols, volcanic sulfuric acid droplets, and type I polar stratospheric clouds: Effects of temperature fluctuations and differences in particle phase. *J. Geophys. Res.* **102**: 3639–3648.

Boucher O. 2010. Stratospheric ozone, ultraviolet radiation and climate change. *Weather* **65**: 105–110.

Bradbury TAM. 1977. The use of wet-bulb potential temperature charts. *Meteorol. Mag.* **106**: 233–251.

Brilliant L. 2007. Climate, Poverty and Health. The John H Chafee Memorial Lecture on Science and the Environment. 7th Nat. Conf. Sci., Policy, Environ., 1 February 2007, Washington DC. http://www.eoearth.org/view/article/51cbed3c7896bb431f690c17/, accessed 16 January 2015.

Brown R. 1973. New indices to locate clear air turbulence. *Meteorol. Mag.* **102**: 347–361.

Browning KA. 1963. The growth of hail within a steady updraught. *Quart. J. R. Meteorol. Soc.* **89**: 490–506.

Browning KA. 1997. The dry intrusion perspective of extra-tropical cyclone development. *Meteorol. Appl.* **4**: 317–324.

Brugge R & Stuttart M. 2003. Back to basics. From Sputnik to Envisat, and beyond: the use of satellite measurements in weather forecasting and research. Part 1 – A history. *Weather* **58**: 107–112.

Bureau of Meteorology. 2014. *Australian climate variability & change – Trend maps.* http://www.bom.gov.au/cgi-bin/climate/change/trendmaps.cgi?map=rain&area=aus&season=0112&period=1910, accessed 25 June 2015.

Burgess ML & Klingaman NP. 2015. Atmospheric circulation patterns associated with extreme cold winters in the UK. *Weather*, **70**: 211–217.

Burkholder JB & Orlando JJ. 1998. Rate coefficient upper limits for the $BrONO_2$ and $ClONO_2 + O_3$ reactions. *Geophys. Res. Lett.* **25**: 3567–3569.

Burroughs WJ. 2005. *Climate change in prehistory.* Cambridge University Press.

Burt S. 1991. Falls of dust rain within the British Isles. *Weather* **46**: 347–353.

Burt SD. 2014. *Synoptic transport and deposition of Saharan dust to the British Isles.* MSc Dissertation. University of Reading, Department of Meteorology.

Butchart N & Austin J. 1996. On the relationship between the QBO, total chlorine and the severity of the Antarctic ozone hole. *Quart. J. R. Meteorol. Soc.* **122**: 183–217.

Bysouth CE. 2000. Clear air turbulence – a reply. *Weather* **55**: 122, 147.

Calhoun RC. 1981. *Typhoon, the other enemy: the Third Fleet and the Pacific storm of December 1944.* Naval Institute Press, Annapolis, MD.

Callen NS & Prescott P. 1982. Forecasting daily maximum temperatures from 1000–850 mb thickness lines and cloud cover. *Meteorol. Mag.* **111**: 51–58.

Camberlin P & Philippon N. 2001. The stationarity of lead-lag teleconnections with East Africa rainfall and its incidence on seasonal predictability, in Brunet India M, Lopez Bonillo D (eds), *Detecting and Modelling Regional Climate Change*. Springer-Verlag, Berlin: 291–307.

Carlson KM, Curran LM, Asner GP et al. 2013. Carbon emissions from forest conversion by Kalimantan oil palm plantations. *Nature Clim. Change* **3**: 283–287.

Chamberlain G, Shah S & Jones S. 2010. Pakistan floods: UN urges world to step up aid efforts. *The Guardian.* http://www.theguardian.co.uk/world/2010/aug/15/pakistan-floods-united-nations-aid, accessed 7 December 2014.

Chang CB. 1993. Impact of desert environment on the genesis of African wave disturbances. *J. Atmos. Sci.* **50**: 2137–2145.

Chaudhuri S & Middey A. 2009. The applicability of bipartite graph model for thunderstorms forecast over Kolkata. *Adv. Meteorol.*: doi:10.1155/2009/270530.

Cohen AJ, Anderson HR, Ostro B et al. 1999. Chapter 17. Urban air pollution. *Comparative Quantification of Health Risks*. Geneva, WHO.

Collins M & Senior CA. 2002. Projections of future climate change. *Weather* **57**: 283–287.

Collins M, Fricker T & Hermanson L. 2011. From observations to forecasts – Part 9: what is decadal forecasting. *Weather* **66**: 160–164.

Connell D. 2005. Managing climate for the Murray–Darling basin (1850–2050). In Sherratt T, Griffiths T, Robin L (eds) *Change in the Weather*. National Museum of Australia Press, Canberra: 82–91.

Cornforth R, Hoskins B & Thorncroft C. 2005. The African easterly jet-waves system: moist physics. *RMetS Conf. 2005 Abs., Plenary Rev. Pap. Weath., Air Quality Forecast.*: 36.

Cornish MM & Ives EE. 2006. *Reed's maritime meteorology*, 3rd edition. Adlard-Coles Nautical, London.

Coronas J. 1912. *Meteorological bulletin for October 1912*. Philippines Weather Bureau, Manila.

Cox PM, Betts RA, Jones CD, Spall SA & Totterdell IJ. 2000. Acceleration of global warming due to carbon-cycle feedbacks in a coupled climate model. *Nature* **408**: 184–187.

Cox PM, Betts RA, Collins M, Harris PP, Huntingford C & Jones CD. 2004. Amazonian forest dieback under climate-carbon cycle projections for the 21st Century. *Theor. Appl. Climatol.* **78**: 137–156.

Dai A. 2013. Increasing drought under global warming in observations and models. *Nature Clim. Change* **3**: 52–58.

de Gruijl FR. 1995. Impacts of a projected depletion of the ozone layer. *Consequences* **1** (2). http://gcrio.org/CONSEQUENCES/summer95/impacts.html, accessed 17 January 2015.

de Villiers MP & van Heerden J. 2007a. Dust storms and dust at Abu Dhabi international airport. *Weather*, **62**: 339–343.

de Villiers MP & van Heerden J. 2007b. Fog at Abu Dhabi international airport. *Weather* **62**: 209-214.

de Villiers M & van Heerden J. 2011. Nashi dust storm over the United Arab Emirates. *Weather* **66**: 79–81.

Delgado Martin L, Garcia Diez A, Rivas Soriano L & Garcia Diez EL. 1997. Meteorology and forest fires: conditions for ignition and conditions for development. *J. Appl. Meteorol.* **36**: 705–710.

Demographia. 2014. *Demographia world urban areas*, 10th annual edition. http://demographia.com/db-worldua.pdf, accessed 25 June 2015.

Desbois M, Pircher V & Pinty B. 1988. Validation and use of Meteosat cloud winds for the study of low-level flows during the WAMEX experiment. *WAMEX Related Res. Tropic. Meteorol. in Africa*: 115–116.

Dhar ON & Nandargi S. 2005. Areas of heavy precipitation in the Nepalese Himalayas. *Weather* **60**: 354–356.

Diaz HF & Anderson CA. 1995. Precipitation trends and water consumption related to population in the southwestern United States – a reassessment. *Water Resources Res.* **31**: 713–720.

Ding Y. 1994. *Monsoons over China*. Kluwer Acad. Sci. Lib., Dordrecht.

Ding Y & Hu J. 1988. The variation of the heat sources in east China in the early summer of 1984 and their effects on the large-scale circulation in east Asia. *Adv. Atmos. Sci.* **5**: 171–180.

Diop M & Grimes DIF. 2003. Satellite-based rainfall estimation for river-flow forecasting in Africa. II: African Easterly Waves, convection and rainfall. *Hydrol. Sci. J.* **48**: 585–599.

Dobson R. 2005. Ozone depletion will bring big rise in number of cataracts. *Brit. Med. J.* **331** (7528): 1292–1295.

Dubey RC & Chandra S. 1991. Application of meteorology for the effective control of the desert locust. *Meteorology for locust control* (WMO/TD-No.404). WMO, Geneva.

Dunlop S. 2004. *Wild guide. Weather.* Collins, London.

Ellis S & Mellor A. 1995. *Soils and environment.* Routledge, London.

Ellrod GP. 1989. A decision-tree approach to clear air turbulence analysis using satellite and upper air data. *NOAA Tech. Mem. NESDIS 23.* Satellite Applications Laboratory, Washington DC.

Ellrod GP & Knapp DI. 1992. An objective clear-air turbulence forecasting technique: verification and operational use. *Weath. Forecast.* **7**: 150–165.

Elsberry RL. 2006. Research to support improved tropical cyclone landfall forecasts and warnings. *WMO Bull.* **55**: 200–209.

Emanuel KA. 1988. Large-scale and mesoscale circulations in convectively adjusted atmospheres. *Proc. Workshop Diabatic Forcing, ECMWF, Reading, 30 Nov.–2 Dec. 1987*: 323–348.

Emanuel K. 2005. *Divine Wind: The History and Science of Hurricanes.* Oxford University Press, New York NY.

Eumetsat. 2014. *Products.* eumetsat.int/website/home/Data/Products/index.html, accessed 16 December 2014.

FAO-UNESCO. 1989. *Soil map of the world: revised legend.* Food and Agriculture Organization/United Nations Educational Social and Cultural Organization, International Soil Reference and Information Centre, Wageningen.

Farman JC, Gardiner BG & Shanklin JD. 1985. Large losses of total ozone in Antarctica reveal seasonal CIOx/NOx interaction. *Nature* **315**: 207–210.

Fears TR, Bird CC, Guerry D et al. 2002. Average midrange ultraviolet radiation flux and time outdoors predict melanoma risk. *Cancer Res.* **62** (14): 3992–3996.

Fedorov AV. 2002. The response of the coupled tropical ocean-atmosphere to westerly wind bursts. *Quart. J. R. Meteorol. Soc.* **128**: 1–23.

Fink AH & Knippertz P. 2003. An extreme precipitation event in southern Morocco in spring 2002 and some hydrological implications. *Weather* **58**: 377–387.

Fink AH & Reiner A. 2003. Spatiotemporal variability of the relation between African Easterly Waves and West African Squall Lines in 1998 and 1999. *J. Geophys. Res.* **108D**: 4332–4348.

Fisher M & Membery DA. 1998. Climate. In Ghanzafar SA, Fisher M (eds) *Vegetation of the Arabian Peninsula*: 5–38.

Fleming RJ. 1986. *The Tropical Ocean and Global Atmosphere programme.* ICSU and WMO International TOGA Project Office, Boulder CO.

Fominyen G. 2009. West Africa's seasonal floods in 2009. *ReliefWeb.* http://reliefweb.int/rw/rwb.nsf/db900SID/SNAA-7VLBLG?OpenDocument, accessed 10 December 2014.

Food and Agriculture Organization. 2009. *Frequently Asked Questions (FAQs) about locusts.* http://www.fao.org/ag/locusts/en/info/info/faq/index.html, accessed 4 Janaury 2015.

Forster PM de F & Shine KP. 1997. Radiative forcing and temperature trends from stratospheric ozone changes. *J. Geophys. Res.* **102**: 10841–10855.

Fothergill A, Berlowitz V, Brownlow M, Cordey H, Keeling J & Linfield M. 2006. *Planet earth – as you've never seen it before.* BBC Books, London.

Francis P, Capacci D & Saunders R. 2006. Improving the Nimrod nowcasting system's satellite precipitation estimates by introducing the new SEVIRI channels. *Proc. 2006 EUMETSAT Meteorol. Satell. Conf., Helsinki, Finland.* http://www.eumetsat.int/home/Main/AboutEUMETSAT/Publications/ConferenceandWorkshopProceedings/2006/groups/cps/documents/document/pdf_conf_p48_s2a_04_francis_v.pdf, accessed 10 December 2014.

Frank WM. 1977. The structure and energetics of the tropical cyclone, Paper I: storm structure. *Mon. Weath. Rev.* **105**: 1119–1135.

Fuller SR. 2004. Recent forecast model improvements. *NWP Gazette*, June 2004: 10.

Gale EL & Saunders MA. 2013. Thailand flood: climate causes and return periods. *Weather* **68**: 223–227.

Galvin JFP. 2004. Lee-slope katabatic winds on the Philippine island of Mindanao. *Weather* **59**: 127–131.

Galvin JFP. 2005. Typhoon *Nida* and its effects in the Philippines. *Weather* **60**: 71–74.

Galvin JFP. 2007. Weather Image: Severe tropical storm over India and South-West Asia. *Weather* **62**: 337–338.

Galvin JFP. 2010. Weather Image: Advancing monsoon. *Weather*, **65**: 195.

Galvin JFP. 2012. Dust events in Cyprus. *Weather* **67**: 283–290.

Galvin JFP. 2014. The development, track and destruction of typhoon Haiyan. *Weather* **69**: 307–309.

Galvin JFP & Lakshminarayanan R. 2006. Weather image: The South-West Monsoon and the Equatorial Easterly Jet. *Weather* **61**: 296.

Galvin JFP & Walker JM. 2007. Weather image: Cloudy South-East Asia. *Weather* **62**: 55–56.

Galvin JFP, Black AI & Priestley D. 2011. Mesoscale features over the Mediterranean, Part 2. *Weather* **66**: 87–94.

Gass I et al. 1986. *The Earth's physical resources (S238), Block 4. Water resources*. Open University, Milton Keynes.

Gates WL & Newson RL. 2006. World Climate Research Programme: a history. *WMO Bull.* **55**: 210–216.

Gaye A, Viltard A & de Felice P. 2005. Squall lines and rainfall over western Africa during summer 1986 and 87. *Meteorol. Atmos. Phys.* **90**: 215–224.

George JJ. 1960. *Weather Forecasting for Aeronautics*. Academic Press, New York.

Ghulam A & Dorling S. 2005. Synoptic study on mechanisms producing spring thunderstorms in Saudi Arabia. *RMetS Conf. 2005 Abs., Plenary Rev. Pap. Weather Air Quality Forecast.*: 35.

Giles BD. 2011. The Queensland floods of December 2010/early January 2011 – and the media. *Weather* **66**: 55.

Global Atmospheric Research Programme. 1969. *Report of the First Session of the Study Group on Tropical Disturbances*. Joint Organizing Cttee. GARP, Madison, WN.

GDACS. 2009. *Green Flood Alert in Senegal*. Global Disaster Alert and Coordination System. webcitation.org/5jdQQGq59, accessed 10 December 2014.

Godbole RV & Shukla J. 1981. Global analysis of mean sea-level pressure. *NASA Tech. Memo. 82097*. Goddard Space Flight Centre, Greenbelt MD.

Gordon AH. 1973. The great Philippine floods of 1972. *Weather* **28**: 404–415.

Graham R. 2009. *Met Office tropical storm forecast for the North Atlantic, July to November 2009*. http://www.metoffice.gov.uk/media/pdf/t/p/forecast2009.pdf, accessed 25 June 2015.

Grahame NS & Davies P. 2008. Forecasting the exceptional rainfall events of summer 2007 and communication of key messages to Met Office customers. *Weather* **63**: 268–273.

Grahame N, Page A, Hickman A & Pearson C. 2015. An unusual thunderstorm event overnight 13/14 June 2014. *Weather* **70**: 167–172.

Grant K. 1995. The British 'dry line' and its role in the genesis of severe local storms. *J. Meteorol.* **20**: 241–259.

Gray L & Lahoz W. 2002. *The impact of the stratosphere on tropospheric climate*. http://www.ugamp.nerc.ac.uk/sr9798/wl.htm, accessed 7 January 2015.

Grimes DIF. 2003. Satellite-based rainfall monitoring for food security in Africa. In *Crop and Rangeland Monitoring in Eastern Africa – for Early Warning and Food Security*, Proc. Int. Works. Crop Rangeland Monit. East Africa, Nairobi, Jan. 2003. European Commission, Strasbourg.

Grimes DIF & Diop M. 2003: Satellite-based rainfall estimation for river flow forecasting in Africa. Part I. Rainfall estimates and hydrological forecasts. *Hydrol. Sci. J.* **48**: 567–584.

Groeningen KJ, Van Kessel C & Hungate BA. 2013. Increased greenhouse-gas intensity of rice production under future atmospheric conditions. *Nature Clim. Change* **3**: 288–291.

Grosvenor G, Choularton T, Coe H & Held G. 2005. Cloud resolving model studies of deep tropical convection observed during HIBISCUS 2004. *RMetS Conf. 2005 Abs., Plenary Rev. Pap. Weath., Air Quality Forecast.*: 25.

Gu G & Adler RF. 2003. Seasonal rainfall variability within the West African monsoon system. *CLIVAR Exchanges* **8**: 11–15.

Guhathakurta P. 2007. Highest recorded point rainfall over India. *Weather* **62**: 349.

Gullison RE, Frumhoff PC, Canadell JG, et al. 2007. Tropical forests and climate policy. *Science* **316**: 985–986.

Guo Q & Wang J. 1981. The distribution of precipitation in China during the summer monsoon period for 30 years. *Acta Geogr. Sin.* **36**: 187–195.

Häder D-P, Worrest RC & Kumar HD. 1991. Aquatic ecosystems. *Environmental Effects of Ozone Depletion: 1991 update*. United Nations Environment Programme, Nairobi.

Hall JD, Matthews AJ & Karoly DJ. 2001. The modulation of tropical cyclone activity in the Australian region by the Madden–Julian Oscillation. *Mon. Weath. Rev.* **129**: 2970–2982.

Hall JW, Brown S, Nicholls RJ, Pidgeon NF & Watson RT. 2012. Proportionate adaptation. *Nature Clim. Change* **2**: 833–834.

Hamilton MG. 1975. Some characteristics of cloud clusters over the Indian Ocean. *Weather* **30**: 2–13.

Hastenrath S. 1991. *Climate Dynamics of the Tropics*. Kluwer Academic Publishers, Dordrecht.

Hawkins HF & Imbembo SM. 1976. The structure of a small, intense hurricane, Inez 1966. *Mon. Weath. Rev.* **104**: 418–442.

Heaps A, Lahoz W & O'Neill A. 1999. *The quasi-biennial zonal wind oscillation (QBO)* http://www.ugamp.nerc.ac.uk/hot/ajh/qbo.htm, accessed 18 January 2015.

Heming J. 1999. More cyclonic success in the tropics. *NWP Gazette*, September 1999: 7.

Hewitt C, Mason S & Walland D. 2012. The global framework for climate services. *Nature Clim. Change* **2**: 831–832.

Higgins RW & Shi W. 2001. Intercomparison of the principal modes of interannual and intraseasonal variability of the North American monsoon system. *J. Clim.* **14**: 403–417.

Hofmann DJ, Deshler T, Aimedieu P et al. 1989. Stratospheric clouds and ozone depletion in the Arctic during January 1989. *Nature* **340**: 117–121.

Hole R. 2006. *Cyclone Larry at Tolga on the 20-3-06*. weather.org.au/tolga/larry.htm, accessed 7 December 2014.

Hollermann PW. 1993. Fire ecology in the Canary Islands and central California – a comparative outline. *Erdkunde* **47**: 177–184.

Holton JR. 1979. *An Introduction to Dynamic Meteorology*, 4th edition. Academic Press, San Diego, CA.

Holton JR. 1990. On the global exchange of mass between the stratosphere and troposphere. *J. Atmos. Sci.* **47**: 392–395.

Howard L. 2011. *Essay on the Modifications of Clouds*, 3rd edition. John Churchill, London.

International Rice Research Institute 2014. *Rice science for a better world*. Los Baños. http:/wsww.irri.org, accessed 15 December 2014.

IPCC. 2000. *Special Report on Emissions Scenarios. Summary for policymakers*. WMO and UNEP, Geneva.

IPCC. 2014. *Climate change 2014. Synthesis report*. WMO and UNEP, Geneva.

IRIN 2009a. *Burkina Faso-Ghana: One country's dam, another's flood*. Integrated Regional Information Networks. http://www.irinnews.org/Report.aspx?ReportId=86015, accessed 10 December 2014.

IRIN 2009b. *Niger: Desert flooding wipes out electricity, homes, livestock*. Integrated Regional Information Networks. irinnews.org/Report.aspx?ReportId=85996, accessed 10 December 2014.

IRIN. 2009c. *West Africa: Seasonal rains, seasonal misery*. Integrated Regional Information Networks. http:/www.irinnews.org/Report.aspx?ReportId=85942, accessed 10 December 2014.

Jackson CCE. 1988. Sea temperature in the Arabian Gulf. *Weather* **43**: 429–439.

Jackson GE, Smith RK & Spengler T. 2002. The prediction of low-level mesoscale convergence lines over northeastern Australia. *Austral. Meteorol. Mag.* **51**: 13–23.

James IN. 1994. *Introduction to circulating atmospheres*. Cambridge University Press.

Jefferson GJ. 1963. A further development of the instability index. *Meteorol. Mag.* **92**: 313–316.

Jia J-Y, Wang Y, Sun Z-B, Liu Y & Chen H-S. 2010. Leading spatial patterns of summer rainfall quasibiennial oscillation and characteristics of meteorological backgrounds over eastern China. *Chin. J. Geophys.* **53**: 740–749.

John S. 2006. MOGREPS: Met Office ensemble prediction system for short-range weather forecasting. *NWP Gazette*, February 2006: 2–5.

Jones C. 2000. Occurrence of extreme precipitation events in California and relationships with the Madden–Julian Oscillation. *J. Clim.* **13**: 3576–3587.

Jones CG & Thorncroft CD. 1998. The role of El Niño in Atlantic tropical cyclone activity. *Weather* **53**: 324–336.

Kelly J. 2011. First death confirmed after Cyclone Yasi. *ABC News*. http://www.abc.net.au/lateline/content/2011/s3130675.htm, accessed 7 December 2014.

Kendrew WG. 1937. *Climates of the continents*. 3rd edition. Oxford University Press.

Kendrew WG. 1961. *The climates of the continents*, 5th edition. Clarendon Press, Oxford.

Kennedy J, Hood S, Titchner H, Palmer M & Parker D. 2009. Global and regional climate in 2008. *Weather* **64**: 288–297.

Kenworthy JM. 2000. The use of the word 'tornado' in West Africa and the eastern tropical Atlantic. *Weather* **55**: 60–62.

Kiangi PMR. 1989. The monsoons of east Africa and the associated rainfall deficiency. *Lectures presented at the seminar on tropical meteorology,*

Erice, 26 September – 4 October 1986. WMO, Geneva: 181–227.

Kiehl J & Trenberth K. 1997. Earth's annual mean energy budget. *Bull. Amer. Meteorol. Soc.* **78**: 197–206.

Kington T. 2007. Climate change brings malaria back to Italy. *The Guardian*, 6 January.

Knippertz P. 2007. Tropical–extratropical interactions related to upper-level troughs at low latitudes. *Dyn. Atmos. Oceans* **43**: 36–62.

Knippertz P & Stuut J-B W. 2014. *Mineral Dust: A Key Player in the Earth System.* Springer-Verlag, Amsterdam.

Koech KV. 2015. Atmospheric divergence over equatorial East Africa and its influence on distribution of rainfall. *Weather* **70**: 158–162.

Krishnamurti TN. 1961. The subtropical jetstream of winter. *J. Meteorol.* **18**: 172–191.

Laing AG & Fritsch JM. 1997. The global population of mesoscale convective complexes. *Quart. J. R. Meteorol. Soc.* **123**: 389–405.

Lalli CM & Parsons TR. 1993. *Biological oceanography – an introduction.* Pergamon Press, Oxford.

Landsea CW. 1993. A climatology of intense (or major) Atlantic hurricanes. *Mon. Weath. Rev.* **121**: 1703–1713.

Landsea C & Delgado S. 2014. *What are the average, most, and least tropical cyclones occurring in each basin?* http://www.aoml.noaa.gov/hrd/tcfaq/E10.html, accessed 25 June 2015.

Landsea CW & Gray WM. 1992. The strong association between western Sahelian monsoon rainfall and intense Atlantic hurricanes. *J. Climate* **121**: 1703–1713.

Lapworth A. 2009. Diurnal variation of the dry overland boundary layer vertical temperature profile. *Weather* **64**: 337–339.

Le Treut H, Somerville R, Cubasch U et al. 2007. Historical Overview of Climate Change. *Climate Change 2007: The Physical Science Basis. Contribution of Working Group I to the Fourth Assessment Report of the Intergovernmental Panel on Climate Change.* Cambridge University Press.

Leroux M. 1970. La dynamique des précipitations en Afrique occidentale. *Publ. Dir. Expl. Météo, No. 23.* ASECNA, Dakar.

Leroux M. 1972. Climatologie dynamique de l'Afrique occidentale. *Atlas Int. de l'Ouest Africain.* IFAN, Dakar.

Leroux M. 2001. *The meteorology and climate of tropical Africa.* Springer/Praxis, Chichester.

Levi M. 1963. Local winds around the Mediterranean Sea. *Bull. Meteorol. Serv. Israel* **4**: 1–5.

Li L, Yang Y, Zong R, Mao X, Jiang Y & Zhuang H. 2015. The first tropospheric wind profiler observations of a severe typhoon over a coastal area in South China. *Weather* **70**: 9–13.

Liebmann B, Kiladis GN, Vera CS, Saulo AC & Carvalho LMV. 2004. Seasonal variations of rainfall in the vicinity of the South American low-level jet stream and comparison to those in the South Atlantic Convergence Zone. *J. Clim.* **17**: 3289–3842.

Lindsey R. 2009. Climate and Earth's energy budget. *NASA Earth Observatory.* www.earthobservatory.nasa.gov/Features/EnergyBalance, accessed 16 June 2014.

Lorenc AC. 2006. 4D-Var and the butterfly effect. Statistical 4D-Var for a wide range of scales. *Forecasting Research Technical Report, No. 481.* Met Office, Exeter.

Lowe J, Pope V & Smith F (eds). 2005. *Climate change, rivers and rainfall.* Met Office/Defra, Exeter.

Lu J & Lin C. 1982. The relationship between the retreat of the summer monsoon in East Asia and seasonal variation of circulation over East Asia. *Proc. Symp. Summer Monsoon S and E Asia, 10–15 Oct. 1982, Kunming, China*: 80–96.

Lucas R, McMichael T, Smith W & Armstrong B. 2006. Solar ultraviolet radiation: Global disease burden from solar ultraviolet radiation. *Environmental Burden of Disease Series, No. 13.* Geneva, WHO.

Ludlam FH. 1980. *Clouds and storms.* Pennsylvania State University Press, University Park, PA.

Lumb FE. 1970. Topographic influences on thunderstorm activity near Lake Victoria. *Weather* **25**: 404–410.

Madden RA & Julian PR. 1971. Detection of a 40–50 day oscillation in the zonal wind in the tropical Pacific. *J. Atmos. Sci.* **28**: 702–708.

Madden RA & Julian PR 1972. Description of global-scale circulation cells in the tropics with a 40–50 day period. *J. Atmos. Sci.* **29**: 1109–1123.

Madden RA & Julian PR 1994. Observations of the 40–50 day tropical oscillation: a review. *J. Atmos. Sci.* **112**: 814–837.

Maddox RA. 1980. Mesoscale convective complexes. *Bull. Amer. Meteorol. Soc.* **61**: 1374–1387.

Maddox RA. 1986. Mesoscale convective complexes. *Satellite imagery interpretation for forecasters.* Vol. 2: *Precipitation, convection.* National Weather Association, USA: 4-J-1/14.

Maddox RA, Perkey DJ & Fritsch JM. 1981. Evolution of upper tropospheric features during the development of a mesoscale convective complex. *J. Atmos. Sci.* **38**: 1664–1674.

Maloney ED & Hartmann DL. 2000. Modulation of eastern North Pacific hurricanes by the Madden–Julian Oscillation. *J. Clim.* **13**: 1451–1460.

Marshall M. 2011. Pacific shouldn't amplify climate change. *New Sci.* http://newscientist.com/article/dn20509-pacific-shouldnt-amplify-climate-change.html, accessed 11 December 2014.

Martyn D. 1992. *Climates of the World.* Elsevier, Amsterdam.

Mathews JH. 1982. The sea breeze – forecasting aspects. *Austral. Meteorol. Mag.* **30**: 205–209.

Mathur VK & WMO Secretariat. 2006. Sustainable development through integrated flood management. *WMO Bull.* **55**: 164–169.

Matthews AJ. 2004. Intraseasonal variability over tropical Africa during northern summer. *J. Clim.* **17**: 2427–2440.

Mayes JC. 2012. From Observations to Forecasts – concluding article (Part 15): Opportunities and challenges for today's operational weather forecasters. *Weather* **67**: 100–107.

Mayes JC & Perry AH. 1989. *Winds and local climates.* Weather Watchers Network Scotland, Laurieston.

McCallum E & Heming JT. 2006. Hurricane Katrina – an environmental perspective. *Phil. Trans. R. Soc. London* **364A**: 2099–2115.

McIlveen JFR. 1992. *Fundamentals of weather and climate.* Stanley Thornes, Cheltenham.

McKnight TL & Hess D. 2000. Climate zones and types: The Köppen system. *Physical Geography: A Landscape Appreciation.* Prentice Hall, Upper Saddle River, NJ: 200–240.

McNab A. 1994. *Bravo-two-zero. The true story of an SAS patrol behind enemy lines in Iraq.* Corgi Adult, London.

McNab A. 2008. *Seven troop.* Bantam Books, London: 327–328.

Medicalecology.org. 2006. *Stratospheric ozone depletion.* http//:www.medicalecology.org/atmosphere/print_a_app_strat.htm, accessed 16 January 2015.

Membery DA. 1982. A documented example of strong wind shear. *Weather* **37**: 19–22.

Membery DA. 1983a. Severe storm in Dubai on 30 April 1981. *Weather* **38**: 62.

Membery DA. 1983b. Low-level wind profiles during the Gulf Shamal. *Weather* **38**: 18–24.

Membery DA. 1998. Famous for 15 minutes: An investigation into the causes and effects of the tropical storm that struck southern Arabia in June 1996. *Weather* **53**: 102–110.

Membery DA. 2001. Monsoon tropical cyclones, Part 1. *Weather* **56**: 431–436.

Membery DA. 2002. Monsoon tropical cyclones, Part 2. *Weather* **57**: 246–255.

Met Office. 2004a. *Uncertainty, risk and dangerous climate change.* Hadley Centre for Climate Prediction and Research, Exeter.

Met Office. 2004b. On-screen field modification. *NWP Gazette,* June 2004: 4–6.

Met Office. 2006. *Cloud types for observers: reading the sky.* Exeter.

Met Office. 2011a. *How we forecast the behaviour of our oceans.* metoffice.gov.uk/learning/science/days-ahead/oceans, accessed 11 December 2014.

Met Office. 2011b. *Fact sheet 2 – Thunderstorms.* National Meteorological Library and Archive, Exeter.

Met Office. 2011c. *Past tropical cyclones,* metoffice.gov.uk/data/tropicalcyclone/tctracks/ause10-11.gif, accessed 11 December 2014.

Met Office. 2012. *Expertise. Making a difference to aviation services with our specialist knowledge.* Exeter.

Met Office. 2013a. *Continually improving our forecasts.* http://www.metoffice.gov.uk/services/accuracy, accessed 16 June 2014.

Met Office. 2013b. The power of climate science. *Barometer* **24**: 7–12.

Met Office. 2014a. *North Atlantic tropical storm seasonal forecast 2014.* http://www.metoffice.gov.uk/weather/tropicalcyclone/seasonal/northatlantic2014.

Met Office. 2014b. *Past tropical cyclones.* http://www.metoffice.gov.uk/weather/tropicalcyclone/observations, accessed 7 December 2014.

Met Office. 2014c. *Met Office climate prediction model: HadGEM2 family.* http://www.metoffice.gov.uk/research/modelling-systems/unified-model/climate-models/hadgem2, accessed 15 December 2014.

Met Office. 2014d. *Weather and your health: UV and sun health.* http://www.metoffice.gov.uk/health/

yourhealth/uv-and-sun-health, accessed 28 December 2014.

Met Office. 2014e. *Forecast verification*. http://www.metoffice.gov.uk/weather/tropicalcyclone/verification, accessed 7 December 2014.

Met Office. 2014f. *Unified Model*. http://www.metoffice.gov.uk/research/modelling-systems/unified-model, accessed 28 December 2014.

Met Office. 2014g. *International role*. http://www.metoffice.gov.uk/about-us/what/international, accessed 28 December 2014.

Met Office. 2014h. *Unified Model collaboration*. http://www.metoffice.gov.uk/research/collaboration/um-collaboration, accessed 28 December 2014.

Met Office. 2014i. *International climate services*. http://www.metoffice.gov.uk/services/climate-services/international, accessed 2 January 2015.

Met Office. 2014j. *DFID-Met Office Climate Science Research Partnership (CSRP), CSRP Phase 1 – Final report*. metoffice.gov.uk/media/pdf/5/i/CSRP1_Report.pdf.

Met Office. 2014k. *Aviation international responsibilities*. http://www.metoffice.gov.uk/aviation/international_responsibilities, accessed 28 December 2014.

Meteorological Office. 1975. *Handbook of weather forecasting*. HMSO, London.

Meteorological Office. 1991. *Meteorological glossary*, 6th edition. HMSO, London.

Meteorological Office. 1994. *Handbook of Aviation Meteorology*, 3rd edition. HMSO, London.

Meteorological Office. 1997. *Source book to the Forecasters' Reference Book*. Shinfield Park, Reading.

Miles MK. 1959. Factors leading to the meridional extension of thermal troughs and some forecasting criteria derived from them. *Meteorol. Mag.* **88**: 193–203.

Millington S. 2006. Weather image. Satellite imagery of Saharan mineral dust. *Weather* **61**: 60.

Mo K & Higgins RW. 1998. Tropical convection and precipitation regimes in the western United States. *J. Clim.* **10**: 3028–3046.

Mooley DA & Parthasarathy B. 1983. Indian summer monsoon and El Nino. *Pure Appl. Geophys.* **121**: 339–352.

Mukhopadhay P, Singh HAK & Singh SS. 2005. Two severe nor'westers in April 2003 over Kolkata, India, using Doppler radar observations and satellite imagery. *Weather* **60**: 343–353.

Myneni RB, Hall FG, Sellers PJ & Marshak AL. 1995. The interpretation of spectral vegetation indexes, *IEEE Trans. Geosci. Remote Sens.* **33**: 481–486.

NASA-GSFC. 2013. *Deadly tropical cyclone Mahasen comes ashore*. National Aeronautics and Space Administration, Goddard Space Flight Center. http://trmm.gsfc.nasa.gov/publications_dir/mahasen_may_2013.html, accessed 7 December 2014.

NASA-GSFC. 2014. *Ozone hole watch*. http://ozonewatch.gsfc.nasa.gov, accessed 16 January 2015.

National Centre for Ocean Forecasting. 2009. *Operational sea surface temperature and sea ice analysis (OSTIA)*. http://ghrsst-pp.metoffice.com/pages/latest_analysis/ostia.html, accessed 16 January 2015.

National Hurricane Research Laboratory. 1970. *Project STORMFURY annual report 1969*. Coral Gables, FL.

National Oceanic and Atmospheric Administration. 2010. *NASA/NOAA study finds El Niños growing stronger*. http://www.noaanews.noaa.gov/stories2010/20100825_elnino.html, accessed 1 June 2013.

National Oceanic and Atmospheric Administration. 2013. *State of the Climate: Global Analysis: Annual 2012*. Washington, DC.

Natural Resources Conservation Service. 1999. *Soil taxonomy: a basic system of soil classification for making and interpreting soil surveys*, 2nd edition. United States Department of Agriculture, Washington DC.

Natural Resources Conservation Service. 2006. *Keys to soil taxonomy*, 10th edition. United States Department of Agriculture, Washington, DC.

Naujokat B. 1986: An update of the observed quasi-biennial oscillation of the stratospheric winds over the tropics. *J. Atmos. Sci.*, **43**: 1873–1877.

Newell RE. 1979. Climate and the ocean. *Am. Sci.* **67**: 405–416.

Ng'ongolo HK & Smyshlyaev SP. 2010. The statistical prediction of East African rainfalls using quasi-biennial oscillation phases information. *Sci. Res.* **2**: 1407–1416.

Nieto Ferreira R, Schubert WH & Hack JJ. 1996. Dynamical aspects of twin tropical cyclones associated with the Madden–Julian Oscillation. *J. Atmos. Sci.* **53**: 929–945.

Normand Sir Charles. 1946. Energy in the atmosphere. (G J Symons memorial lecture delivered

20.3.1946). *Quart. J. R. Meteorol. Soc.* **72**: 145–167.

Oke TR. 1987. *Boundary layer climates*, 2nd edition. Routledge, New York.

Ologunorisa TE & Chinago A. 2007. The diurnal variation of thunderstorm activity over Nigeria. *Int. J. Meteorol.* **32**: 19–29.

Oltmans SJ & Hofmann DJ. 1995: Increase in lower stratospheric water vapour at a midlatitude Northern Hemisphere site from 1981 to 1994. *Nature* **374**: 146–149.

Oña IL & DiCarlo G. 2011. *Climate change in the Galápagos Islands*. Conservation International/World-Wide Fund for Nature, Quito.

Overton A & Galvin JFP. 2005. Drizzle from various clouds. *Weather* **60**: 357–358.

Overton A & Strangeways I. 2007. Meeting report. Summer visit 2007 Centre for Ecology and Hydrology, Wallingford. *Weather* **62**: 306.

Owusu K & Waylen P. 2009. Trends in spatio-temporal variability in annual rainfall in Ghana (1951–2000). *Weather* **94**: 115–120.

Paegle JN, Byerle LA & Mo KC. 2000. Intraseasonal modulation of South American summer precipitation. *Mon. Weath. Rev.* **128**: 837–850.

Parker D & Diop-Kane M. 2015. *Meteorology and forecasting for tropical West Africa: the forecasters' handbook*. J Wiley & Sons, Oxford

Patwardhan SK & Asnani GC. 2000. Mesoscale distribution of summer monsoon rainfall near the Western Ghats (India). *Int. J. Climatol.* **20**: 575–581.

Pearce EA & Smith CG. 1984. *World weather guide*. Hutchinson, London.

Perkins J. 2010. Green Revolution. *Encyclopedia of Earth*. eoearth.org/view/article/153125/, accessed 12 December 2014.

Persson A. 2000. Back to basics: Coriolis, Part 3. The Coriolis force on the physical earth. *Weather* **55**: 234–239.

Petr J (ed). 1991. *Weather and yield*. Elsevier, Amsterdam.

Pidwirny M. 2013. Energy balance of Earth. *Encyclopedia of Earth*. http://www.eoearth.org/view/article/152458/, accessed 11 December 2014.

Prüss-Üstün A & Corvalán C. 2006. *Towards an estimate of the environmental burden of disease*. Geneva, WHO.

Rabier F, Thepaut J-N & Coutier P. 1998. Four dimensional variational assimilation at ECMWF.

Proc. ECMWF Semin. Data Assim., 2–6 September 1996: 213–249.

Ramage CS. 1971. *Monsoon meteorology*. International Geophysics Series, 15. Academic Press, London.

Rao YP. 1976. Southwest monsoon. *Meteorological Monograph No. I.* India Meteorological Department, New Delhi.

Raschke E, Ohmura A, Rossow WB et al. 2005. Cloud effects on the radiation budget based on ISCCP data (1991 to 1995). *Int. J. Climatol.* **25**: 1103–1125.

ReliefWeb. 2009a. *Floods across Western Sahel (as of 08 Sep 2009)*. http://reliefweb.int/rw/rwb.nsf/db900SID/LPAA-7VQAVF?OpenDocument, accessed 10 December 2014.

ReliefWeb. 2009b. *600,000 people affected by floods in West Africa*. http://reliefweb.int/rw/rwb.nsf/db900SID/LSGZ-7VPHBH?OpenDocument&emid=FL-2009-000172-BFA, accessed 10 December 2014.

Republic of Cyprus, Department of Labour Inspection. 2015. *Air quality in Cyprus: Health effects*. http://www.airquality.dli.mlsi.gov.cy/Default.aspx?pageid=647, accessed 5 January 2015.

Reynolds R. 2000. *Philip's Guide to Weather*. George Philip, London.

Richardson D. 2011. From Observations to Forecasts – Part 11: Ensemble products for weather forecasters. *Weather* **66**: 235–241.

Riehl H. 1974. Hot-tower precipitation. *Weather* **29**: 196.

Riehl H. 1979. *The Climate and Weather of the Tropics*. Academic Press, London.

Roach WT & Bysouth CE. 2002. How often does severe clear air turbulence occur over tropical oceans? *Weather* **57**: 8–19.

Roberts H. 2014. The weather of 2013. *Weather* **69**: 302–303.

Roca R, Lafore J-P, Piriou C & Redelsperger J-L. 2005. Extratropical dry-air intrusions into the West African monsoon midtroposphere: an important factor for the convective activity over the Sahel. *J. Atmos. Sci.* **62**: 390–407.

Rodda HL. 1983. Severe storm in Dubai on 30 April 1981. *Weather* **38**: 61.

Sarkies JW. 1967. Dust and the incidence of severe trachoma. *Br. J. Ophthalmol.* **51**: 97–100.

Saunders MA & Lea AS. 2008. Large contribution of sea surface warming to recent increase in Atlantic hurricane activity. *Nature* **451**: 557–560.

Saunders MA & Rockett P. 2001. Improving typhoon predictions, *Global Reinsurance Mag., East Asia Spec. Rep.*: 26–29.

Scaife AA, Butchart N, Warner CD & Swinbank R. 2002. Impact of a spectral gravity wave parameterization on the stratosphere in the Met Office Unified Model. *J. Atmos. Sci.* **59**: 1473–1489.

Schamp H. 1964. *Die Winde der Erde und ihre Namen. Ragelmassige, periodische und lokale Winde Klimaelemente.* Franz Steiner Verlage, Wiesbaden (translated into English by C Long).

Schipper JW & Mühr B. 2010. Observation of a complex halo phenomenon in central Europe. *Weather* **65**: 335–338.

Schlein L. 2009. West Africa Hit by Devastating Floods. *Voice of America.* http:/www.voanews.com/content/a-13-2009-09-05-voa15-68807527/412584.html, accessed 10 December 2014.

Schmetz J, Borde R, König M & Lutz H-J. 2004. Upper tropospheric divergence fields in a tropical convective system observed with Meteosat-8. *Proc. Fifth Int. Winds Workshop*. Helsinki.

Scorer RS & Verkaik A. 1989. *Spacious skies*. David and Charles, Newton Abbot.

Sellers PJ. 1985. Canopy reflectance, photosynthesis, and transpiration. *Int. J. Remote Sens.* **6**: 1335–1372.

Setlow RB, Grist E, Thompson K & Woodhead AD. 1993. Wavelengths effective in induction of malignant melanoma. *Proc. Natl. Acad. Sci. U.S.A.* **90** (14): 6666–6670.

Sheldon WR, Benbrook JR & Aimedieu P. 1997. Ozone depletion in the upper stratosphere at the dawn terminator, *J. Atmos. Solar Terr. Phys.* **59**: 1–7.

Sheldrick J. 2005. Goyder's Line: the unreliable history of the line of reliable rainfall. In Sherratt T, Griffiths T, Robin L (eds) *Change in the weather: climate and culture in Australia*. National Museum of Australia Press: Canberra: 56–65.

Sherwin R. 2014. Meeting report: Understanding the weather of 2013. *Weather* **69**: 306.

Shutts G. 2005. Numerical simulations of large-scale tropical flow with a cloud-resolving model. *RMetS Conf. 2005 Abs., Plenary Rev. Pap. Weath. Air Quality Forecast.*: 18.

Simmons AJ & Gibson JK. 2000. The ERA-40 Project Plan. *ERA-40 Project Report Series No. 1.* ECMWF, Reading.

Sinha RP, Singh SC & Häder D-P. 1999. Photoecophysiology of cyanobacteria. *J. Photochem. Photobiol.* **3**: 91–101.

Slingo A, Ackerman TP, Allan RP et al. 2006. Observations of the impact of a major Saharan dust storm on the atmospheric radiation balance. *Geophys. Res. Lett.* **33**: L24817.

Smith MA. 2005. Palaeoclimates: an archaeology of climate change. In Sherratt T, Griffiths T, Robin L (eds) *Change in the weather: climate and culture in Australia*. National Museum of Australia Press, Canberra: 176–186.

Smith PM & Warr K. 1991. *Global environmental issues*. Hodder and Stoughton/Open Univ. Press.

Smith DM, Eade R, Dunstone NJ et al. 2010. Skilful multi-year predictions of Atlantic hurricane frequency. *Nature Geosci.* **3**, 846–849.

Solomon S, Rosenlof KH, Portmann RW et al. 2010. Contributions of stratospheric water vapour to decadal changes in the rate of global warming. *Science* **327**: 1219–1223.

Sow B, Viltard A, de Felice P, Deme A & Adamou G. 2005. Are squall lines detected by NCEP-NCAR reanalyses? *Meteorol. Atmos. Phys.* **90**: 209–214.

Stansel JW & Fries RE. 1980. A conceptual agromet rice yield model. *Proc. Sympos. Agromet. Rice Crop*: 201–212.

Strangeways I. 2011. The greenhouse effect: a closer look. *Weather* **66**: 44–48.

Suzuki S-I & Hoskins BJ. 2005. The circulation changes associated with the end of the Baiu season in Japan. *RMetS Conf. 2005 Abs., Plenary Rev. Pap. Weath. Air Quality Forecast*: 37.

Thomas A. 2003. Soaring the wave. *Scribbly Gum*, August 2003. abc.net.au/science/scribblygum/august2003/, accessed 6 December 2014.

Thorncroft CD & Blackburn M. 1999. Maintenance of the African easterly jet. *Quart. J. R. Meteorol. Soc.* **125** A: 763–786.

Thorncroft CD, Hall NMJ & Kiladis GN. 2008. Three-dimensional structure and dynamics of African easterly waves. Part III. Genesis. *J. Atmos. Sci.* **65**: 3596–3607.

Tilev-Tanriover S & Kahraman A. 2015. Saharan Dust Transport by Mediterranean Cyclones Causing Mud Rain in Istanbul. *Weather* **70**: 145–149.

Todd MC & Washington R. 2005. Variability in rainfall over central equatorial Africa at synoptic timescales. *RMetS Conf. 2005 Abs., Plenary Rev. Pap. Weath., Air Quality Forecast.*: 38.

Tomita T, Yoshikane T & Yasunari T. 2003. *Interannual variability of Baiu in Japan uncovered – a mechanism of air–sea interaction in the Western Pacific affecting the interannual variability of the Baiu front*. www.jamstec.go.jp/frsgc/eng/press/040831/index.html, accessed 25 October 2007.

Turner AG & Annamalai H. 2012. Climate change and the South Asian summer monsoon. *Nature Clim. Change* **2**: 587–595.

Turner JA & Bysouth CE. 1999. Automated systems for predicting clear air turbulence in global aviation forecasts. *Proc. 8th Conf. Aviation, Range Aerosp. Meteorol., Dallas, Texas*. American Meteorological Society: 368–372.

Turner BL & McCandless SR. 2004. How humankind came to rival nature: a brief history of the human-environment condition and the lessons learned. *Earth system analysis for sustainability*: 227–243.

United Kingdom Stratospheric Ozone Review Group. 1999. *Stratospheric ozone 1999*. HMSO, London.

United Nations, Economic and Social Affairs Division. 2012. *World population prospects, the 2012 revision*. Geneva.

University of Illinois, Department of Atmospheric Sciences. 2010. El Niño. *WW2010*. http://ww2010.atmos.uiuc.edu/(Gh)/guides/mtr/eln/home.rxml, accessed 11 December 2014.

van Ormondt M. 2013. *Storm surge of Super Typhoon Haiyan at Tacloban City, The Philippines*, using data from the Deltares Institute, Delft. https://www.youtube.com/watch?v=l6ht1JoRv_A, accessed 12 July 2014.

Van der Wal J, Murphy HT, Kutt AS et al. 2013. Focus on poleward shifts in species' distribution underestimates the fingerprint of climate change. *Nature Clim. Change* **3**: 239–243.

Varga G, Cserháti C, Kovács J, Szeberényi J & Bradák B. 2014. Unusual Saharan dust events in the Carpathian Basin (Central Europe) in 2013 and early 2014. *Weather* **69**: 309–313.

Vedel H. 1978. *Trees and shrubs of the Mediterranean*. Penguin, Harmondsworth.

Verbickas S. 1998. Westerly wind bursts in the tropical Pacific. *Weather* **53**: 282–284.

Villarini G & Vecchi GA. 2012. Twenty-first-century projections of North Atlantic tropical storms from CMIP5 models. *Nature Clim. Change* **2**: 604–607.

Voluntary Aid Societies. 1997. *First aid manual*, 7th edition. Dorling Kindersley, London.

Walker GT. 1923. Correlation in seasonal variations of weather, VIII. A preliminary study of world weather. *Ind. Meteorol. Mem.* **24**: 75–132.

Walker GT. 1924a. Correlation in seasonal variations of weather, IX. A further study of world weather. *Ind. Meteorol. Mem.* **24**: 275–332.

Walker GT. 1924b. Correlation in seasonal variations of weather, X. Applications to seasonal forecasting in India. *Ind. Meteorol. Mem.* **24**: 333–346.

Wallace JM & Hobbs PV. 2006. *Atmospheric Science. An introductory survey*, 2nd edition. Academic Press, Amsterdam.

Warr K. 1991. The ozone layer. In *Global Environmental Issues*, PM Smith & K Warr (eds). Hodder and Stoughton/Open Univ. Press: 121–171.

Washington R, Todd M, Middleton NJ & Goudie AS. 2003. Dust-storm source areas determined by the Total Ozone Measuring Spectrometer and surface observations. *Annal. Assoc. Amer. Geog.* **93**: 297–313.

Weight N. 2001. *The local winds of the world. A comprehensive review of local and regional winds, their causes, general characteristics and tips for forecasters*. Met Office, High Wycombe.

Werner M, Cranston M, Harrison T, Whitfield D & Schellekens J. 2009. Recent developments in operational flood forecasting in England, Wales and Scotland. *Meteorol. Appl.* **16**: 13–22.

Wheeler TR, Challinor AJ, Crauford PQ, Slingo JM & Grimes DIF. 2005. Forecasting the harvest – from proverbs to PCs. *Biologist* **52**: 1, 45–49.

Wikipedia. 2013. *Tropical cyclone*. http://en.wikipedia.org/wiki/Tropical_cyclone, accessed 7 December 2014.

Wikipedia. 2014a. *Cloud forest*. http://en.wikipedia.org/wiki/Cloud_forest, accessed 17 January 2015.

Wikipedia. 2014b. *Bergeron process*. http://en.wikipedia.org/wiki/Bergeron_process, accessed 18 January 2015.

Wikipedia. 2014c. *2004 Africa locust infestation*. http://en.wikipedia.org/wiki/2004_locust_infestation, accessed 4 January 2015.

Wikipedia. 2014d. *World population*. http://en.wikipedia.org/wiki/World_population, accessed 28 December 2014.

Wilby RL & Dessai S. 2010. Robust adaptation to climate change. *Weather* **65**: 180–185.

Willett S & Milton S. 2005. Modelling the transition from suppressed to deep tropical convection:

comparison of the Met Office Unified Model with TOGA-COARE (GCSS WG4 Case 5). *RMetS Conf. 2005 Abs., Plenary Rev. Pap. Weath., Air Quality Forecast.*: 22.

Williams D (ed). 1983. *Geology, Block 4: Surface processes*. Open University, Milton Keynes.

WHO. 2002. *Global solar UV Index: a practical guide*. Geneva.

WHO. 2013. *TDR Research Report 2013*. http://www.who.int/tdr/publications/about-tdr/ar2013_research_report.pdf.

WHO. 2014. *Ultraviolet radiation and the INTERSUN Programme*. http://www.who.int/uv/intersunprogramme/activities/uv_index/en/index1.html.

WHO & WMO. 2012. *Atlas of health and climate*. WMO No. 1098. Geneva.

WMO. 1956. *International cloud atlas*. Abridged atlas. Geneva.

WMO. 1982. *Manual on the Global Data-Processing System*. Geneva.

WMO. 1999. Scientific assessment of ozone depletion, 1998. *Global Ozone Research and Monitoring Project – Report No. 44*. Geneva.

WMO. 2009a. *WMO Technical Progress Report on the Global Data-Processing and Forecasting System (GDPFS) and Numerical Weather Prediction (NWP) Research*. http://www.wmo.int/pages/prog/www/DPFS/ProgressReports/GDPFS-NWP_Progressreports.html, accessed 2 January 2015.

WMO. 2009b. *Manual on codes, Vol. I.1. International codes. Part A. Alphanumeric codes*. WMO No. 306. Geneva.

WMO. 2011. Scientific assessment of ozone depletion, 2010. *Global Ozone Research and Monitoring Project – Report No. 52*. Geneva.

WMO. 2013. *Watching the weather to protect life and property: celebrating 50 years of World Weather Watch*. WMO No. 1107. Geneva.

WMO. 2012. *Tropical Cyclone Operational Plan: 2012*. RA I Tropical Cyclone Committee, Geneva.

Yihui D & Chan CL. 2005. The East Asian summer monsoon: an overview. *Meteorol. Atmos. Phys.* **89**: 117–142.

Young MV. 1995. Severe thunderstorms over southeast England on 24 June 1994: a forecasting perspective. *Weather* **50**: 250–256.

Zhang C. 2005. The Madden–Julian oscillation. *Rev. Geophys.* **43**: RG2003.

Zhang C & Dong M. 2004. Seasonality of the Madden–Julian Oscillation. *J. Clim.* **17**: 3169–3180.

Zipser E. 1969. The role of organised unsaturated convective downdrafts in the structure and rapid decay of an equatorial disturbance. *J. Appl. Meteorol.* **8**: 799–814.

Zipser EJ. 1981. Life cycle of mesoscale convective systems. Nowcasting: mesoscale observations and short-range prediction. *Proc. Int. Sympos., 25–28 Aug. 1981, Hamburg, Germany* (part of the IAMAP Third Scientific Assembly): 381–386.

Zipser EJ, Twohy CH, Tsay S-C et al. 2009. The Saharan air layer and the fate of African easterly waves. *Bull. Amer. Meteorol. Soc.* **90**: 1137–1156.

Zobel RF & Cornford SG. 1966. Cloud tops over Malaya during the south-west Monsoon season. *Meteorol. Mag.* **95**: 65–68.

Glossary

Many of these terms are defined more completely in the *Meteorological glossary* (Meteorological Office 1991).

Absoroption Removal of radiation from an incident beam with conversion to another form of energy – in meteorology, this is usually heat.

Acceleration Rate of change of velocity (V) with respect to time. In meteorology, the acceleration of air is almost always measured relative to a set of axes fixed in the earth. However, because the earth is rotating, Newton's second law of motion cannot be applied and both the angular velocity of the earth and centripetal acceleration must be considered, giving:

$$a = dV / dt + 2\Omega \times V + g$$

Accretion In meteorology, the growth of ice crystals by collision and coalescence with water drops.

Adiabatic An adiabatic thermodynamic process is one in which heat does not enter or leave the system. Because the atmosphere is compressible and pressure varies with height, adiabatic processes play a fundamental role in meteorology. Thus, if a parcel of air rises, it expands against its lower environmental pressure, the energy to expand it coming from its internal energy. Its temperature falls, despite the fact that no heat leaves the parcel. Conversely, the internal energy of a parcel increases as it descends and its temperature rises as it is compressed. Processes in the troposphere may be regarded at adiabatic. Adiabatic temperature changes occur at a definite rate. For unsaturated air, the rate is $-0.98°C$ per 100 m altitude gain (see Appendix 4). The rate of change of temperature for saturated air varies according to temperature and pressure.

Advection The transfer of an air-mass property by motion. The term is almost always used to signify only horizontal transfer (although advection may also be in the vertical). In the tropics, advection is seen mainly in association with motion of tropical upper tropospheric troughs at the boundaries of the tropical air mass.

Advection fog Fog formed by the passage of relatively warm moist stable air over a cool surface. It is associated mainly with cool sea areas and may affect adjacent coasts. The term is also used to describe the motion of pre-existing fog transferred from a distant source.

Aerological diagram See Appendix 4.

Aerosol In meteorology, an aggregation of minute particles (solid or liquid) suspended in the atmosphere.

African easterly waves (AEWs) Short-wavelength mid-tropospheric waves observed close to the equator in West and Central Africa

An Introduction to the Meteorology and Climate of the Tropics, First Edition. J F P Galvin.
© 2016 John Wiley & Sons, Ltd. Published 2016 by John Wiley & Sons, Ltd.

during the summer monsoon. (Easterly waves may develop elsewhere, but these are usually of lesser importance.)

Ageostrophic wind The vector difference between the actual wind and the geostrophic (balanced) wind due to atmospheric convergence, divergence or vertical motion in the atmosphere:

actual wind vector = geostrophic wind vector + ageostrophic wind vector

Agglomeration The growth of cloud drops by collision to become sufficiently heavy to descend, thus collecting more cloud drops and growing to fall as rain.

Agriculture The cultivation of land on a relatively large scale to grow crops or rear livestock. (Agrometeorology is the scientific study of weather and climate in relation to agriculture.)

Air The mixture of gases which form the earth's atmosphere. In the absence of water vapour and dust, the composition of air is approximately 78% nitrogen, 21% oxygen, 1% argon and 0.03% carbon dioxide (by volume) with very small amounts of other (trace) gases. (See also water vapour.)

Air mass An extensive body of air in which the horizontal gradients of temperature and humidity are relatively slight and which is separated from the adjacent body of air by a relatively sharp transition zone. (In the tropics, there is a single air mass and the transition zone is found at its northern and southern edges, which may also be marked by the subtropical jet stream. However, there are significant variations in humidity within the tropics.)

Albedo A measure of the reflecting power of a surface, being that fraction of incident radiation which is reflected by that surface.

Altitude In meteorology, altitude generally signifies height above mean sea level, but in dynamical meteorology altitude (height) is usually expressed in geopotential metres (or feet).

Altocumulus cloud See Appendix 3.

Altostratus cloud See Appendix 3.

Analysis In synoptic meteorology, the plotting of isopleths or representative symbols using observed data (usually for a single point in time).

Angular velocity A vector quantity which is usually measured in either revolutions per unit time or radians per unit time. Its units are s^{-1}.

Anti-cyclone The atmospheric pressure system in which there is a high central pressure relative to the surroundings enclosed by a number of isobars. In the tropics, anti-cyclones are generally of the 'warm' type, having warmth in depth (although this is sometimes not present close to the surface), with a high (and so cold) tropopause, associated with a cold stratosphere. Anti-cyclones are characterized by divergence in the lowest tropospheric levels with convergence above.

Anvil cloud See Appendix 3.

Arid climate A climate in which the rainfall is insufficient to support vegetation (see Chapters 1, 3, 6 and 7).

Baiu front See mei-yu front.

Barchan A sand dune formed into a crescent shape by the wind. It possesses two 'horns' that point downwind with a steep slip face also on the downwind side. These dunes may be 90 m tall and 350 m in the crosswind direction.

Baroclinic An atmosphere in which there is an intersection of pressure and density surfaces at some level. Where there is strong baroclinicity, dynamical process may be generated on gradients of temperature and humidity (in the presence of strong thermal winds). In the tropics, there is strong baroclinicity only near the poleward boundaries, usually associated with tropical upper-tropospheric troughs.

Barotropic An atmosphere in which surfaces of pressure and density coincide at all levels. Barotropic processes do not involve steep gradients of temperature or humidity and are usually driven by convection. (These processes are relatively common in the tropics, in particular in association with easterly waves in the middle troposphere.)

Boundary layer The layer of air in contact with the earth's surface within which mixing occurs. (The boundary layer may be thin at night over land and is usually capped by a temperature inversion within tropical anticyclones. Over the sea and through much of the day it commonly reaches an altitude of at least 1500 m.)

Carbon dioxide (CO_2) A trace atmospheric gas, formed as a result of respiration or burning, essential to photosynthesising plant life, but commonly associated with climate change.

Chlorofluorocarbons (CFCs) Trace (anthropogenic) atmospheric gases combining chlorine, fluorine and carbon produced as refrigerants (now banned), usually associated with the destruction of ozone in the stratosphere, but which are also greenhouse gases.

Cirrocumulus cloud See Appendix 3.

Cirrostratus cloud See Appendix 3.

Cirrus cloud See Appendix 3.

Clear-air turbulence (CAT) Rapidly changing winds of the upper troposphere that may affect aircraft, usually associated with jet streams, but also seen in stable airflows over mountains.

Climate A zone of characteristic weather types, temperature and humidity variations, synthesized over a long period (usually at least 30 years) and often with an annual cycle.

Coalescence See accretion.

Congo Air Boundary The line along which warm moist air from the Indian Ocean meets air of lower humidity with a less well-defined origin over the tropical Atlantic Ocean. This line is oriented north–south and has little zonal movement near the range of mountains along the western edge of the East African Rift Valley.

Continentality The extent to which distance from the sea affects climate, the effect of this isolation increasing temperature range, reducing humidity and in some areas tending to reduce rainfall.

Confluence A pattern of decreasing separation between adjacent wind streamlines in the direction of flow.

Convection In general, the buoyant ascent of parcels of air that transfers heat upwards in an unstable troposphere where the temperature of the environment is lower than that of the ascending air. (Convection may also be mechanically forced to allow the release of potential instability.)

Convergence See divergence.

Coriolis force The apparent acceleration of air due to the rotation of the earth. In balanced flow the Coriolis acceleration matches the pressure gradient force on air.

Cumulonimbus cloud See Appendix 3.

Cumulus cloud See Appendix 3.

Current, ocean A general motion, of a permanent or periodic nature, of the surface water of the ocean. Because water has a high specific heat and its temperature changes only slowly, currents are particularly important in the climatic system, the advection of relatively cool water into warm areas bringing stability and lower temperatures in the boundary layer, and the advection of warm water assisting the development of deep convection where air temperatures are relatively low.

Cyclogenesis The initiation or strengthening of cyclonic circulation to form or deepen a depression.

Cyclone The name for tropical revolving storms in the northern Indian Ocean.

Depression The atmospheric pressure system on a large scale in which there is a low central pressure relative to the surroundings and enclosed by a number of isobars. In the tropics, depressions generally represent anomalously cool air near the surface, but there may be anomalously warm air aloft, in particular in mature tropical depressions (due to convective entrainment) or in the special case of a tropical revolving storm. If the air is cool in depth, the tropopause and lower stratosphere are warm. Depressions are characterized by convergence in the lower troposphere and divergence in the upper troposphere.

Diffluence A pattern of increasing separation between adjacent wind streamlines in the direction of flow.

Divergence In meteorology, we are mainly concerned with the horizontal divergence of the wind velocity ($\mathrm{div}_{H}V$ or ∇V) which expresses the time-rate of horizontal expansion of the air. Values are small numbers. Negative values indicate convergence. Divergence in the upper troposphere or convergence in the lower troposphere assist in the development of deep convection.

Dobson–Brewer circulation The motion of air away from the near-equatorial zone by vertical and meridional motion. This circulation is associated with the motion of ozone, produced in the tropics, into the high latitudes.

Drizzle Liquid precipitation, conventionally with drop size between 0.2 and 0.5 mm, produced from relatively shallow clouds near the surface of the earth.

Dry adiabatic lapse rate A constant rate of cooling of an unsaturated parcel of rising air of approximately $-10°C\ km^{-1}$ due to its expansion.

Easterly waves See African easterly waves.

El Niño A periodic warming of the eastern tropical Pacific Ocean (most often seen around Christmastide) associated with increased convection and rainfall, as well as global climatic variation. (See also Southern Oscillation.)

Ensemble forecasting The use of a range of numerical model output based on initializing data that has been adjusted to cover the range of observational and model errors to generate a representative sample of possible future states of the atmosphere. Resolution tends to be low, but use allows a level of confidence to be assigned, in particular to medium-range forecasts, depending on the divergence or similarity of predicted states. A 'poor man's' version involves the comparison of the output of a range of numerical models from different centres.

Equatorial easterly jet stream (EEJ) A wind close to the equator in the upper troposphere associated with the development of monsoon circulations. Its speed is related to the development of an upper-tropospheric high-pressure system on its poleward side.

Fog Obscurity in the layer of air at the surface due to the condensation of water, causing the horizontal visibility to fall below 1000 m.

Föhn A warm dry wind that forms to the leeward of mountains (originally the Alps) due to the condensation of moisture on the windward slopes (see Appendix 2).

Forecast A statement (sometimes in graphical form) of the anticipated meteorological conditions for a place or area. Short-period forecasts are for part or the whole of the following 48 hours, medium-range forecasts cover the period up to about 10 days ahead and long-range forecasts cover monthly to seasonal periods.

Freezing level The altitude (usually above the surface) at which the air temperature is 0°C.

Friction velocity (u_*) The wind speed at the surface resulting from frictional effects.

Front A sloping zone separating air masses, the cooler, denser air lying below the warmer, moist air. Fronts are mainly a feature of mid-latitude weather systems, but may extend into the tropical zone, particularly where there are tropical upper-tropospheric troughs, where they tend to have a modified convective form.

Geostrophic wind Wind in equilibrium between the pressure-gradient force and Coriolis force, thus blowing parallel to isobars or height contours, but neglecting the effect of curvature.

Gradient wind The equilibrium horizontal wind blowing parallel to curved isobars or contours, representing balance between the Coriolis force, the pressure gradient force and the acceleration due to curvature.

Greenhouse gases (GHGs) A range of gases that, due to their absorption and emissivity of radiation, tend to warm the atmosphere. Examples include carbon dioxide, methane, water vapour and chlorofluorocarbons.

Hail Pellets or spheres of ice, sometimes in the form of layers, that fall from convective clouds. Large hail may cause significant damage to crops, buildings, vehicles and aircraft.

Halons Trace (anthropogenic) atmospheric components. A group of unreactive gaseous compounds of carbon with bromine and other halogens used in fire extinguishers or as refrigerants, now banned due to the damage they cause in the ozone layer.

Humidity mixing ratio The proportion of gas in the air that is water vapour, normally given in units of g kg^{-1}.

Hurricane Tropical revolving storms in the North Atlantic, north-east Pacific, south-west Pacific and southern Indian Oceans.

Hygroscopic mineral A mineral compound (usually clay) that has an affinity for water and which therefore promotes condensation to form clouds.

Instability See stability.

Inter-tropical convergence zone (ITCZ) A relatively narrow zone of deep convection, usually found up to about 10° from the equator, but rarely on it, formed where the trade winds from each hemisphere converge.

Inter-tropical front (ITF) The name sometimes given to the humidity discontinuity associated with the movement of monsoon weather systems across West Africa. (Unlike the fronts of the mid-latitudes, it is not a true air-mass discontinuity.)

Jet stream A wind of the upper troposphere with speed greater than 40 m s^{-1} generated by gradients of temperature (and thus thickness).

Katabatic wind A local wind which blows down a slope that cools under clear skies. Its converse is the anabatic wind.

Kelvin wave Fluctuations in wind speed and direction at the ocean surface near the equator result in these eastward propagating waves associated with enhanced convection. They affect the ocean thermocline and cause variations in sea-surface temperature, a key element in the MJO.

Kinetic energy The energy associated with motion.

Köppen classification The classification of climate according to the way in which the fauna and flora of regions are dependent on variations of temperature, winds and rainfall. Each climate is described by a short group of capital letters with subdivisions indicated by small letters, as listed below for climates found in the tropical zone:

A	Tropical rain climates
B	Arid climates
BS	Steppe grassland climates
BW	Tropical-desert climates
C	Temperate rain climates
H	Mountain climates (generally at altitudes above at least 2000 m in the tropics)
a	warmer temperate regimes (mean summer temperatures above 22°C)
b	cooler temperate regimes (mean summer temperatures 10–22°C), often higher ground near the edges of the tropical zone
f	no marked dry season
h	desert climates in which mean minimum temperatures are above 0°C

m   monsoon climates with a short dry winter

n   arid climates with foggy coastal zones

s   dry summer season

w   dry winter season

La Niña   A periodic warming of the western Pacific Ocean (the opposite of El Niño) associated with generally lower temperatures around the globe and unusually dry weather in the eastern Pacific.

Land breeze   An offshore wind generated when temperatures on land fall below those of a neighbouring sea surface, usually commencing around midnight. Speeds are generally low, requiring pre-existing gradient winds to be light. Land breezes are most often seen in association with drainage from high ground near the coast.

Leaching   The loss of water-soluble plant nutrients from the soil due to washout, mainly due to high rainfall or rainfall rates.

Low-level jet   A strong wind (but not a true jet stream), associated with cooling under clear skies, usually observed in arid areas. It forms as wind speeds fall at the surface in response to developing stability and represents the preservation of momentum in the boundary layer.

Madden–Julian Oscillation (MJO)   The main inter-annual fluctuation in the humid tropics. A significant component of the variability of weather in the ITCZ that moves east around the globe over a period of 30–60 days. The MJO moves as a combined Kelvin–Rossby wave affecting the entire troposphere and is most evident in the Indian and western Pacific Oceans. Monitored using the variability in the amount of radiation emitted from the tops of cloud, the MJO involves wind reversals, sea-surface temperature changes and enhanced cloud development, bringing a significant increase in precipitation.

Mei-yu front   An air-mass discontinuity associated with the advance of the moist monsoon flow across East Asia, formed at the right entrance of the sub-tropical jet stream. Its baroclinic characteristics differentiate it from the ITF over Africa.

Meridional   The northward component of motion on the rotating earth.

Mesoscale convective complex (MCC)   A mesoscale convective system defined by a minimum cold-cloud area of 15×10^4 km^2.

Mesoscale convective system (MCS)   An area of showers and thunderstorms which becomes organized on a scale larger than the individual cumulonimbus clouds and which normally persists for several hours or more. MCSs frequently bring large hail and heavy precipitation, and are often seen to develop in association with an area of low pressure or a mid-tropospheric trough.

Meteorological equator (ME)   The line of convergence of air with northward and southward components, associated with the ITCZ over the oceans, but often poleward of the areas of deep convection in summer monsoon flows.

Methane   A trace atmospheric gas, produced naturally, sometimes enterically, or as a result of combustion, or by enteric fermentation. It is a powerful greenhouse gas.

Moist air   A mixture of dry air and water vapour, usually an area of relatively high humidity, signifying the potential or presence of cloud and rain.

Moist climate   A climate in which rainfall is generally sufficient to support agriculture (see Chapters 1, 3, 6 and 8).

Nimbostratus cloud   See Appendix 3.

Numerical modelling   The representation of the atmosphere (or the ocean) by mathematical equations, allowing the atmospheric (or oceanic) state to be forecast into the future.

Numerical weather prediction (NWP)   The objective forecasting of the expected state of the atmosphere using mathematical equations.

Nutrients The nutritional components of foods that are used by organisms to survive and grow.

Ozone A form of oxygen containing three atoms in each molecule (O_3). The lower stratosphere contains a relatively high concentration of ozone which absorbs harmful solar ultraviolet radiation. In recent years the ozone layer over high latitudes has lost a significant amount of ozone and these areas are referred to as 'ozone holes'.

Phytoplankton Microscopic organisms, the autotrophic (plant) component that forms a key part at the base of the food chain of water bodies, particularly the oceans.

Pineapple Express A meteorological phenomenon characterized by a strong persistent flow of moisture and associated with heavy precipitation in the area close to the Hawaiian Islands to any part of the Pacific coast of North America. The upper-tropospheric wind system that forms in association with it includes an equatorward branch of the STJ. The Pineapple Express is a significant source of precipitation in these relatively dry areas.

Polar-front jet stream (PFJ) The upper-tropospheric wind system that carries and develops mid-latitude frontal systems.

Polar stratospheric clouds (PSCs) Clouds formed from ice, nitric acid, nitric acid trinitride or sulphuric acid that form in the lower stratosphere at a temperature below $-78°C$. Chemical reactions with chlorine or bromine associated with these clouds cause the depletion of the ozone layer.

Potential temperature (θ) The temperature to which unsaturated air will rise if compressed to a particular pressure (usually 1000 hPa).

Potential vorticity (PV) This plays an important role in the generation of vorticity in cyclogenesis. It is also very useful in tracing dry intrusions of stratospheric air deep into the troposphere in the vicinity of jet streams.

Where there are positive values of PV in the troposphere, upward motion is supported. PV values are usually relatively modest in the tropics, although significant positive values may be found in upper troughs.

Potentially unstable troposphere A temperature profile that will permit (or promote) deep convection in the presence of sufficient moisture. Classically, this is a profile where wet-bulb potential temperature (θ_w) falls with height (see stability).

Precipitation In meteorology, the production of ice crystals (snow), water drops (rain or drizzle) or hail of sufficient size to fall from the base of a cloud.

Pressure-gradient force The force induced by a difference in pressure (or geopotential height). Wind speed is inversely proportional to this difference.

Pressure tendency The character and amount of atmospheric pressure change during a specified period of time, usually a 3-hour period preceding an observation. The pressure tendency is used to assist in the prediction of the development and motion of weather systems.

Quasi-biennial oscillation (QBO) A well-defined reversal of the zonal wind component in the equatorial stratosphere with a period of about 27 months.

Radiation The transmission of energy by electromagnetic waves. In meteorology we are mainly concerned with (i) the short-wave emissions from the sun that peak in the visible spectrum and that warm the earth's surface and sustain life, part of which is reflected from the surface, and (ii) the long-wave emissions from the surface, clouds or water vapour in the atmosphere found mainly in the infra-red spectrum.

Radiation balance The net flux of radiation through a horizontal surface, considered positive if the downward flux exceeds the upward flux. The radiation balance is generally positive by day as a result of insolation and negative by night as a result of long-wave radiation from the ground. However, incoming radiation generally exceeds the outgoing flux in the tropics, the excess resultant atmospheric heat being exported to higher latitudes, where the outgoing flux generally exceeds the incoming radiation from the sun. Two complications are the effect of clouds, which prevent some incoming radiation from reaching the ground, but also reflect outgoing radiation back to the surface, and the reflection of incoming radiation from the surface. Considering the earth as a whole through a year, the outgoing radiation equals that from the sun, although a very small component of outgoing radiation is now known to be retained by GHGs, causing a gradual rise in global temperatures.

Radiation fog Fog that forms in situ mainly by loss of infra-red radiation from the ground under clear skies.

Radiosonde A device suspended beneath a balloon (or dropped on a parachute) that sends readings of temperature, humidity and wind velocity through the troposphere by radio to a ground station.

Rain Liquid precipitation in the form of water drops at least 0.5 mm across. Drops produced by convective cloud tend to be large (often more than 1 mm across), tending to enhance rainfall rates in many tropical environments. This is the most common form of precipitation observed at the earth's surface.

Ridge An area of anti-cyclonic curvature and relatively high pressure within which there is general subsidence, suppressing the formation of areas of cloud.

Rossby wave A large-scale horizontal atmospheric wave in the middle and upper troposphere associated with the propagation of weather systems. Rossby waves are associated with the MJO.

Saturated adiabatic lapse rate The rate at which a parcel of saturated air cools as it rises.

This varies with temperature and pressure, the cooling smallest at high temperatures and pressures. At 850 hPa and a temperature of 20°C the rate is approximately $-4.5°C\ km^{-1}$ (see Appendix 4).

Sea breeze An onshore wind that develops around midday against light winds early in the day that have an offshore component. The sea breeze forms due to heating of the land surface several degrees above that of the adjacent sea surface. Cooler air of higher humidity is brought inland and in some situations may be carried 100 km or more from the sea.

Seif dune A longitudinal sand dune that forms almost parallel to the prevailing wind, growing in height due to a small cross-wind component, which also results in a steeper downwind side. Some dunes reach a height of 90 m, with a width of 550 m and a length of 100 km. They are particularly common in the Sahara.

Southern Oscillation Variations in the pressure difference across the Pacific Ocean between Tahiti and Darwin used to measure the strength of the El Niño phenomenon.

Specific heat The heat required to raise the temperature of a unit mass of a substance by 1K.

Squall line An extended line of cloud, most of which is formed by deep convection, associated with heavy rain and strong gusty winds, often with hail and thunder.

Stability Three forms of stability may be recognized: (i) absolute stability exists where air cannot rise due to convection, even if saturated with moisture, typically the situation where θ_w increases with height or where a layer of warm air overlies cooler air; (ii) conditional instability, when the air will rise by convection only if it is saturated, typically where θ_w falls with height at some level above the surface; (iii) instability in dry air, the temperature of which falls at a rate greater than the dry adiabatic lapse rate. Such instability is only maintained through the whole depth of the troposphere if it becomes saturated and θ_w falls with height.

Stellate dune   A large sand dune of approximately pyramidal shape with several radial sinuous arms, formed due to winds blowing from several directions. They may be 150 m high and 1 km across.

Stratocumulus cloud   See Appendix 3.

Stratus cloud   See Appendix 3.

Streamlines   Lines parallel to the wind at all directions along them, usually at a particular level (but may be used in the vertical sense).

Sub-tropical jet stream (STJ)   A wind of more than 40 m s^{-1} near the top of the tropical troposphere at its poleward edge driven by the temperature difference.

Subsidence   The general descent of air in a stable atmospheric regime, typically associated with areas of high pressure.

Subsistence crops   Food crops, typically grown by family groups for local consumption or exchange and not generally sold.

Teleconnections   The links between weather and climate systems separated by large distance and typically in different masses of air or different zones.

Temperature inversion   A layer of warmer air overlying cooler air.

Thermal wind   An apparent wind representing advection, the vector difference between wind velocities at two levels in the atmosphere. These winds are parallel to lines of equal atmospheric thickness.

Thickness   In meteorology, the difference in height between two (standard) pressure levels. The height difference is directly proportional to the mean temperature of the layer. So-called 'total' thickness is the difference between the 1000 and 500 hPa pressure levels; so-called 'partial' thickness is the height difference between the 1000 and 850 hPa levels.

Transverse dune   A large asymmetrical sand dune that forms at right angles to the prevailing wind. These dunes have a steep leeward side and are typically found on beaches. They may develop to a height of about 100 m and may be 10 km long.

Tropical revolving storm   These storms develop over open ocean when sea-surface temperatures are at least 27°C and winds through the troposphere support deep organized convection. Circulation is cyclonic in the lower troposphere and anti-cyclonic above. They are defined as having surface wind speeds of gale force or more. Tropical revolving storms develop from tropical depressions.

Tropical upper-tropospheric troughs (TUTTs)   Troughs of relatively low pressure found near the poleward boundaries of the tropical air mass. These features are semi-permanent and are found most often in particular areas, but are usually most developed in winter.

Tropopause   The level marking transition from the troposphere below to the stratosphere above. In the lower stratosphere temperatures are approximately constant, thus forming a cap to convection or layer-cloud development in the troposphere.

Troposphere   The lowest layer of the atmosphere in which, in general, temperatures fall with height and that contain weather systems, including all but the deepest of convective clouds.

Trough   An elongated area of cyclonic curvature and relatively low pressure which, if present in depth and in otherwise conducive conditions, will promote deep convection and shower formation as a result of convergence. Troughs are also associated with fronts.

Turbulence   Flow in which there is irregular random oscillation, often in the form of eddies, due to interaction with a rough surface or rapid changes in wind speed.

Typhoon Tropical revolving storms in the north-west Pacific Ocean.

Upwelling The dynamic uplift of water from moderate depths to the sea or ocean surface. Upwelling is relatively common on coasts where there are long-shore or offshore winds. The resultant cooling of the sea surface may prevent convection.

Velocity The vector quantity of speed and direction (of wind).

Vorticity The vector quantity of rotation (expressed as the change of velocity with distance). Positive values indicate areas of upward dynamical motion.

Water vapour The gaseous form of water in the atmosphere, the condensation of which forms weather systems. Amounts are usually small, rarely exceeding 25 grams per kilogram of dry air, even in the humid tropics.

Weather systems In general, depressions or anticyclones, but also the related trough and ridge. Sometimes a term applied to areas of unsettled weather.

Westerly wind bursts Local areas of wind at low levels that has crossed the equator and that originally had an easterly component but becomes westerly as a result of the reversal of the Coliolis force as the equator is crossed.

Wet-bulb potential temperature (θ_w) The degree to which the wet-bulb temperature of air will rise by compression to a standard level (usually 1000 hPa from 850 hPa).

Wind gust Rapid fluctuations in the 10-m wind speed, significant when there is a variation of 10 knots or more between peaks and the mean speed. The speed of the gust is the maximum instantaneous wind speed.

Wind profiler A system using radio or sound waves to track winds in the troposphere by sensing variations in water vapour.

Zonal The eastward component of air motion on the rotating earth.

Zone In meteorology, a portion of the earth oriented east–west, named according to its prevailing temperature and wind regime.

Zooplankton Microscopic animal life of the oceans and other water bodies; an important part of the food chain, generally above phytoplankton and below crustaceans or fish.

Index

An Introduction to the Meteorology and Climate of the Tropics, First Edition. J F P Galvin.
© 2016 John Wiley & Sons, Ltd. Published 2016 by John Wiley & Sons, Ltd.

Printed and bound by CPI Group (UK) Ltd, Croydon, CR0 4YY

09/10/2024

14571432-0005